AF676945

Aesthetic Fatigue

Aesthetic Fatigue: Modernity and the Language of Waste

Edited by

John Scanlan and John F. M. Clark

CAMBRIDGE SCHOLARS

PUBLISHING

Aesthetic Fatigue: Modernity and the Language of Waste,
Edited by John Scanlan and John F. M. Clark

This book first published 2013

Cambridge Scholars Publishing

12 Back Chapman Street, Newcastle upon Tyne, NE6 2XX, UK

British Library Cataloguing in Publication Data
A catalogue record for this book is available from the British Library

Copyright © 2013 by John Scanlan, John F. M. Clark and contributors

All rights for this book reserved. No part of this book may be reproduced, stored in a retrieval system, or transmitted, in any form or by any means, electronic, mechanical, photocopying, recording or otherwise, without the prior permission of the copyright owner.

ISBN (10): 1-4438-4912-X, ISBN (13): 978-1-4438-4912-8

Contents

Preface vii

Introduction 1
Aesthetic Fatigue, Modernity and Waste
John Scanlan

Chapter One 17
Wasteland Mania: The Alternative Modernity of Peat Moss
Fredrik Albritton Jonsson

Chapter Two 32
Landscape and Waste: Hydroelectric Development in the Highlands of Scotland
Jill Payne

Chapter Three 45
Scenes from a Desert Wasteland
John Scanlan

Chapter Four 71
Exhaust: On Aerial Rejectamenta
Steven Connor

Chapter Five 81
Debating Waste, Nature and Justice: Spencer, Huxley and Morris
Piers J. Hale

Chapter Six 99
H. G. Wells's Taxonomy of Waste
William Kupinse

Chapter Seven 119
The Future as a Virus in the Midst of our Waste
Alberto Duman

Chapter Eight 138
The Future of Ruins
William Viney

Chapter Nine 159
Cézanne and Van Gogh's Quarries: Wasted Landscapes as Modern Art
Maura Coughlin

Chapter Ten 173
Arman's Spectacles of Waste
Jaimey Hamilton

Chapter Eleven 190
Retread America: Postwar Re-Adjustment, boredom, and Life in the "Lonely Crowd"
Edward J.K. Gitre

Chapter Twelve 206
Garbage and Gatsby
Rex Ferguson

Chapter Thirteen 220
Exteriority: Boredom, Disgust and the Margins of Humanity
Harvie Ferguson

Chapter Fourteen 251
The "Litter Nuisance" in Twentieth-century Britain
Timothy Cooper

Chapter Fifteen 274
Shipping and Receiving: the Social Death of Electronics
Jennifer Gabrys

Chapter Sixteen 298
Managing Household Refuse: Towards a History of the Municipal Waste-Handling Business
Ray Stokes and Stephen Sambrook

Contributors 319

Index 322

PREFACE

The majority of the essays in this volume were originally presented as papers at a conference on "Modernity and Waste," organised and chaired by John Scanlan, and held in July 2006 at the Arts and Humanities Research Council (AHRC) Centre for Environmental History at the University of St Andrews. It was the final of three annual international events sponsored by the AHRC Centre, whose work was themed around studies of *waste* and *wastelands*; in previous years—2003 and 2004—similar gatherings had taken the form of invited workshops whose primary purpose was to bring together a diverse body of international scholars from a variety of academic disciplines in the hope that they might benefit from sharing their thoughts and ideas with others similarly intrigued to understand what could be learned from the study of "waste."

The origins of the AHRC Centre lay in a collaboration between the universities of St Andrews and Stirling, dating back to the 1990s, and at that time led by the historians Chris Smout, the Historiographer Royal since 1993 (St Andrews) and Fiona Watson (Stirling). John Clark arrived at the University of St Andrews in autumn 2000 to take over the running of its Institute of Environmental History, established by Smout, and proposed to build on their success by submitting a collaborative bid to the Arts and Humanities Research Centres Scheme on the historically rich potential of "waste" as a subject around which a series of related projects could be themed. As uncultivated or disused land, and as rejectamenta, or discarded material objects, waste encompassed rural and urban environmental history. The bid was successful, and resulted in a substantial grant to form a joint AHRC (then AHRB) Centre for Environmental History that was established in October 2002.

Our principal remit was to draw upon waste to establish environmental history as an intellectually urgent field of enquiry within mainstream history, and cognate areas in the humanities and social sciences. By design, our research was interdisciplinary and multidisciplinary from the outset, and this volume aptly embodies our research agenda. In 2003, John Scanlan joined the Centre to undertake a project on "The Language of Waste," and soon after became the Centre's Project Officer.

During the lifetime of the Centre, over eighty academics visited St Andrews to present their research, and their number included historians, geographers, art and architectural theorists, sociologists, philosophers, literary scholars, artists, and a film-maker—Michael Chanan, whose 2005 documentary, *Detroit: Ruin of a City*, co-directed with American sociologist George Steinmetz, was given a first British showing at the Centre.

Most of the papers gathered here have been substantially revised in light of discussion at the "Modernity and Waste" conference and subsequent editorial and peer review. Several of the essays had other origins and were, in a manner of speaking, "commissioned" in its aftermath, including the essay by Stokes and Sambrook, which gives a hint of the work that will soon emerge from their recent ESRC study on the management of household waste. The essays by Piers J. Hale and Jill R. Payne arise out of their prior association with the Centre at St Andrews—Piers Hale had been a speaker at the Centre in 2003, and the environmental historian Jill Payne (as well as being a speaker at earlier workshops) completed her PhD at the Centre under the supervision of Chris Smout and, latterly, John Clark; the essays by Scanlan and Viney were, respectively, written and selected especially for this volume.

John Scanlan
John F.M. Clark

INTRODUCTION

AESTHETIC FATIGUE, MODERNITY AND WASTE

JOHN SCANLAN

It is difficult not to arrive at the view that a spectre of waste hangs over modern western society, and its culture and history. The extent to which we seem to be willing to acknowledge this suggests that waste actually haunts our *consciousness*—and, specifically, consciousness as a kind of temporal awareness or sense of finitude.[1] If we reflect on the discourse of sustainability that we are now so familiar with, after all, it irrefutably rests on a language of waste reduction that is itself shaped by the social desirability of fixing our collective gaze on a horizon that surpasses merely the immediate or close at hand. It recognizes that the consequences of the lives we live and the choices we make today reverberate for longer than we—finite and human—can perhaps properly comprehend.

WHEN THE WORLD IS RUNNING DOWN ...

Before the idea of working towards a sustainable future had become a common feature of everyday life (in the last two or three decades) - and more specifically attached to concerns about the fate of life on this planet - the threat of mere waste stuff, it may surprise us to learn, loomed large enough that it could promise the reversal of human efforts to fashion life, human society and the world in accordance with the apparent mastery human reason had attained over nature since the emergence of a modern world shaped by Enlightenment ideals. And despite a contemporary worldview that is increasingly attentive to the sustainability of the resources and energy we use, there still lingers, in the background, a *language* of waste that reveals itself upon closer scrutiny. Sometimes it is found in a kind of moral discourse that cautions against the dangers of profligacy; at other times we see it beneath technocratic visions of efficiency and innovation, or even in artistic gestures that illustrate how easily our waste stuff gives away an essential truth about what it is to be human today. In the media, and in culture more generally, the dangers of all kinds of waste are revealed through visions and representations that hint at societies in terminal decline, wallowing in filth, enslaved to the temporality—and ideology—of consumer society and technological innovation, and where artists and museum curators, through their ability to disarm us, often seem to offer the most powerful visions of our situation and the complicity of each of us within it.[2] Waste also always seems to be the product of an over-zealous or exuberant expenditure of energy, as we see if we look at the recent history of Wall Street and the City of London, the citadels of capital whose own waste products brought western economies low—"toxic" loans and trading in the suspiciously *leftover*-sounding "derivatives" superseded the earlier "junk" bonds of the 1980s and 1990s.[3] This language of pollution and negative value is not accidental: where human interests are at risk of erosion, it is a terminology of waste that comes ready-to-hand to flag up what is at stake. At this very moment, as I write, my own bank—the news announces—has been downgraded to "junk" status.

Waste illustrates the deadly effect of time, and begins to loom where we lose control of events and places, or become negligent of the consequences of our actions. The language we use to come to terms with it characterises modern life's pursuit of order, efficiency and perfection as itself the source of a kind of *aesthetic fatigue*.

In recent years, waste has continued to appear in many of its familiar guises, but it also lurks behind some of the most forward-looking attempts to develop an eco-architecture, and encourage so-called "cradle-to-cradle" product design, and a variety of smart technologies that would reshape our relationship to the environment.[4] The familiar is relatively easy to identify: we need only point to the repeated recurrence of garbage strikes and the force with which they can remind a community that whatever other advances in waste reduction and disposal have been made, the residues of daily life are permanently present and can quickly become "active" in ways that unmake the fabric of society. On one such occasion, authorities on the Italian island of Capri had to take steps to prevent an influx of tourists as the summer approached, thereby jeopardizing the local economy, simply because the closure of a municipal dump had left the island awash with rubbish. "We just cannot manage to get rid of the waste generated by 13,500 residents and 3,500 guests," the mayor of a village on the island told *The Independent* newspaper. "Imagine what it will be like when we have to deal with the rubbish produced by tens of thousands of tourists who arrive at the heart of the summer."[5] Similar fears about being overrun with material wastes followed New York City mayor Rudy Giuliani's decision to close the Fresh Kills landfill in 2001 before legislation had been approved to allow for the exporting of waste to other states.[6] One risk—and a fear expressed by some—was that if public authorities ceded control of waste management it could all-too-easily fall into the hands of the kind of unscrupulous operators portrayed in the hit TV series, *The Sopranos*—about a mobster for whom the trafficking of waste offers a convenient disguise and a non-suspicious form of business respectability.[7]

The mountains of trash depicted in the Oscar-winning 2010 documentary *Waste Land,* set around Jardim Gramacho near the city of Rio de Janeiro—the world's largest landfill site—may suggest a strong equivalence between poverty and waste, yet it merely echoes what has seemingly always been the case within modern societies geared around production and consumption. Not only do the most desperate always find a use for what the better off deemed to be either worthless or of negligible use, but they usually found themselves condemned to live in "wastelands" of one kind or another.[8] If we travel back in time to mid-Victorian London, we will find an economy of waste and poverty that was the main source of subsistence for many: those in pursuit of wealth created vast amounts of waste, and the poor absorbed it and extended its life.[9] A link between poverty and waste has been identified also in more contemporary examples of what Lindsey Dillon calls the "socio-ecological transformations" of former urban wastelands (often termed "brownfield" sites). The transformations of these sites have become common in recent decades but, Dillon suggests, still seem to be unable to offer the hope that there is a way out of a wasteland that would be defined according to recognised indices of poverty, and often end up merely remaking existing inequalities around race, health and their links to the quality of habitable space.[10]

But perhaps less apparent to a casual observer is the impact of the accumulated wisdom gained from centuries of struggles against waste, pollution and the effects of profligacy. It is a historical knowledge and sensibility that fundamentally grounds the ideas of waste-minimisation, or "zero waste," that seem to be driving technological innovation today, remaking the product world and the homes and buildings we live and work in. And alongside a world that is, in aesthetic terms, refashioned, it is hoped that a new eco-awareness will emerge in the form of attitudes and actions that recognise the impossibility of separating the present from a distant future where the effects of how we live today will be felt. We might point to developments in architecture, such as the Eastgate Building in Harare, Zimbabwe, which minimizes energy waste by mimicking the heating and cooling principles of a termite mound in order to maintain a constant temperature and reduce energy use—but this seems a long way from the more makeshift kind of buildings and austere dwellings celebrated in Martin Pawley's 1975 book *Garbage Housing.*[11] The former embodies the rise of a design movement that has sought to learn from natural systems, while the examples revealed by Pawley—a generation ago—worked towards minimizing energy use through the quite basic method of recycling waste products (such as bottles and cans) as building materials.

Some of the most celebrated names in the world of architecture and design have for some time now been trying to meet the challenge of working towards futures that are, for instance, carbon neutral. Foster + Partners' Masdar development in Abu Dhabi, led by Sir Norman Foster, claims to combine "state-of-the-art technologies with the planning principals of traditional Arab settlements to create a desert community that aims to be carbon neutral and zero waste":

> A mixed-use, low-rise, high-density development, Masdar City includes the headquarters for the International Renewable Energy Agency and the recently completed Masdar Institute. Strategically located for Abu Dhabi's transport infrastructure, Masdar is linked to neighbouring communities and the international airport by existing road and rail routes. The city itself will be the first modern community in the world to operate without fossil-fuelled vehicles at street level. With a maximum distance of 200 metres to the nearest rapid transport links and amenities, the city is designed to encourage walking, while its shaded streets and courtyards offer an attractive pedestrian environment, sheltered from climatic extremes. The land surrounding the city will contain wind and photovoltaic farms, research fields and plantations, allowing the community to be entirely energy self-sufficient.
>
> The development is divided into two sectors, bridged by a linear park, and is being constructed in phases, beginning with the larger sector. The masterplan is designed to be highly flexible, to allow it to benefit from emergent technologies and to respond to lessons learnt during the implementation of the initial phases. Expansion has been anticipated from the outset, allowing for growth while avoiding the sprawl that besets so many cities. While, Masdar's design represents a specific response to its location and climate, the underlying principles are applicable anywhere the world. In that sense it offers a blueprint for the sustainable city of the future.[12]

Early in the new century, the "intelligent design" of such large-scale projects was matched at the everyday level by a drive towards "intelligent homes," as illustrated by MIT's *House_n,* one of a number of examples which, according to Lynn Spigel, offer "a curious inversion of the relation between people and their things," resulting, we might say, in a situation in which the active subject is "dispersed" in systems and technologies:

> Intelligent agents become more lifelike (as they take on a series of cognitive and motor tasks) while humans become more "thing-like" (as they submit their bodies to checkups, judgments and repairs.[13]

As we slipped into what a few generations ago was conceived as a distant sci-fi kind of existence, such visions of the future were as much driven by ideas about how to manufacture environments for sustainable living as they were by the blurring of distinctions between the human and the machine in the Posthuman world to come. Thus, researchers on the *House_n* project, it was reported, were "also exploring ways that a home can use sensors to monitor the home's performance to help people understand the implications of their behaviour on energy consumption."[14] That such intelligence now extends to the mobile technologies we increasingly carry around in our pockets means that anyone can, in theory, monitor their own home energy use, and remotely control it via smartphone applications. Indeed, you might want to want to raise your environmental awareness through "Apps" that monitor, in real-time, the sources of energy supplying our National Grid. In the terms of Martin Heidegger's well-known essay "The Question Concerning Technology" this is yet a further technological "revealing" of natural forces:

> The energy concealed in nature is unlocked, what is unlocked is transferred, what is transferred is stored, what is stored up is, in turn distributed, and what is distributed is switched about ever anew. Unlocking, transforming, storing, distributing, and switching about are ways of revealing.[15]

In terms of working towards a goal of saving energy and eliminating waste (whether that is wasted resources, or wasted human effort), these are all arguably extensions of the same logic that produced commonplace technologies such as refrigeration, and even the kind of automation evident in sink units for food waste disposal; that is to say, they are extensions of a technological reason that continually remakes, refines and hopefully makes more sustainable our relationship to nature.[16]

But off-loading tasks onto intelligent systems and devices also runs the risk that we are simply freer to indulge other distractions that reduce our level of engagement with how our actions affect the world around us. The expansion of digital memory provides a useful analogy here: the increasing processing power of personal computers and the advent of apparently limitless storage capacity that drives revolutions in home entertainment media remakes the ways that we may understand "leisure time": it allows for a surfeit of possible future experiences. That might seem to many of us as an undeniably positive development—yet, we need to ask also what it does to our sense of the world that we inhabit in "real" time. It surely illustrates one possible challenge to living with an awareness of our finite limits. Reflecting on the challenges faced in simply remaining aware of the dangers of un-checked energy use, for instance, the environmentalist Bill McKibben saw that it was something akin to what I have described elsewhere as a kind of "surfacing" that warps our awareness:

> We are fatally confused about time and space. Though we know that our culture has placed our own lives on a demonic fast-forward, we imagine that the earth must work on some other timescale. The long slow accretion of epochs—the Jurassic, the Cretaceous, the Pleistocene—lulls us into imagining that the physical world offers us an essentially stable background against which we can run our race. Humbly, we believe that the world is big and that we are small. This humility is attractive, but also historic and no longer useful. In the world as we have made it, the opposite is true. Each of us is big enough, for example, to produce our own cloud of carbon dioxide.[17]

Energy, in a sense, is what the language of waste (in its positive expression—efficiency, good resource management) is all about.

Modernity, Consciousness, Waste

All of this is to say that increasingly, our sense of waste is conditioned by the ascendency of a sustainability agenda that reaches from global politics to the design of systems of energy conservation. Even if we don't acknowledge that in encouraging us to think positively about how we shape the future we have diverted attention away from representations of ruin, depletion, pollution and decline, any serious reflection on what drives such developments must come back to the relationship between modernity (the culture, beliefs and ways of organising modern societies) and waste. Studies of the impact of capitalist modernity on nature, for instance, don't have to look far to find evidence of its inherent destructiveness.[18] Teresa Brennan points to a truism when she says that "to satisfy the demands of large-scale production, more and more of nature" was destroyed:

> In this sense production under capitalism is consumption, not production; it gobbles that which is already there, gives nothing back but waste [...] Because so many of the commodities it produces cannot be recycled (in general) or because, if they are agricultural, they tend to exhaust the conditions of their own production, then, given that nature is the source of value, capital can only profit by continuing to exploit every available natural source.[19]

But I would stress also that waste works against the future, and against sustainable living in more subtle ways: it is not just because the stuff of waste may pollute, or that it illustrates the profligate use of scarce resources that we have to be concerned about it. Aesthetic sensibility, consciousness, and time-awareness are all bound up in "experience," and are equally deserving of our attention, precisely because waste creates a space—an interior life, a sensibility, one might say—where normal attentiveness to present or future concerns become disengaged. Hence, however we might define a wasteland, it is not merely an abandoned, depleted or polluted space. Here Michel Serres elaboration of the ways in which a variety of "hard" and "soft" wastes also points toward a broader phenomenological consideration of our subject:

> Let us define two things and clearly distinguish them from one another: first the hard, and second the soft. By the first I mean on the one hand solid residues, liquids, and gases, emitted through the atmosphere by big industrial companies or gigantic garbage dumps, the shameful signature of big cities. By the second, tsunamis of writing, signs, images, and logos flooding rural, civic, public and natural spaces as well as landscapes with their advertising. Even though different in terms of energy, garbage and marks [of the latter kind] nevertheless result from the same soiling gesture, from the same intention to appropriate […][20]

Wastelands are also imaginatively constructed zones—indeterminate, unregulated, un-policed or supervised—that seem to exert an almost magnetic pull as location or dumping ground for other "wastes."[21] Peter Wynn Kirby, an anthropologist of contemporary Japan, gives an interesting illustration of how desolate areas around Tokyo attract unusual acts of dumping, including the "abandonment of pets such as cats and dogs on the periphery […] leading to an abrupt rustication of these domesticated pets in a simultaneous dilution of 'the wild'."[22] Such examples illustrate in one way how waste takes up residence in the mobile and the indeterminate, of how it applies to almost everything in transit between accepted categories of value and understanding.[23] Kirby thereafter makes a persuasive case for the importance of seeing waste within a context—in his terms, an *ethnographic* one—that embraces a nuanced understanding of how environmental practices in different societies produce their own highly variable kind of waste consciousness:

> Attitudes toward waste are not universal; they modulate with reference to history, differences of language, notions of productivity, constructions of thrift, regulatory landscapes, classifications of vermin and outcasts, exposure to and participation in local, regional, and global environmental dis-

> courses, and so on. Even within a given society there are marked variations.[24]

As a student of western modernity who indeed tends to view phenomena like waste within a broader context, I recognise that it is still all too easy to generalise—I am probably guilty of it to some extent, and realise that "western modernity" could be said to incorporate many modernities; but in thinking *historically*—thinking of modernity as a kind of consciousness that has roots in certain philosophical assumptions about the world—it is useful to make general observations that can sum up the epochal shifts in experience that otherwise might be lost. One conviction I share with Kirby is in the need to expand our understanding of what an *environment*, and an environmental problem, is. My own more recent work on the interrelationship of human-environmental-technological factors in sustaining memory, for instance, rests on a view of human consciousness and action that stresses the primacy of the various kinds of *ecologies* we might be said to occupy in contemporary western societies—specifically ecologies of remembering and forgetting—and which alter our environmental consciousness and our time-place awareness.[25] These, I suggest, ought to take precedence over assumptions that historically drove the development of my own nominal discipline of sociology, such as "agency" and "subjectivity," but which today can't begin to account for the importance to society of the broader ecological fate of life.[26] The realisation that has to be made is summed up by the French Sinologist François Jullien, whose work has elaborated on the differences between western philosophy and Chinese thinking, and stresses the many ways in which the latter places the individual *within* nature, as opposed to in the position of "subject" viewing the natural world as a separate object. Here he is reflecting on how he responds to an image of himself undergoing the "silent transformation" of aging, but we might equally apply these words to the way that knowledge of our own wasting activities often comes to us as a shock:

> I thought I was a *subject*: a subject of initiative, one who conceives and desires, is active or passive, but one which always retains the sense of its being and self-possession. Admittedly it is a self that knows itself to be caught in a totality of external as well as internal interactions which hem it in, but it still considers itself to be "generated from within," according to the expression dear to metaphysics, *causa sui*. And then, right in front of me, this perspective is suddenly violently shaken, capsizing into this otherness: that of a cause or a continuum whose sole consistency stems from a mutual correlation of factors—mutually maintained and without regard to "me" […].[27]

What we might learn from looking beyond traditional western ideas of "subject" and "object" is how individuals in contemporary societies come to act in the kinds of thoughtless and disengaged ways that Peter Kirby finds also in his ethnography of Japan's "wastescapes."[28] One reason that a focus on consciousness seems to be desirable is that the phenomenon under observation is then not limited to the easily defined and generally accepted; it is not limited to the "wastes" we can point to in recognition, but allows for our understanding to be open to its constantly mutating presence. To consider the importance of consciousness extends our view to the whole repertoire of means through which waste is manifested, ignored or otherwise takes on some kind of monstrous quality.

In historical terms, our attempts to preserve and conserve, to save rather than squander and, indeed, to fashion a more sustainable future points to an unease with the pace of untrammelled change. But from the dawn of modernity the danger of progress tumbling into reverse—manifested in many battles against a variety of wastes—has been evident in a number of spheres of life: from public health to economic management, the regulation of land use, and so on. This collection of essays examines a strange and seemingly paradoxical relationship: between progress and decline, novelty and obsolescence, modernity and waste. And, as much as language can reveal aspects of this relationship to us, one aim of this volume is to show how modernity, with its desire to colonise unknown terrain and turn nature into use-value, implicates history—and a historical consciousness—in defining what it is to be human, and what the social organisation of human affairs, needs and interests does to our relationship with nature.

In broad terms "modernity" is shorthand for the historical break with "traditional" ways of life that gave rise to a particular kind of experience.[29] In this sense, modernity institutes a new kind of time-space that takes effect as the re-orientation of the human to the natural, and of the modern individual to its environment (which includes the human-made, or built, environment). Equally, modernity might be thought to represent a leap into the unknown: a condition of self-estrangement that splits individuals and communities from a familiar and seemingly self-sustaining habitat. In enacting a rupture with the past, the modern impulse sought a decisive break with the shortcomings of the old world (encompassing, for example, the political, the religious and the philosophical): it was fixed upon a clean, unspoiled, and more perfect future. Yet, paradoxically, even the most infinitesimal of imperfections may reveal how central waste and waste stuff is to merely being alive; and even though we may become cleaner and healthier, a world fashioned from bright surfaces and infused with light may just show up the irreducible and filthy reality of life.[30]

The Utopian impulse, which is one version of a modern desire to begin anew, places memory, forgetting and imagination together. Amongst the many artistic and literary explorations of this impulse one particularly interesting case is offered by the 1997 movie, *Gattaca*.[31] The society of the future we are presented with in this film illustrates the forward leap as an attempt to elude—if not to entirely erase—the reach of the past and its perceived contaminants, which are represented in the human form of a sub-class of human beings, so-called "invalids," whose genetic imperfections are illustrated against the background of the chosen ones; those who are known as "valids" and who constitute the perceived future of a society that sees its viability in terms of the genetic manipulation of life. The invalids are those possessed of relatively negligible flaws—projected illnesses associated with aging—but who have been conceived "naturally" and without having been selected on the basis of in utero DNA testing, and are thus seen as condemned to a fate that can no longer be controlled.

The story follows the progress of Jerome, an "invalid" whose aim is to succeed in the elite space programme. His problem is that at birth he was given a projected life expectancy of 30 years, an age that coincides with the peak of his likely suitability as an astronaut. But with the aid of subtle means of genetic subterfuge, he manages to take on the identity of a once perfect, but now wheelchair-bound, man. To exist under this disguise he has to carry around bodily waste stuff—urine, hair, and loose skin from the one-time "valid" who has thus donated his identity. As an "imperfect" Jerome represents the real. At one point, when his identity becomes known to a woman he has confided in, he is referred to as one of "God's children," thus identifying chance and imperfection—cosmic luck—with the once perfect of all imagined entities, the Almighty, who now falls short of the miracles that genetic engineering has delivered for the wealthy in this utopian world.

But Jerome, like a futuristic Odysseus fighting free of the gods, overcomes his fate; he scrubs and cleans the loose skin cells from his body and incinerates contaminated clothes because the testing of elite astronauts is so widespread that mere traces of bodily matter can reveal the true identity of its owner. In one scene, a single hair believed to be from the under suspicion Jerome's head is taken to a lab for testing—but the subterfuge works; the hair was planted by Jerome, and it produces a long computer printout, a profile, complete with a photograph and the verdict, "valid." The message is clear: the cleaner and more perfect the world is, the more its imperfections can be shown up. Technology turns the waste of the material world into information, heritability into code—into a thing that speaks; that will not forget. Waste, whatever its forms, reveals us for who we are because it is always *of* us.

The Essays in this Volume

George Kubler in his book *The Shape of Time* elaborates an idea of "aesthetic fatigue" that sums up much of what, in this volume, seems to be characteristic of waste—its relation to time, to energy, to modernity's desire for evermore perfect designs and forms of living—and to why we discard the past and the stuff that once adorned our world.[32] Overfamiliarity, he suggests, makes us tired of the way the world looks; it makes the world itself look tired. "Waste," as this book aims to show, is a multifaceted phenomenon—a by-product of technological progress, of efforts at landscape improvement and rational efficiency, of self-consciousness and of a jaded familiarity with the way things are. But it is also simply a by-product of living, which is essentially the expenditure of energy that is fed by the use we make of our environment.

What *Aesthetic Fatigue* therefore aims to do is look at how waste has been represented, imagined, narrated, analysed, and how, in its various manifestations—as material stuff, moral lack, or other spectral presence—it has transformed the nature of everyday life in modernity. While the essays that are presented in this volume represent a variety of academic traditions and disciplinary specialisms, they have in common shared themes, conceptual unities, and a concern with the very language of waste. *Aesthetic Fatigue* begins with considerations of early modern ideas of waste in the understanding of landscape, and works through literary, artistic, philosophical, social and economic considerations of waste as a key to understanding modern life.

Fredrik Albritton Jonsson's essay on eighteenth-century wasteland "improvement" in the Scottish Highlands focuses on the role of marginal lands as the medium through which a hybrid modernity of agricultural productivity and traditional political loyalties was developed, and in which the socially redemptive powers of peat moss and a "wasteland mania" acted as a critique of the industrialism and revolutionary radicalism that was gripping societies elsewhere at the time. Jill Payne's essay moves the discussion of the Highland "wasteland" as an under-used resource into the twentieth century when the economic and geographic potential for large-scale development in the Scottish Highlands was arguably at its greatest. But, as she argues, the twentieth-century veneration of the region's outstanding natural beauty is the product of an eighteenth- and nineteenth-century stress on the importance of seeing in such natural environments an alternative to industrialization, and a world that seemed to exist beyond the utilitarian emphasis on "progress." This examination of hydroelectric development explores how the potential for improvement contradicted and threatened visions of the Highland landscape that had taken root in the aftermath of the rise of Romantic idealism, laying bare competing twentieth-century ideas about the "best use" of non-urban space, and the associated language of "waste" and "wasteland."

My own essay (chapter 4) looks at the example of the Mojave Desert as, variously, a space of liminal encounters, military research, and play, in order to explore how cultural and phenomenological dimensions of forgetting can inform a more nuanced understanding of the ways in which "waste" and "wastelands" are manifested in contemporary life, yet at the same time obscured in their real dimensions and serious consequences. Steven Connor's consideration of air pollution focuses on how the very idea of air has changed as a result of an awareness that the atmosphere itself had, in modernity, become the target of waste dumping in the form of a variety of aerial rejectamenta.

The next three essays have as a focus ideas about waste that drove ideas about political economy and sanitation in the nineteenth and early twentieth centuries. Piers Hale's essay looks at the relationship between waste, nature and social justice by focusing on the emergent "green politics" of the nineteenth-century English writer William Morris—who saw the wasteful organization of labour in society as a means through which socialist ideas about the inequities of capitalism could be revealed and addressed—and the contrasting ideas of political economy popular at that time (in, for example, the writings of Herbert Spencer), which were underpinned by a belief in a form of social Darwinism that rested significantly on an association between human nature and the natural environment. William Kupinse's essay examines the widespread use of waste as rhetorical device in the writing of H. G. Wells, and argues that closer examination reveals that two related issues inform the body of Wells's oeuvre: the utopian impulse that manifests itself in Wells's various programs of political and social reform, and Wells's concept of "waste" as a negative motive force driving his utopianism. It is a reading that reveals Wells's systematic exploration of the cultural construction of value, and aims to sketch the epistemological underpinnings of waste in the early twentieth century. Alberto Duman's "The Future as a Virus in the Midst of our Waste" is concerned with the management of waste matter and hygienic practices in utopian projections of nineteenth century British urban planning, which he ties into concerns about the future of waste management and urban development through a description of his own 2006 artistic project— the

"Lamby Way Time Capsule"—which proposed the burial of a time capsule within the imminently closing landfill facilities at Lamby Way in Cardiff. The purpose of the time capsule, Duman explains, was to function as a *reversed* viral presence, keeping alive a particular discourse of the landfill, and thus waste, in the face of the planned aestheticization of the site which, of course, would continue a modern trend of obscuring the reality of waste.

The following three essays by William Viney, Maura Coughlin and Jaimey Hamilton examine how artists have made use of a variety of wastes to comment upon aspects of modernity. Viney's essay looks at the figure of the ruin in modern culture and as represented in the work of Hubert Robert, Joseph Gandy and other artists. It is the *vision* of ruin, he suggests, that has often allowed us to imagine the future as something that might have a different temporal shape than the past and present. In contrast to the ruins of architecture, Maura Coughlin's essay draws our attention to modern ruins of another kind—in the work of Vincent Van Gogh and Paul Cézanne whose paintings of quarries dwell on what she terms "the feral remains of capitalist exploitation of natural resources." As she illustrates, theirs was not an entirely escapist construction of *nature* (considered, for instance, as the pristine, prelapsarian other to urban modernity) but rather an engagement with the awkward, ugly remains of modern exploitation. Jaimey Hamilton's study of French *Nouveau Réaliste* artist Arman looks at works such as 1973's *La Grande Bouffe* (*The Big Feast*)—a Plexiglas box filled with Seven-Up bottles, milk cartons, canned tomato tins, and detergent boxes embedded in clear polyester—as an example of how art reflected the activity of "wasting" in contemporary consumer culture. And, following Georges Bataille analysis of the role that the "accursed share" plays in culture, for Hamilton, Arman's *spectacles* of waste (glass vitrines, like a shop window, that transform something as abject as trash into image, and then image into a luxury art object) are seen as useful allegories for the paradoxical relationship between commodities and waste.

What we might term the affective dimensions of "waste" occupy the authors of the next three essays. Edward Gitre looks at how the Second World War opened a new chapter on the history of boredom, a social malaise that he suggests became a common feature of an American culture and society overly acquainted with the nervous exhaustion and overstimulation occasioned by modern warfare, as the "retreads" of war (returning soldiers struggling to adjust to normality) illustrate some of the unintended consequences of war's destructive acceleration of human experience. Next to the suburban unease of ex-servicemen that Gitre identifies, Rex Ferguson's essay on F. Scott Fitzgerald's *The Great Gatsby* explores how that novel's grand settings of East and West Egg, as well as the debonair personality of Jay Gatsby and the phantasmagoric imagery of his extravagant parties, provides the context for one of the most telling depictions of the relationship between modernity and waste in literature. He suggests that the major themes of the novel—such as purity, transience and idealism—are all, in fact, underpinned by the spectre of waste. Finally, Harvie Ferguson returns to the question of boredom with a more philosophical probing of the experiential conditions of modernity. His essay plots the ways in which lethargy, a lack of interest in life, and an engulfing sense of indifference to, and of, the world are bound together in a peculiar unity that is distinctive of modern society. He further suggests that while repulsion, defilement, and fear of contagion are possible experiences in any society, for modern society, they constitute the peculiar aesthetic-moral unity of disgust. In the case of both boredom and disgust, he argues, the significance of the peculiarity of their modern dimensions has to be seen through a phenomenology of waste.

The final three essays develop approaches to understanding material wastes and how we dispose of them. Tim Cooper's essay observes that while capitalist modernity may depend upon waste, it nonetheless claims to be able to reabsorb its own leftovers, and to reincorporate its own excretions. In a survey of the idea and application of recycling in twentieth century Britain, and its relationship to modern conceptions of waste, he ponders its role in the reinforcement of capitalist ideology. Jennifer Gabrys's essay examines the consequences of the explosion in electronics and products such as personal computers (including their components) to reveal the often invisible nature of its wastes and contaminants, and also how these particular wastes spread into distant "spaces of remainder" that extend the waste ecologies of western societies. By investigating these spaces of electronic waste, she argues, we can begin to re-map technology through its remainders. In the final essay, Ray Stokes and Stephen Sambrook take a rather different approach to understanding how modernity has dealt with the unwanted stuff of everyday life by identifying trends in the developing business and economy of waste management—that is, through the ways that waste, at the level of local level, is collected, processed and disposed of—by focusing on the British public sector governance of the waste business with specific reference to the city of Glasgow in Scotland, which saw itself as a pioneer in waste management.

It is hoped that these essays and the issues they raise not only enliven the reader's fascination with those parts of life, those spaces and behaviours we ordinarily consider to be negligible, but that through the diversity of its approaches it can stimulate further intellectual engagement about the lessons to be learned from modernity in relation to issues of environmental consciousness, time-awareness, and how we might live in less wasteful and more sustainable ways.

NOTES

[1] This is explored at length in John Scanlan, *On Garbage* (London: Reaktion, 2004).

[2] Recent examples include the Wellcome Collection's major London exhibition, "Dirt," March-August 2011; Michael Landy's exhibition / performance piece, "Art Bin," South London Gallery, January-March 2010; and Anthony Gormley's 2006 figure, "Waste Man," made from 30 tonnes of waste materials, which was temporarily erected next to the sea at Margate before being burned down, as intended, as part of Penny Woolcock's 2007 film, *Exodus*.

[3] On the rise and fall of the junk bonds market in the 1980s and 1990s see, Charles P. Kindleberger and Robert Z. Aliber, *Manias, Panics and Crashes: A History of Financial Crises,* 5th edition (Houndmills, Basingstoke: Palgrave, 2005), pp. 63-64.

[4] See, for example, Jonathan Chapman and Nick Grant, *Designers Visionaries and Other Stories: A Collection of Sustainable Design Essays* (London and Sterling, VA: Earthscan, 2007); Stuart Walker, *Sustainable by Design: Explorations in Theory and Practice* (London and Sterling, VA: Earthscan, 2006); Michael Pawlyn, *Biomimicry in Architecture* (London: RIBA, 2011).

[5] John Phillips, "Capri, the Island Drowning in a Sea of Rubbish," *The Independent,* 07 July 2005, p. 30.

[6] See Benjamin Miller, *Fat of the Land: Garbage in New York, The Last Two Hundred Years* (New York: Four Walls Eight Windows, 2000), pp. 279-287.

[7] On the metaphor of "waste" in *The Sopranos,* see William G. Little, *The Waste Fix: Seizures of the Sacred from Upton Sinclair to The Sopranos* (New York and London: Routledge, 2002), pp. 141-155.

[8] Lucy Walker, dir., *Waste Land* (London & São Paulo: Almega Projects / O2 Filmes).

[9] For an analysis of life within London's mid-Victorian economy of waste as seen through the writings of Henry Mayhew, see John Scanlan, "In Deadly Time: the 'Lasting-on' of Waste in Mayhew's London," in *Time & Society,* 16: 2/3, pp. 189-206.

[10] Lindsey Dillon, "Race, Waste, and Space: Brownfield Redevelopment and Environmental Justice at the Hunters Point Shipyard," in *Antipode* (2013), pp. 1-17.

[11] On Harare's Eastgate Building, see Karen Chung, "The Age of Techneco," in *Wallpaper,* 92 (2006), pp. 135-140; Martin Pawley, *Garbage Housing* (London: The Architectural Press, 1975).

[12] Foster + Partners, "Masdar Development. Abu Dhabi, United Arab Emirates, 2007-2008" (available at: www.fosterandpartners.com/projects).

[13] Lynn Spigel, "Designing the Smarthouse: Posthuman Domesticity and Conspicuous Production," in *Electronic Elsewheres: Media, Technology, and the Experi-*

ence of Social Space, eds. Chris Berry, Soyoung Kim and Lynn Spigel, (Minneapolis, MN: University of Minnesota Press, 2010), p. 62.

[14] "Industry Briefs," in *Building Systems Magazine* (March 2000), p. 10.

[15] Martin Heidegger, *The Question Concerning Technology and Other Essays,* trans. William Lovitt (New York: HarperPerennial, 1977), p. 16.

[16] The relationship between technology and waste is explored in Scanlan, *On Garbage,* pp. 80-88.

[17] On the idea of "surfacing," see John Scanlan, *Memory: Encounters with the Strange and the Familiar* (London: Reaktion, 2013), pp. 130-140; Bill McKibben, "Worried? Us?" in *Granta 83: This Overheating World* (Autumn, 2003), p. 8.

[18] See, for example, Herbert Marcuse, *One-Dimensional Man: Studies in the Ideology of Advanced Industrial Society* (Boston, MA: Beacon Press, 1964).

[19] Teresa Brennan, *Exhausting Modernity: Grounds for a New Economy* (London and New York: Routledge, 2000), p. 115.

[20] Michel Serres, *Malfeasance: Appropriation through Pollution* (Stanford, CA: Stanford University Press, 2011), p. 41.

[21] See, for instance, Zygmunt Bauman, *Wasted Lives: Modernity and its Outcasts* (Cambridge: Polity, 2004), Ch. 3, on the wastes of globalization.

[22] Peter Wynn Kirby, *Troubled Natures: Waste, Environment, Japan* (Honolulu: University of Hawai'i Press, 2011), p. 7.

[23] See the extensive discussion in Scanlan, *On Garbage,* pp. 22-33.

[24] Kirby, *Troubled Natures,* p. 4.

[25] See Scanlan, *Memory*, Ch. 3, "Ecologies."

[26] *Ibid.*, and my discussion in Scanlan, *On Garbage,* around the Greek idea of "techne," pp. 80-83. Exceptions to my general statement here would include recent work in sociology and cognate areas that questions the continuing usefulness of uncomplicated ideas of "agency" and "subjectivity," including: Harvie Ferguson, *Self-Identity and Everyday Life* (London and New York: Routledge, 2009); John Urry, *Mobilities* (Cambridge: Polity, 2007); Nigel Thrift, *Non-Representational Theory* (London: Routledge, 2007); and Bruno Latour, *Reassembling the Social: An Introduction to Actor-Network Theory* (Oxford: Oxford University Press, 2005).

[27] François Jullien, *The Silent Transformations,* trans. Krzysztof Fijalkowski and Michael Richardson (London, New York and Calcutta: Seagull Books, 2011), p. 6.

[28] Kirby, *ibid.*, pp. 1-27.

[29] See, for instance, Marshall Berman, *All That Is Solid Melts Into Air: The Experience of Modernity* (London: Verso, 1983).

[30] New standards of cleanliness not only make us aware of dirt and dust, but seem to give rise to new forms of waste. See Joseph A. Amato, *Dust: A History of the Small and the Invisible* (Berkeley, CA: University of California Press, 2000), pp. 80-81.

[31] Andrew Niccol, *Gattaca* (Los Angeles: Columbia Pictures, 1997).

[32] George Kubler, *The Shape of Time: Remarks on the History of Things* (New Haven, CT: Yale University Press, 1962), pp. 80-82.

Chapter One

Wasteland Mania: The Alternative Modernity of Peat Moss

Fredrik Albritton Jonsson

During the Scottish Enlightenment, agricultural improvers discovered a providential economy in the blanket peat bogs of the uplands. Where cursory observers saw only sterility and irredeemable waste, these naturalists and modernizing landowners claimed to uncover a secret abundance granted by divine mercy. Such sentiments peaked in the decades around the turn of the nineteenth century, when the northern and western edges of Scotland took on the guise of a promised land to many savants. Naturalists such as John Lightfoot and John Walker led the vanguard of improvement into the Highlands on voyages of internal discovery, cataloguing native species and compiling inventories of the natural advantages of the region. Curiously, much of their excitement about peat moss did not focus on the traditional use of peat as a fuel. Rather, they saw in blanket bogs a neglected source of soil fertility, an environmental fount of Gaelic virtue, or a removable cover hiding internal riches of fertile land. Such expectations played a crucial role in shaping the emerging new social and economic order of the Highlands and Hebrides at the end of the eighteenth century. The purported virtues of peat served to justify the specific system of land reclamation through spade agriculture which underpinned the spread of crofting settlements along the northwestern sea coast. Indeed, peat reclamation was fostered as a conservative form of population politics by landlords and intellectuals during the Scottish Enlightenment. This strategy set out to preserve and maintain a large traditional rural population by resettling Gaelic tenants and subtenants from the Highland interior to tiny plots of marginal soil along the Atlantic. The wastelands of the Atlantic coast were to serve as privileged sites of progress and virtue. For the defenders of the rural order, peat moss seemed to offer an alternative path to modernization, which would reconcile commercial profits with a traditional social structure.

Wasteland Theodicy

Peat bogs are the product of waterlogged and acidic conditions which prevent decaying vegetation from fully decomposing. The rock formations of the British uplands tend to be deficient in calcium, making local soils acidic. Such low pH levels, combined with an oceanic climate pattern of high rainfall in the west, favour the accumulation of peat bogs. Blanket bogs (mire) form in the uplands when soil conditions become waterlogged, depleting the oxygen and inhibiting decomposition. On average, blanket bogs grow to a depth of about two to four meters. At present, blanket bogs cover approximately ten percent of the land surface of Great Britain. The corresponding figure for the eighteenth century is not known but was presumably larger, because the conditions of the Little Ice Age are believed to have promoted the expansion of peat bogs in the British uplands. William Aiton's treatise on peat bogs asserted that the moors, wastes and bogs of Scotland amounted to approximately 14,200,000 (Scottish) acres around the turn of the nineteenth century. In this peculiar and difficult environment, a variety of *Sphagnum* bog mosses are among the few species to thrive. These surface plants in turn feed the underlying mire when they decompose, compacting gradually into peat bogs. Over time, the mass of peat tends towards a steady state where accumulation growth approaches zero when losses at the bottom balance additions on the surface.[1] From this perspective, one might describe peat as a form of ecological waste: organic matter which remains suspended in an intermediate stage of decomposition over centuries. Though such a state does not last indefinitely, the changes occur at a pace so slow that they seem utterly divorced from the cycles of husbandry practised in agriculture. Compared with the nutrient cycle of traditional farming, peat bogs appear quite stagnant and sterile.

Perhaps unsurprisingly, many early modern writers regarded peat bogs as irredeemable wastelands. They resorted to theology to explain the origins of such sterility. In texts such as Thomas Burnet's *Sacred Theory of the Earth* (1684) and John Woodward's *An Essay Towards a Natural History of the Earth* (1695), the wastelands of the world served as savage reminders of the Fall and the Deluge. After the expulsion from the Garden of Eden, humanity had been condemned to live on an Earth which refused to yield fruit without labour. The only path to the redemption of the soil lay through the arduous work of husbandry, but some regions of the Earth resisted the restoration of agriculture. In this sense, peat bogs constituted material reminders of moral pollution and divine condemnation. Even the coming of Christ had not redeemed them from God's wrath.[2]

Economic practice, of course, often diverged from the condemnations of the natural philosophers. Notwithstanding the spiritual stigma, peat was highly valued by the population of the British uplands. In a land long denuded of its original forests, peat served as a cheap substitute for wood fuel. Since the late middle ages, peat cutting had been conducted on a large scale by tenants and landlords across Britain. Such practices were in fact carried out across Europe wherever conditions favoured peat accumulation. Dutch historians have shown that peat served as a crucial energy source in the early modern Dutch Republic, sustaining a large urban economy in northern Europe. Later, peat cutting became one of the pillars of the crofting system in Highland Scotland and the Hebrides. Fraser Darling calculated that a crofting family had to cut 15,000 peats per year to sustain its meagre existence.[3] In the postwar Irish Republic, raised bogs became the site of industrial-scale extraction of peat, supplementing energy needs during the oil crisis of the 1970s. Currently, *Sphagnum* moss has become a horticultural mainstay, serving as a soil additive which increases the capacity of the soil to retain moisture and nutrients.[4]

The long-standing recognition of the practical value of peat spread to the intellectual elite in the seventeenth and eighteenth centuries. For a number of leading savants of the Scottish Enlightenment, the virtues of peat moss became a consuming fixation. Alongside the early modern identification of infertility with original sin, another powerful view of wastelands gradually emerged. This was the patriotic vision of a merciful God who had planted within every nation all the blessings necessary for its prosperity and power.[5] Rather than projecting loss and sin on the natural order, it emphasized the agricultural and horticultural dimension of God's covenant with his chosen people. In the writings of John Walker and his fellow improvers, infertility was reduced to a mirage. Wastelands were only superficially barren. The role of the natural historian was to detect the secret blessings hidden on the periphery. In this way, religion, nationalism, and natural history converged in a movement of land reclamation. In so far as Scotland was a chosen nation, Isaiah's promise to Israel directly applied to the peat bogs of the north and west:

> For the LORD shall comfort Zion: he will comfort all her waste places; and he will make her wilderness like Eden, and her desert like the garden of the LORD; joy and gladness shall be found therein, thanksgiving, and the voice of melody. Hearken unto me, my people; and give ear unto me, O

> my nation: for a law shall proceed from me, and I will make my judgment to rest for a light of the people.[6]

In the first flora of Scotland, the *Flora Scotica* (1777), John Lightfoot described the virtues of *Spagnum palustre* in terms of such a providential economy:

> It is generally believed that the roots and decay'd stalks of this moss constitute a principal part of that useful bituminous substance call'd peat, which is the chief fuel of the northern regions. So that those parts of the creation, which, to the inconsiderate minds of man, appear the most trifling and insignificant, will be often found, upon mature enquiry, to be ordain'd by Providence for the wisest and most gracious purposes.[7]

For Scottish naturalists and their landowning patrons, the mercy of providence was simultaneously material and moral. By linking virtue to the effects of a harsh environment, they were able to posit the reclaimed wasteland as a place of moral regeneration: this was an ecological theodicy. Through the discovery of numerous uses of peat, the apparent evil of the wasteland was eliminated, or rather, subsumed into a providential plan. The recognition of such a theodicy begins to explain why the Highland improvers went beyond the obvious use of peat as a fuel and attributed to it so many far grander powers. In addition, the Scottish fascination with the multiple virtues of peat bogs was sustained by wider cross currents of population politics, counter-revolutionary ideology, and anti-imperial internal colonization.

Three case studies highlight the material context of peat bog colonization. First, Henry Home, Lord Kames, sought to reclaim the great peat bog of Blair Drummond in the 1760s and 1770s. For Kames, peat excavation promised not only to uncover a massive tract of fertile land, but also a means to preserve what he took to be the dying traditions of Gaelic society. A generation later, the chemist Archibald Cochrane made the apparent fertility of peat a central component in his anti-radical ideology of rural modernization. Peat was quite literally supposed to prop up the landed order in an age of revolution and imperial crisis by regenerating exhausted and marginal soils in the Highlands. Finally, John Walker's essay on peat from 1803 completed the dialectic by fusing chemistry with the programme of cultural preservation. Walker explained the physical and moral superiority of the Gael in terms of the purifying environment of the northern peat bogs. By resettling the native population on the bogs, these virtues would be preserved and emigration prevented, even as new commercial pressures were brought to bear on the Highland economy.

These three cases were representative of a larger movement, which involved substantial portions of the intellectual and agrarian elite of Scotland, from the aftermath of the Jacobite rebellion of 1745 to the end of the Napoleonic Wars in 1815. These projects included Mr. Graeme's experiments in moss husbandry in the middle of the 1750s, the soldier settlements of the Annexed Estates in the next decade, Smith of Swinridgemuir's reclamation projects, David Young's 1785 treatise on internal colonization through spade culture, and the premiums for wasteland cultivation awarded by the Highland Society of Scotland in 1790s and early 1800s. Theoretical interest in peat moss reclamation peaked between 1794 and 1826 with important works on the subject by James Headrick, James Anderson, John Walker, Archibald Cochrane (Lord Dundonald), William Aiton, Robert Rennie, Allan Maconochie (Lord Meadowbank) and Andrew Steele.[8] Clearly this wave of treatises was spurred by the peculiar social, economic, and ideological pressures of the French and Napoleonic Wars. More generally, the wasteland mania after 1750 constituted the ideological origin of the crofting system which eventually came to dominate much of the northwest lands of Scotland in the nineteenth century.[9]

Peat and Agrarian Modernization

In current histories of the Scottish Enlightenment, Henry Home, Lord Kames (1696-1782), occupies a distinctly second-rate place as a defender of the polygenesis of races, a theoretician of aesthetics, and, perhaps most happily, a patron of Adam Smith and David Hume. His *The Gentleman Farmer* (1776) elicits little attention, despite its widespread appeal in the late Enlightenment. The subsequent significance of that other book, published in 1776 by Kames's client Smith, has, of course, done much to eclipse the reputation of *The Gentleman Farmer*. Yet for anyone interested in the wider currents of enlightened thought, Kames's agricultural theory and practice ought to be a central field of study. Two major themes of his involvement with the agricultural improvers stand out: on the one hand, his defense of the harmony of science and agrarian capitalism; on the other, his practical attempt to combat the problem of rural flight.

Kames was eager to render agriculture a philosophical pursuit. This meant above all the need to examine and integrate into farming practice the known principles of soil fertility and plant physiology. Here, Kames turned to the chemical and botanical expertise of his fellow Scotsmen Joseph Black (1728-1799) and John Walker (1731-1803). By grounding agriculture in the enlightened sciences, Kames sought to underscore the centrality of agricultural productivity to the social and political order.[10] *The*

Gentleman Farmer must be read as a defense of agrarian capitalism and the landed interest against the mercantile, urban, and industrial sectors of the economy. Crucially, this conservative agenda involved harnessing the fruits of modern science as a weapon of social entrenchment. Yet such an alliance between agriculture and philosophy raised a fundamental problem. How could the landed interest be safeguarded if modern agriculture involved the destruction of the old feudal order and the ruthless reduction of the rural population through enclosure and labour-saving techniques? For Adam Smith, such a structural transformation was no cause for sentimental regret. Higher profits in agriculture were simply predicated on the decrease of rural population: "the diminution of the number of cottagers and other small occupiers of land" acted "in every part of Europe" as "the immediate fore-runner of improvement and better cultivation."[11] Yet Smith's clinical analysis held little appeal for men like Kames, Walker, and the leading Highland landowners in the late Enlightenment. Instead they embarked on a series of experiments to balance the priorities of rural modernization against what they regarded as the moral, social, and military imperative of a large and stable Gaelic population in the Highlands.

In a pioneering effort to resolve this tension, Kames turned from theory to practice. As early as 1764, he began to organize a Herculean scheme of drainage and removal on his estate of Blair Drummond in the Scottish Lowlands. His primary objective was to uncover the fertile carse land underneath a 2,000-acre peat bog. This enterprise took more than thirty years of labour to complete and was hailed by John Walker as "the most singular and considerable piece of improvement that has yet been executed in any parish in Scotland." Walker and other admirers were particularly pleased with Kames's recruitment of Gaelic Highlanders to colonize the Blair Drummond peat bog. They praised this recruitment as a humanitarian project of cultural and moral preservation. In the 1790s, the Highland Society of Scotland—the leading body of agricultural improvement in the country—recognized Kames' achievement with a posthumous gold medal. According to the article on moss by Walker in the *Statistical Account of Scotland*, the resettled Gaels were men of "sober, frugal, and industrious" character "inured to hardships in their own country" and therefore "peculiarly qualified to encounter so arduous an undertaking." A few years later, William Aiton—another leading bog expert—compared the floating of Blair Drummond moss favourably to David Dale's industrial village at New Lanark outside Glasgow. Whereas Dale had recruited Highlanders to do manufacturing work, Kames's strategy had retained Gaels in the agrarian order and was therefore a far superior project. Significantly, peat bog reclamation was done through backbreaking manual labour with the spade (though Kames's son eventually installed a water pump to aid the drainage procedure). In other words, the brutal physical drudgery of the bog was here vested with a moral character found entirely lacking in factory labour. Walker emphasized how the distinct habitat of the peat bog reinforced traditional virtues:

> From their first settlement of the present day, not a single instance has occurred amongst them of theft, bad neighbourhood, or of any other misdemeanour, that required the intervention of the civil magistrate. Nor, however poor in circumstances, has any one of them ever stooped to solicit assistance from the funds of the parish appropriated to that purpose.

The lesson to the Highland improvers was obvious: peat bogs formed a viable basis for the retention and strengthening of a traditional rural order, even in the midst of a commercializing society.[12]

Counter-Revolutionary Peat

The eccentric ninth Earl of Dundonald, Archibald Cochrane (1748-1831), helped fashion a chemical justification for the centrality of peat moss to the social order in the 1790s. In the winter of 1795, Cochrane published *A Treatise, showing the Intimate connexion that subsists between Agriculture and Chemistry*. It was addressed to the "Cultivators of the soil" and the "Proprietors of Fens and Mosses" across the British Isles. Writing in the shadow of national grain shortages after 1791, which reached a dismal peak in the winter of 1794-95, Dundonald called for a fundamental shift in the attitude of both the government and the great proprietors of the nation. He insisted that the high price of provisions ought to rouse proprietors from "the golden dreams of manufactures and of commerce" to the realization that only agriculture could guarantee "the permanent prosperity of [the] nation." The problem was moral and political as much as economic. Because of the sensitivity of manufactures to the "fluctuation in demand" and the "total suspension of trade by war or other causes," workers were tempted by the promise of fundamental "political change." The dense concentration of labourers in manufactures and cities also exposed them to the "dissemination of pernicious doctrines, by a few profligate persons" who planted seeds of sedition among them. Dundonald's conclusion was startling but logical: the only proper remedy was to give preference to agriculture over all "other pursuits." Those "branches of manufacture" deemed "wise to encourage" should "be promoted [solely] in scattered villages, resembling the townships in America." By severing the link between cities and manufacture, "the diseases of the body and the mind would be rendered less contagious." The new policy also recovered the connection between the worker and the soil. By giving each operative a "sufficient extent of ground . . . requisite for raising potatoes, and other vegetables cultivated at his leisure by the spade," the manufacturing labourer would find "an agreeable and healthful change to his confined and sedentary occupation."[13]

Chemistry was the ally of the spade in curing sedition. Dundonald had pursued experiments in salt manufacture and the extraction of tar from coal on his estate in Culross for more than a decade before he published his treatise on agriculture. By the middle of the 1790s, his private affairs were in disarray and the prospects of his previous projects were precarious. He needed to raise his status among fellow landowners which might earn him new fame and better connections: hence the appeal to a neglected resource as the solution to wartime pressures of sedition and grain shortage. Moreover, his approach to soil fertility was permeated by a firm belief in the power of increasing returns. Whereas Adam Smith and many English agricultural improvers of the period believed that the supply of cattle manure set an absolute limit to soil fertility, Dundonald emphasized a rich variety of organic substances which might serve as substitute fertilizers. A crucial part of his survey focused on the relation between peat and topsoil. He claimed that they were harmoniously symbiotic in relation, such that peat moss could become the fertilizer of soil and vice versa. Following a line of chemical thinking which had been suggested a generation earlier by Francis Home, he rejected the fixed limit to cereal production which Smith and others had stipulated. Instead, Dundonald suggested that the peat mosses of Great Britain and Ireland opened up an "inexhaustible" supply of fertilizer. The discovery promised to alter fundamentally the economic order of all peripheries blessed with peat bogs. Here was a vision of environmental engineering which dwarfed the achievement of Kames on Blair Drummond moss.[14]

Dundonald's fascination with peat was, perhaps, an attempt to discover an agricultural equivalent to the revolutionary power of coal and steam in transforming output. In the same spirit, William Aiton's work on moss earth from 1805 and 1811, endorsed by the Highland Society of Scotland, pointed not only to the massive reservoir of fertile soils hidden beneath the Scottish peat bogs, but also boldly asserted that the removal of these bogs would fundamentally alter the climate of Scotland for the better. For both of these agricultural improvers, peat seems to have figured as a counterbalance to coal in the conservative imagination. Within the idealized rural system depicted in their treatises, peat served as the guarantor of the growth and modernity of the landed interest and therefore as a means of containing the challenge to the social order posed by the new mineral-industrial economy.[15]

Antiseptic Population Politics

During the Napoleonic Wars, the interest in the redemptive powers of peat moss reached a peak. Perhaps the most influential expression of this faith came with the natural historian John Walker's "Essay on Peat" from 1803. Writing as the Professor of Natural History at the University of Edinburgh, Walker offered a chemical justification not only for the fertility of peat moss, but also for its physiological effects on human populations. Indeed, he pointed to the antiseptic quality of moss water as the cause behind Gaelic racial particularities. Rather than defining peat moss colonization as a labour of removal, Walker, like Dundonald before him, suggested that peat could serve as a valuable element in cultivation. In this way, he presented a chemical and medical guide to peat bog settlement at the very moment when Highland landowners were beginning to introduce potato cultivation on marginal soil as a preferred mode of land use. In a letter to his friend, the chemist Joseph Black, Walker emphasized that his essay was designed for wide consumption by "Country Gentlemen and intelligent farmers" and therefore must avoid "as much as possible, all the Technical Terms in Natural History and Chemistry." Walker completed and published the text during the height of the emigration panic of 1802-03, when Scottish elites worried that a massive number of Highlanders was on the verge of emigration to America. This certainly seems to explain why Walker framed the habitat of the peat bog as a substitute New World. Tellingly, the "Essay on Peat" was embraced by the Highland Society of Scotland and received a hefty premium as the best guide written on the subject.[16]

Walker began the essay by considering the chemical composition of peat moss. Contrary to common opinion, he found peat bogs to be a healthy environment and marveled at the purifying properties of moss in water. Even though this "dark brown infusion" was stagnant "during the heats of summer," it remained "perfectly sweet" when torpid water elsewhere by necessity grew putrid. Captain Cook had discovered this antiseptic quality of peat infusions during his circumnavigation and made good use of such water to maintain his crew in health. From these observations, Walker arrived at a chemical explanation for the health and vigour of the Highlanders. Whereas stagnant water without peat inevitably bred noxious miasma in the carse countries and fenlands of the plains, the "neighbouring uplands" were exempt from contagion. Among the "moors and mosses" of the Highlands, putrid fever, intermittent fever, and "putrid soar throat" were all unknown. Evidently, moss water had a "powerful influence in preserving vegetable and animal substances from corruption," whether animate or inanimate. The same "bitumen" which preserved peat "from the ultimate stage of putrefaction" also "rendered [it] capable of preserving other bodies." Clearly, reports about how corpses resisted decomposition in peat bogs (an effect of tannins in sphagnum moss) served as an inspiration for Walker's analysis. He simply extrapolated that the same restorative effect would also influence living bodies.[17]

Next, Walker turned to the question of the fertility of peat bogs. His approach was more circumspect and cautious than that of Dundonald, perhaps conditioned by the habits of the botanical taxonomist rather than the chemical projector. The essay compiled a typology of seven kinds of peat, and an inventory of species native to these habitats. Although he affirmed the widespread belief that peat bogs had their origin in decaying timber from ancient woods, he also hinted at the succession of plants necessary for the formation of peat. Among the eighty-four "most copious and important Plants" identified by Walker's inventory, more than a dozen of the species had been rated as useful in Lightfoot's *Flora Scotica*. In Walker's opinion, the best prospects lay in harvesting indigenous grasses like *Aira coerulea* (Fly Bent) as hay, or in introducing "exotic cultivated plants" such as turnips, beans, Friesland oats, and potatoes.

To any proprietor still uncertain about the merits of spade cultivation, Walker urged the case of Lord Kames' moss excavation. He recommended that every landowner should follow Kames' example and gather a "colony" of "moss tenants" to settle peat bogs on his estate. These settlers would at the same time advance their own interest in finding subsistence, improve the long-term value of the property, and increase the food supply of the nation in a time of grain shortages. Bog labour was particularly suited for the poorest class of people, since it required "no stock, no cattle, nor any implement but [the] spade." It was also a manner of living "highly conducive to the population of the country" since it allowed "the lower people" to marry and form families. Finally, "such an opportunity" would banish "all thoughts of emigration . . . especially . . . in the Highlands." For all of these reasons, peat bog cultivation was the most patriotic strategy of estate management possible:

> Such a sudden and striking transformation in the soil of any such district, is not to be met with in Britain, nor probably any where in Europe. This is domestic conquest, far preferable to one of a foreign nature. If this is not patriotism, one may be at a loss to know what patriotism is.[18]

Yet Walker's preferred method of moss culture was plebeian rather than patrician in origin. He had noticed the practice of lazy bed cultivation on the island of Islay and elsewhere on his naturalist voyage through the Hebrides in 1764 (commissioned by Lord Kames and the Board for the Annexed Estates). This form of cultivation raised a "bed" perhaps six feet wide, composed of layers of seaweed and peat moss, with trenches on both sides to drain off water. Walker noted that the Irish had grown large crops of potatoes in lazy beds on bogs for many years and found several advantages to this system. The labour of the spade in the "soft peat" was "very inconsiderable" by comparison with other forms of digging. It also produced a superior quality and quantity of potatoes. For this reason, Walker recommended that even gentlemen farmers should cultivate them with the spade rather than the plough. In addition, he observed that the lazy bed destroyed heather and "matted turf" by increasing the rate of their putrefaction. He recommended the use of lazy beds as "the first and most effectual method of culture in reclaiming many sorts of wild land." Once "completely rotted," the underlying moss would form a fertilizer in its own right, far superior to the ashes produced by burning the top layer.[19] Lazy beds thus seemed to form the first step in an extended process of wasteland reclamation. This meant that Walker's manual provided landlords with the rationale of long-term soil improvement even as they confined Gaelic tenants to marginal soils. By reformulating the rhetoric of internal colonization in a minimalist vein, Walker thus dressed up the new crofting system as a form of ethnic conservation and moral idealism. Whereas Dundonald and Aiton imagined a world of resurgent agrarian capitalism based on bog removal, Walker embraced peat as a source of virtuous poverty. The riches of the Highlands were to be found not beneath the bogs but in the simultaneous conservation of the native population and its boggy habitat.

Conclusion

Though the crofting movement pioneered by Kames, Walker and their followers might seem a peculiar aberration born out of strange local circumstances, it was in fact part of a wider pattern in enlightened culture. Across Europe and its peripheries, improvers appealed to the authority of nature to explain the workings of the economy and to justify programs of reform. While particular policies were contingent on local conditions, a common core of principles connected Scottish naturalists, Swedish cameralists, French physiocrats, agrarian republicans in Virginia, and colonial administrators from Bengal to Mauritius. In each case, agriculture was held to offer the fundamental source of economic growth. [20] The lens of nature was also applied to understand patterns of circulation, the division of labour, population growth, and modes of subsistence. Natural, political and divine economies seemed inextricably intertwined. In part, such intellectual uniformity may have been a response to greater environmental pressures within the advanced organic economy before industrialization.[21] In part, one might also understand the enlightenment penchant for natural economies as an attempt to rearticulate and reanimate the ideology of agrarian capitalism by framing it as a science of nature and human nature. Notwithstanding the humanitarian rhetoric of the improvers, natural economies were realized only through backbreaking labour. Lazy beds were hardly the first choice of the poor. Bullying threats of eviction prompted the exodus to the peat bogs. The redemption of peat was predicated on a system of organized coercion. Those among us who are interested in exploring the ecological basis of economic life do well to remember that dreams of waste easily beget violence and oppression.

Notes

[1] Alan H. Fielding and Paul F. Haworth, *Upland Habitats* (London: Routledge, 1999), pp. 30-33; R. S. Clymo, "The Limits to Peat Bog Growth," in *Philosophical Transactions of the Royal Society of London. Series B, Biological Sciences*, 303, 1117 (January 30, 1984), pp. 606-07; I.G. Simmons, *An Environmental History of Great Britain* (Edinburgh: Edinburgh University Press, 2001), pp. 39-40, 114; William Aiton, *A Treatise on the Origin, Qualities, and Cultivation of Moss-Earth* ... (Air, 1811), pp. xxi; T. C. Smout, *Nature Contested: Environmental History in Scotland and Northern England Since 1600* (Edinburgh: Edinburgh University Press, 2000), pp. 52-3. One Scottish acre was the equivalent of 1.3 English acres.

[2] Simon Schaffer, "The Earth's Fertility as a Social Fact in Early Modern England," in *Nature and Society in Historical Context*, ed. Mikuláš Teich, Roy Porter and Bo Gustafsson (Cambridge: Cambridge University Press, 1997); Clarence Glacken, *Traces on the Rhodian Shore: Nature and Culture in Western Thought from Ancient Times to the End of the Eighteenth Century* (Berkeley, ca: University of California, 1967), pp. 379-82, 406-15.

[3] The exact quantity of "a peat" as a unit of measure remains difficult to ascertain precisely. Darling, for instance, claims "that a strong man could cut 1,000 peats in a day." See Fraser Darling, *Crofting agriculture: Its practice in the West Highlands and Islands* (Edinburgh, 1945), p. 5.

[4] I. G. Simmons, *The Moorlands of England and Wales; An Environmental History 8000 bc to ad 2000* (Edinburgh: Edinburgh University Press, 2003), p. 66, 97; J. W. De Zeeuw, "Peat and the Dutch golden age: the historical meaning of energy attainability," *A.A.G. Bijdragen*, 12 (1976), pp. 3-31; Richard W. Unger, "Energy Sources for the Dutch Golden Age: Peat, Wind, and Coal," *Research in Economic*

History, 9 (1984), pp. 221-253; Kevin C. Kearns, "Development of the Irish Peat Fuel Industry," *American Journal of Economics and Sociology*, 37, 2 (April 1978), pp. 184-85. For additional traditional uses of peat in the Scottish economy, see William Milliken and Sam Bridgewater, *Flora Celtica; Plants and People in Scotland* (Edinburgh: Birlinn, 2004), pp. 271-73.

[5] Lisbet Koerner, *Linnaeus: Nature and Nation* (Cambridge, ma: Harvard University Press, 1999), pp. 92-93; Alix Cooper, *Inventing the Indigenous: Local Knowledge and Natural History in Early Modern Europe* (Cambridge: Cambridge University Press, 2007) p. 24; E. C. Spary, "'Peaches Which the Patriarchs Lacked': Natural History, Natural Resources, and the Natural Economy in France," in Margaret Schabas and Neil De Marchi, eds, *Oeconomies in the Age of Newton: Annual Supplement to Volume 35, History of Political Economy* (Durham, nc: Duke University Press, 2003), pp. 25-27; Smout, *Nature Contested*, p. 20; John Scanlan, *On Garbage* (London: Reaktion, 2005), pp. 22-25; T.C. Smout, "The Improvers and the Scottish Environment: Soils, Bogs and Woods," in T. M. Devine and J. R. Young, eds, *Eighteenth Century Scotland* (East Linton: Tuckwell Press, 1999), pp. 217-19.

[6] Isaiah 51: 3-4.

[7] Lightfoot, John, *Flora Scotica: or, a Systematic Arrangement in the Linnaean Method of the Native Plants of Scotland and the Hebrides*, 2nd ed., 2 vols (London: R. Faulder, 1789), 2, p. 692. Compare James Headrick, *Essay on the Various Modes of Bringing Waste Lands into a State Fit for Cultivation* (Dublin, 1801), p. 105.

[8] Christopher Tait, *An Account of the Peat Mosses of Kincardine and Flanders in Perthshire: from the Transactions of the Royal Society of Edinburgh* (Edinburgh, 1792); James Anderson, *A practical treatise on peat moss: considered as in its natural state fitted for affording fuel, or as susceptible of being converted into mold capable of yielding abundant crops of useful produce: with full directions for converting it from the state of peat into that of mold, and afterwards cultivating it as a soil* (Edinburgh, 1794); Archibald Cochrane (Earl of Dundonald), *A treatise, shewing the intimate connection that subsists between agriculture and chemistry: addressed to the cultivators of the soil, to the proprietors of fens and mosses, in Great Britain and Ireland, and to the proprietors of West India estates* (London, 1795); Anon, *An Account of the improvement of moss, &c. in a letter to a friend,* 2nd ed. (Dunse, 1797); Headrick, *Essay on the various modes ...;* Lord Meadowbank (Allan Maconochie), *Directions for making compost dunghills from peat-moss* (Edinburgh, 1802); John Walker, "An Essay on Peat ... " in Henry Mackenzie, ed. *Prize Essays and Transactions of the Highland Society of Scotland*, vol. ii (Edinburgh, 1803), pp. 1-137; William Aiton, *A treatise on the origin, qualities, and cultivation of moss-earth* (1805; Glasgow, 1811); Robert Rennie, *Essays on the Natural History and Origin of Peat Moss* (Edinburgh, 1807); John Naismith, "An Essay on Peat, its Properties and Uses" in *Prize Essays and Transactions of the Highland Society of Scotland*, vol. iii, ed. Henry Mackenzie (Edinburgh, 1807), pp. 17-85; Anon (Allan Maconochie), *Directions for preparing manure from peat: instructions for foresters* (Edinburgh, 1815); Andrew Steele, *The Natural and Agricultural History of Peat-moss or Turf-bog* (Edinburgh, 1826).

[9] For the economic and social structures of crofting, see James Hunter, *The Making of the Crofting Community* (Edinburgh: John Donald, 1976); T. M. Devine, *The Transformation of Rural Scotland: Social Change and the Agrarian Economy, 1660-1815* (Edinburgh: Edinburgh University Press, 1994), pp. 64-5; Robert A. Dodgshon, *From Chiefs to Landlords: Social and Economic Change in the Western Highlands and Islands, c. 1493-1820* (Edinburgh: Edinburgh University Press, 1998). On the ideological roots of crofting in the Scottish Enlightenment, see Andrew Mackillop, *"More Fruitful than the Soil": Army, Empire and the Scottish Highlands, 1715-1815* (East Linton: Tuckwell, 2000), pp. 80-85, 89-91; as well as my thesis, "The Enlightenment in the Highlands: Natural History and Internal Colonization in the Scottish Enlightenment, 1760-1830," University of Chicago, 2005.
[10] Henry Home, *The Gentleman Farmer* (Edinburgh: 1776), p. xii, xvii; Schaffer, "The Earth's Fertility as a Social Fact in Early Modern England"; Jan Golinski, *Science as Public Culture; Chemistry and Enlightenment in Britain, 1760-1820* (Cambridge: Cambridge University Press, 1992), pp. 31-33, 35, 39; Sarah Wilmot, *The Business of Improvement; Agriculture and Scientific Culture in Britain c. 1700-c. 1870* (Historical Geography Research Series, 1996), pp. 36-38.
[11] Adam Smith, *An Inquiry into the Nature and Causes of the Wealth of Nations*, 2 vols, eds, R. H. Campbell, A. S. Skinner, W. B. Todd (Oxford: Oxford University Press, 1976), vol. i, p. 133, 243.
[12] John Walker, "Account of the Moss of Kincardine," in Sir John Sinclair, *The statistical account of Scotland. Drawn up from the communications of the ministers of the different parishes, …*, 21 vols (Edinburgh, 1791-99), vol. xxi, Appendix, pp. 173-4; Aiton, *A Treatise on the Origin, Qualities and Cultivation of Moss Earth …*, pp. 341-42, quoted in T. C. Smout, *Nature Contested*, p. 20; Edinburgh University Library (EUL) la iii 352/1b 153. For a longer discussion of Kames's project in Blair Drummond moss, see chapter three of my thesis.
[13] Archibald Cochrane, *A Treatise, shewing the Intimate Connection that subsists between Agriculture and Chemistry*, pp. 2-5.
[14] Cochrane, *A Treatise …*, p. 3, 7, 178; Francis Home, *The Principles of Agriculture and Vegetation* (Edinburgh, 1756), p. 23.
[15] William Aiton, *A Treatise on the Origin, Qualities and Cultivation of Moss Earth,* p. xxiv. Wrigley has argued that the "advanced organic economy" blinkered the imagination of the classical political economists and blinded them to the power of coal and steam. My interpretation of Dundonald's thought suggests that at least some agrarian writers may have been looking actively for ways to propel agriculture towards the growth rates of the "mineral energy economy." See E. A. Wrigley, *Continuity, Chance and Change: The Character of the Industrial Revolution in England* (Cambridge: Cambridge University Press, 1988).
[16] Walker, "An Essay on Peat," pp. 3-137; National Archives of Scotland, rh4/188/1 Sederunt Book 3, 412; Edinburgh University Library (EUL), GEN 874/iv 51-2; EUL LA iii 352/1b, 181.
[17] Walker, "An Essay on Peat," pp. 48-55; compare with Nicholas Turner, *An Essay on Draining and Improving Peat Bogs* (London, 1784), pp. 38-40. For bodies preserved in peat, see also James Headrick, *Essay on the Various Modes …*, p. 73. There was widespread interest in putrid fevers and antiseptic cures among eight-

eenth-century doctors such as Charles Alston, William Wright, William Alexander, William Falconer, William Fordyce, and John Pringle. See for example Charles Alston's *A Dissertation on Quick-Lime and Lime-Water* (Edinburgh, 1752), pp. 31-33.

[18] Walker, "An Essay on Peat," p. 83, 102-3.

[19] Walker, "An Essay on Peat," pp. 62-72, 90-94. On lazy bed cultivation with turf and seaweed, see Fraser Darling, *Crofting Agriculture*, p. 8; William Milliken and Sam Bridgewater, *Flora Celtica*, p. 252.

[20] Emma C. Spary, "Political, Natural and Bodily Economies," in *Cultures of Natural History*, ed. Nicholas Jardine, James A. Secord, and E. C. Spary (Cambridge: Cambridge University Press, 1996), pp. 178-196; Lorraine Daston, "The Nature of Nature in Early Modern Europe," in *Configurations* 6 (1998), pp. 149-172; Margaret Schabas, *The Natural Origin of Economics* (Chicago, IL: University of Chicago Press, 2005); Catherine Larrère, *L'invention de l'économie au XVIIIe siècle* (Paris: Presses Universitaires de France, 1995); Keith Tribe, *Land, Labour and Economic Discourse* (London: Routledge, 1978); Paul B. Wood, "The Natural History of Man in the Scottish Enlightenment," in *History of Science,* 27 (1989), pp. 89-123; Schabas and De Marchi, eds, *Oeconomies in the Age of Newton*; Richard Grove, *Green Imperialism; Colonial Expansion, Tropical Island Edens and the Origins of Environmentalism, 1600-1860* (Cambridge: Cambridge University Press, 1995); John Gascoigne, *Science in the Service of Empire; Joseph Banks, the British State and the Uses of Science in the Age of Revolution* (Cambridge: Cambridge University Press, 1998); Richard Drayon, *Nature's Government; Science, Imperial Britain, and the "Improvement" of the World* (New Haven, CT: Yale University Press, 2000); Lisbet Koerner, *Linnaeus: Nature and Nation* (Cambridge, ma: Harvard University Press, 1999); Drew McCoy, *The Elusive Republic: Political Economy in Jeffersonian America* (Chapel Hill, NC: The University of North Carolina Press, 1980); Ranajit Guha, *A Rule of Property for Bengal: An Essay on the Permanent Settlement* (Durham, NC: Duke University Press, 1996).

[21] Wrigley, *Continuity, Chance and Change*; Kenneth Pomeranz, *The Great Divergence; China, Europe, and the Making of the Modern World Economy* (Princeton, NJ: Princeton University Press: 2000).

Chapter Two

Landscape and Waste: Hydroelectric Development in the Highlands of Scotland[1]

Jill Payne

Landscape and Wasteland

The state-sponsored Scottish hydroelectric programme that emerged from Britain's post-war reconstruction drive reflects the extent to which the application of ideas of "waste" influenced landscape change in the twentieth century. As the landscape architect Mira Engler suggests, "waste is a key element in the ways in which we order and shape our environments, and it is inseparably intertwined and evolving with those environments."[2] The nineteenth-century Romantic movement, in determining a newly-positive attitude towards "natural" and "wild" landscapes, fuelled an increasingly entrenched idea of landscape "development" as landscape "degradation." The concept of development transforming "landscape" into "wasteland," in addition to emergent concerns regarding waste as a by-product of development, stands in stark contrast to the concurrent, post-Industrial Revolution, utilitarian emphasis on development-as-progress and the explicit equation of non-development with stagnation and a sense of opportunity lost. The Scottish Highlands was one of a number of European and North American regions which became, as technology initiatives extended beyond nineteenth-century confines, vulnerable to bitterly contested, land-hungry energy development proposals. The utilitarian, pro-hydroelectric development conviction that unregulated Highland water systems were "running to waste" was challenged by landscape protectionists who argued that artificial constructions would inevitably detract from the aesthetics of significant "natural" Highland landscapes, turning them into "waste dumps." The incompatibility of twentieth-century energy requirements with post-Romantic landscape aesthetic ideals highlights tensions and conflicts not only between land-use and landscape appreciation, but between landscape appreciation and landscape protection, drawing attention to the inflexibility of landscape ideals constructed in, and still reflective of, the values and priorities of the nineteenth century.

The post-war proposals for large-scale hydroelectric development in the Highlands gained impetus from socio-economic concerns stretching back to the eighteenth century. Scotland had experienced a protracted period of economic decline and the Highlands in particular lagged behind the remainder of Britain in terms of basic living standards, employment levels and industrial development. The growing power of Scotland's political left during, and subsequent to, the inter-war years focused attention on this imbalance. As the Second World War drew to a close, Scottish interests within Churchill's war cabinet ensured that the Highland region was incorporated within country-wide post-war reconstruction initiatives. Hydroelectric development utilised two resources abundant in the Highlands, open space and water, and, since the turn of the century, had appeared to represent a viable solution to a swathe of problems. The advantages, laid out in utilitarian terms, were clear-cut: the electrification of rural communities; the expansion of energy-intensive industrial initiatives and related benefits of increased employment opportunities; and an end to population decline.

However, the implementation of hydroelectric proposals, so clearly advantageous from a utilitarian perspective, proved highly problematic. Many of the Highland sites most suited to hydroelectric development were landscapes valued for their natural beauty and socio-cultural associations. Herein lies the enhanced conundrum attached to the aesthetic appreciation of the Highlands. Landscapes may be categorised as "central" or "marginal" by the groups that prioritise them, and accordingly ascribed positive and negative value.[3] At the same time, David Matless emphasizes the complexities that arise when a specific landscape reflects dual (or multiple) categorisation and evaluation by contradictory interests.[4] Highland landscape protection concerns conflicted with more than contemporary economic expansion. The viability of hydroelectric power development as an antidote to socio-economic decline was complicated to a considerable degree by the post-1745 evolution of an aesthetic appreciation of a humanized landscape in decline. The Highland landscape aesthetic at its most undiluted therefore conflicted with efforts intended to raise the Highland way of life onto a par with other, more developed, areas of Britain.

The "Highland Myth" and the "Highland Problem"

The Highland region had begun to assume wider aesthetic significance in the late eighteenth and nineteenth centuries, when, as part of the backlash against industrialization, Romantics extended their travels in search of non-industrialised spaces representative of the "natural" world, as opposed to the world that had been made "unnatural" by the machine age. Factors such as rural depopulation and a lack of all but the most primary of industries, so central to the economic decline of the region, made for a countryside conforming visually to emergent post-Industrial Revolution "Romantic" and "wilderness" ideals evolving concurrently with and, in a sense, critical to the development of the "Highland Myth."

The "Highland Myth" is the ideological antithesis of the "Highland Problem"; much of what underpinned twentieth-century appreciation of certain significant landscapes in Scotland can be retraced to this.[5] Womack locates the origins of the "Highland Myth" in the years following the final quelling of the Jacobite uprising by the Hanoverian government forces in 1746.[6] With the Highlands ceasing to represent a threat to the state (and with the Clearances reducing the humanized aspects of the landscape) it became possible to celebrate rather than fear the "otherness" of the region, with the growing Romantic movement serving to incline this celebration towards an admiration of real (increasingly de-peopled and re-naturalised) prospects glamorised by quasi-historical events and figures.[7]

Considerations of the original meanings of "waste" reflect negative connotations of "desolation" and "wilderness" that, on the back of Romanticism, began to undergo significant re-evaluation.[8] During the latter half of the eighteenth century the perception—on the part of the English elite—of the Highlands as a region of dreary mountainous wastes gave way to an appreciation of Highland scenery for the "sublime" experiences of the grandeur of "Nature" it afforded.[9] This change, deeply intertwined with the "Highland Myth," marked the start of the positive appreciation of Scotland's landscape by outsiders: an extended evolution that incorporated the later eighteenth- and nineteenth-century admiration for the "picturesque," Sir Walter Scott's romanticized re-workings of the historical landscape and the approbation of the Victorian sporting classes, including Queen Victoria herself.[10] By the twentieth century, this "culture of landscape," to use Matless's expression, had further diversified; outsider appreciation included growing numbers of Scots, national, and international tourists intent on celebrating the visual delights of the region both as relatively passive onlookers or viewers, and as participants in outdoor activities such as hiking, cycling and mountain climbing.[11] The "wild" or "unspoiled"

Highland landscape so appreciated by these visitors (both onlookers and participants) was socio-economically moribund, re-naturalised rather than natural, and one to which economic uplift and repopulation would inevitably bring change.

The inevitably large-scale structures accompanying hydroelectric development—dam walls, power houses, pipelines and transmission lines, not to mention the scars left by the actual construction sites—threatened to stand out in stark contrast to this backdrop.

Hydroelectric Development and Landscape Protection

Aesthetic degradation caused by the first hydroelectric enterprises therefore initiated some of the earliest ideas regarding the need to protect what were by now considered "natural" Highland landscapes. Steered by private enterprise and unfettered by legislative constraints, the British Aluminium Company redirected river systems and left scenic waterfalls dry or submerged, and watercourses severely reduced in flow. Subsequent hydroelectric proposals were countered with calls for the protection of both Scotland's landscape and Scotland's natural resources from commercial exploitation. Here, landscape protection interests merged with those of the Scottish political Left, which emphasised hydroelectric development as a vehicle for social reform and Scottish (as opposed to British) economic fortification. In the 1940s, the metropolitan marginality[12] of the Highlands, simultaneously celebrated and regretted, was fenced off from private utilitarian exploitation when the management of hydroelectric development was ceded to a public corporation, the North of Scotland Hydro-electric Board (NSHEB). Over the following thirty years, the NSHEB became the architect of wide-scale hydroelectric development—and a significant catalyst for the emergence of Scotland's landscape protection movement.

In countering arguments in favour of hydroelectric development, so clearly and confidently expressed in the language of cost-benefit analysis and social upliftment, landscape protectionists' original resort was a difficult-to-articulate conviction regarding the value of "natural' landscape, free of artificial constructions and constraints, to the human condition. In the case of many individuals, this aesthetic conviction was offset by a sincere and realistic appreciation of the then socio-economic climate. The result was that, where hydroelectric expansion in the Highlands was concerned, landscape protectionists compromised their aesthetic ideals. In a number of instances, the compromises were made relatively willingly. Sir John Stirling Maxwell was one of many protectionists who genuinely appreciated the need for socio-economic improvement in the Highlands. Even those who did not necessarily place liberal views of social justice ahead of aesthetic integrity were increasingly anxious not to be seen to stand in the way of a means to an improved standard of living for Scots in general and Highlanders in particular.

On the occasions when aesthetic ideas were prioritized, the standoff between developers and landscape protectionists forced the latter to seek to clarify and articulate their position in a manner that could be considered positively within the confines of a state system struggling to recover from the combined impact of two world wars and a depression. From the 1930s onwards, the landscape protection movement began to connect more fully with the socio-cultural perspective of the emergent welfare state. The Scottish landowning elite, outdoors and sporting enthusiasts, and wealthy outsiders heavily influenced the movement during the early years. It became clear during the Second World War that the social tide was turning against this demographic and it became apparent to those involved that they needed to both widen their appeal and come to terms with the new socio-economic direction taken by post-war Britain. The perception that they were standing in the way of "progress" or development would not have given them any advantage in the eyes of the public. As a result, increasingly "democratic" concerns about landscape change were utilized by the Association for the Protection of Rural Scotland (APRS) and National Trust for Scotland (NTS) as plausible rationales for the protection of what would otherwise have been viewed as an elitist preoccupation with beautiful scenery, in as much that it is largely those whose material needs are sufficiently met who are able to prioritise the aesthetics of their wider surroundings. Direct engagement with a high-profile social issue helped landscape protection to become less peripheral; hydroelectric development did not initiate landscape protection in Scotland, but it played a substantial part in its subsequent evolution.

The discourse reflects a measure of reconciliation between land-use and landscape appreciation interests between 1943 and 1962. There was a distinct improvement in relationship between NSHEB and the protection organisations. The planning process became more transparent, and the NSHEB became more open to suggestion and negotiation. The NSHEB allowed for few compromises where some of its earliest projects were concerned (Sloy and Tummel-Garry), but the castigation it received during public inquiries (Tummel-Garry) meant that it went to some lengths to satisfy concerns about the degradation of Glen Affric, a glen valued for its "picturesque" beauties.

Pressure from landscape protection groups also led to the modification of the visual impact of the hydroelectric structures. Pipelines placed above ground in the early years, as they were in the pre-NSHEB Galloway Scheme and at Sloy, were later painstakingly undergrounded. The same treatment was given to transmission lines in areas judged to be of aesthetic significance. Power stations became more sensitively constructed, with a view to integration within the wider landscape. An architect-designed stone building was created to house the Fasnakyle power station at Glen Affric and to promote what it was hoped might become a revived Scottish stonemasonry industry.

These conciliatory gestures on the part of the NSHEB stand out as cosmetic adjustments that did little to interrupt the momentum of a particularly intense period of hydroelectric expansion. Opposition to hydroelectric development on the grounds of aesthetic diminution seldom actually prevented the NSHEB from achieving its objectives. Schemes such as Sloy, Tummel-Garry and Affric-Farrar-Beauly were rapidly implemented, even though proposals on this scale were held at bay for years before the NSHEB was established. The relatively minimal changes affected at Glen Affric, its most showcased aesthetic compromise, came at a heavy cost to the neighbouring, less renowned Glen Cannich, where the creation of the enormous Loch Mullardoch engendered significant landscape change. The NSHEB was also not prepared to concede to subsequent requests to retain Glen Strathfarrar as the remaining "east-west" glen untouched by hydroelectric development.

State support for hydroelectric development became more muted in the 1960s, signalling a deceleration in hydroelectric expansion at a time when the argument over land-use and landscape protection in the Highlands was focusing on a proposal to develop Glen Nevis. The Macmillan administration's accounting drive gave rise to the understanding that, in terms of economic costing, hydroelectric generation and transmission was inefficient. Energy provision strategies (now taking shape within a Cold War climate preoccupied with nuclear technology) increasingly began to take the new costing measures into account, making it almost inevitable that the Mackenzie Commission, which released its conclusions in 1962, would find in favour of downscaling the NSHEB's operations. However, it is also the case that, by the early 1960s, landscape protection had reached the point where its support base could not be ignored. The hydroelectric threat to the iconic landscape of Glen Nevis, in particular, meant that the movement attracted more backing than it might otherwise have done.

Anti-hydroelectric development protest was arguably at its most successful a decade on from the publication of the Mackenzie Report, when it focused on a hesitantly-publicized NSHEB plan to harness the waters of Loch Lomond and create a pumping station inside Ben Lomond. Concrete evidence of success is difficult to establish. The NSHEB was aware of the controversy that the further utilization of Loch Lomond would engender, with one spokesperson referring to "this can of worms," and another clarifying how "if we could have done it elsewhere, then obviously we should prefer to avoid walking into this particular lion's den."[13] The Craigroyston plans were never published and neither developers nor protestors were called upon to defend their perspectives at a public inquiry. Nonetheless, the 1970s outcry in defence of an undeveloped Loch Lomondside reflected the unresolved issues of previous decades and engaged with these within the context of the new climate of environmental consciousness that was to remain a characteristic of the remaining decades of the century. The backlash against Craigroyston therefore highlights the appreciation of, and the broad will to protect, the Highland landscape at the end of the "NSHEB era."

In 1989, the NSHEB was privatized. It became known as Scottish Hydro Electric, prior to the general amalgamation that formed Scottish and Southern Energy plc in 1998. Although it provides energy wholesale to other suppliers under its new name, it is as Scottish Hydro Electric that it continues to supply domestic consumers. The NSHEB's social responsibilities were removed at privatisation, but the company wished to continue to emphasise its past social and community orientation.[14]

Romanticism, Wasteland and Landscape Appreciation

In terms of language, contemporary landscape appreciation no longer employs strictly Romantic categorization as a means of assessing landscape quality. The term "picturesque" remains in general use as a means of describing aesthetically pleasing landscape.[15] The terms "sublime" and "terrific" are, in their original meanings, largely out of circulation. At the same time, the late twentieth-century appellation "wild land" reflects a similar appreciation of something lying beyond the control of humans, and for this reason awe-inspiring. The appreciation of "wild land" in Britain emerged from the appreciation of "wilderness" that had begun to gel in the later decades of the nineteenth century, particularly in North America. Here, a Romanticized appreciation of the vast forested areas with which European colonists wrestled influenced the transcendentalist thinking of Ralph Waldo Emerson and Henry David Thoreau, and the subsequent celebration of wilderness engendered by John Muir and Aldo Leopold.[16]

"Wilderness" thinking re-echoed back in Muir's native Scotland, but in some respects remained equable to "wasteland," with utilitarian undertones, for much of the twentieth century. Glen Cannich, although appreciated in some interwar circles, and in 1936 included in the APRS's list of landscapes threatened by hydroelectric development, lacked the precise mix of loch and woodland that lent such a "picturesque" quality to Glen Affric. Less wooded, loch-less Cannich was utilized for hydroelectric development in the interests of protecting Romantic Affric. However, by 1978 the Countryside Commission for Scotland was including Cannich alongside Affric and Strathfarrar in its list of Highland glens affected by hydroelectric development.

Towards the 1990s, the Romantic circle closed. Along with the understanding that, since probably no area of Britain could correctly be termed "wilderness," "wild land" was a more accurate descriptor, aesthetic appreciation emerged for the boggy tracts of Rannoch Muir and the Flow Country, set aside as "waste" land suitable for "improvement" through industrialized afforestation for much of the twentieth century. In taking visual pleasure (as opposed to the ecologist's satisfaction in the preservation of biodiversity) in the subtleties of "harsher," less picturesque scenery, late twentieth- and early twenty-first century landscape appreciation has distinct parallels with that of the late eighteenth and early nineteenth centuries.

Some changes in landscape ideals are part of the "democratisation" of landscape, in which the latter is appreciated on an increasingly wide scale for the physical challenge it represents to climbers, hill-walkers, and those who pursue other outdoor activities, as much as for the memories it evokes. Early twentieth-century mountaineering enthusiasts, who found a refuge from the industrialised confines of Scotland's Central Belt in the open spaces of the Highland landscape, formed the vanguard of this army of urban-dwellers who, on a metaphorical level, claimed these places as their own. Appreciation of "wild land" and "the remote" has expanded in line with the greater general utilisation of landscape for outdoor pursuits. In the 1930s "the remote" was the objective of an elite (in terms of numbers and fitness levels) and hardy few. The Danish-born entrepreneur Percy Unna converted his passion for mountaineering in remote Highland places into a national legacy through the anonymous purchase of mountainous areas that he placed under the guardianship of the emergent NTS. By the 1980s, usage of "remote" as a positive landscape descriptor was more widespread, and increased environmental awareness on the part of the general public has also contributed to the perception of "remote" landscape as not only a habitat or refuge for wildlife but as a sanctuary for "wilderness" or "wild land" itself.[17]

Romanticism, Wasteland and Landscape Protection

Romanticism no longer wholly informs the landscape protectionist standpoint. In the 1960s, post-war utilitarianism fed into a steadily growing environmental consciousness that actively rejected modernist values and the perceived obsession with technological innovation and development. This philosophy differed from that of the more traditional protectionist approach in terms of its support base and emphasis on aspects of environmentalism other than landscape integrity. Radicalised, post-*Silent Spring* non-governmental organisations such as Friends of the Earth and Greenpeace were the more youthful, cutting-edge face of later Cold War environmental awareness.[18] This demographic questioned the priorities of protection, conservation and development in terms that differed from traditional protectionist groups such as the APRS and NTS, which nonetheless pragmatically attempted to amend their agendas and perspectives in the face of new directions espoused by the former. Reacting to criticism of its elitist, non-integrationist agenda, the APRS made a renewed attempt to emphasise the resolution of social problems as well as issues of amenity and aesthetics:

> APRS is not a negative thinking or obstructive organisation. Its policy is to promote suitable and harmonious development and encourage the rational enjoyment of the countryside by urban dwellers, while at the same time safeguarding the unequalled beauty of the Scottish landscape.[19]

The hydroelectric debate was robbed of a climax because the Mackenzie Commission reported in economic rather than aesthetic terms. This, together with the fact that public inquiries into development plans for Glen Nevis and Loch Lomond were never held, with the respective proposals being shelved rather than cancelled, prevented landscape protectionists from flexing their increased "muscle" in the shape of growing public support. Consequently, a resolution on the issue of aesthetics was never adequately articulated. An increasing variety of interest groups, such as Friends of the Earth, began to include landscape issues within their spheres of interest.[20] However, a clear consensus as to the best use of Scotland's landscape remains elusive. The result has been that the hydroelectric energy debate, since the early 1980s largely quiescent but far from neutralized, has reemerged with an amended focus on the impact of wind, rather than water, power on the landscape.

As a signatory of the Renewables Obligation, Britain has committed itself to obtaining 10.4 per cent of its energy requirements from renewable sources by 2010.[21] The country has already gone some way towards reaching this objective because hydroelectricity has been classified as renewable energy. But it is estimated that a further 3.7 per cent of total energy will need to be drawn from renewable sources if the target is to be met.[22] As a result, the rival merits of solar, wind and hydroelectric power are currently the subject of much debate. At present it appears likely that future hydroelectric developments will be small, but there is substantial scope for large-scale wind energy production in Scotland, with the result that landscape development/protection priorities remain a highly contested issue.

Conclusion

The debate that unfolded over the integration of hydroelectric development within Scotland's Highland landscape is revealing in terms of what it suggests about the juxtaposition of competing twentieth-century mindsets regarding "best use" of non-urban space and associated interpretations of "wasteland." The Highland landscape ideal is grounded in images and values generated by the Romantic movement, itself a backlash against industrial development. Twentieth-century appreciation of this region is, therefore, to a great extent the product of an eighteenth- and nineteenth-century stress on the importance of engaging with the beauties of nature (the antithesis of industrialisation), as opposed to the utilitarian emphasis on "progress" and, as a matter of course, industrialisation. It has been argued that the language invoked in twentieth-century landscape discourse bears testimony to the legacy of the Romantics. In 1983, the chair of an American foundation fundraising for the NTS invoked "the wild splendor of [Scotland's] countryside."[23] In the ideal landscape to which Elliott's words allude, there is no place for development; anything human-induced or human-made will negate the "wild" element. Government agencies and non-governmental organisations have generated similarly indicative reports. The 1992 Countryside Commission for Scotland report, "Conceptual Review of Wild Land in Scotland," lists those elements impacting on landscape quality in Scotland: roads, electricity transmission pylons, hydroelectric power constructions, mineral exploitation and afforestation.[24] In essence, any component of the industrial environment is seen to detract from the aesthetic appeal of any particular landscape. Such late twentieth-century reports have built on earlier studies, similar in outlook and constructed around the same ideals, as to what constitutes a pleasing landscape. However, while the origins of current attitudes to landscape are undoubtedly rooted in the eighteenth and nineteenth centuries, it is possible to detect certain modifications and diversions. All areas exhibiting signs of change may be sites of future evolution and merit continued observation.

The Highlands became the focus of a significant amount of the increased attention paid to British landscape issues during the twentieth century because it was there that the economic and geographic potential for large-scale development was arguably the greatest. It was also where this development potential most contradicted, and therefore threatened, visions of the Romantic landscape ideal. The resultant dialogue illustrates the shifting interface between ideas of landscape and land-use, emphasising how, towards the end of the twentieth century, "priceless national asset" had become a term used to describe both energy resources, including hydroelectric power, *and* aesthetically significant landscape. Current debate as to the placement of renewable energy mechanisms, such as wind turbines within the landscape, is essentially concerned with whether energy-as-national-asset or landscape-as-national-asset has priority. It also demonstrates that the debate was never adequately resolved during the hydroelectric development period.

The extent to which the superficially opposed interests of development and landscape protection have been historically dependent upon one another requires greater emphasis. Energy development was perhaps the most immediate large-scale "threat" to the integrity of Highland landscapes, but it also lent momentum and focus to the landscape protection movement. Moreover, the increased appreciation and experience of Highland landscapes (particularly in the most democratic context) has been greatly reliant on technological advances in transport and energy; utilitarian innovation paved and lit the tourist's path to the Highlands. Rejection of the visual appeal of humanized land in favour of de-humanised landscape takes the viewer further along the route established by followers of the Romantic movement, searching for the "natural" remnants of pre-industrialised Britain. The distance between this trajectory and that reflecting the increasingly technological requirements of Britain's global community membership may continue to widen. If the pressing demand for increased domestic supplies of renewable energy is to be met, it may be necessary for policymakers, pressure groups and the public to accept that Britain's Romantic heritage is what informs this country's appreciation of "natural," "unspoiled" landscapes: that this appreciation is as much a cultural construct as any other. And if Britain is to remain technologically-driven in a global society lacking fossil fuels and acutely conscious of the spectre of global warming and associated climate change, a greater proportion of the non-urban environment may have to be visibly utilised in support of this.

NOTES

[1] J.F.M. Clark and T. C. Smout have been instrumental in shaping the ideas on which this essay is based. I am profoundly grateful to them both; all errors remain my own.
[2] Mira Engler. *Designing America's Waste Landscapes* (Baltimore, MD: Johns Hopkins University Press, 2004), p. xiii-xiv.
[3] *Ibid.*, p. xvi.
[4] David Matless, *Landscape and Englishness* (London: Reaktion, 1998), pp. 11-13.
[5] Peter Womack, *Improvement and Romance: Constructing the Myth of the Highlands* (Basingstoke: Palgrave Macmillan, 1989).
[6] *Ibid.*
[7] Yi-Fu Tuan, *Topophilia: A Study of Environmental Perception, Attitudes and Values* (1974; New York: Columbia University Press, 1990), p. 94.
[8] See, Engler, *Designing America's Waste Landscapes*, p. 12; Tuan, *Topophilia*, pp. 109-110.
[9] T. C. Smout, "Tours in the Scottish Highlands from the Eighteenth to the Twentieth Centuries," *Northern Scotland*, Vol. 5: 2 (1983), pp. 99-102.
[10] *Ibid.*, pp. 103-112.
[11] Matless, *Landscape and Englishness*, pp. 14-15; 70-95.
[12] Engler, *Designing America's Waste Landscapes*, pp. 30; 33.
[13] "Appeal to heartstrings of Scots abroad to help save Loch Lomond," *The Times*, 12 September 1978.
[14] Malcolm Grant, Estates Surveyor, Scottish and Southern Energy plc, personal communication, Perth, 29 February 2001.
[15] See for example Charles Warren, "Historical Perspectives" in *Scotland's Landscape—A Fixed Asset? Proceedings of a Forum held on 8 May 2003 at SNH Battleby Centre, Perth, Scotland*, ed. Helen Talbot and Mike Talbot, (Scottish Natural Heritage, 2003), p. 13.
[16] Leo Marx, *The Machine in the Garden: Technology and the Pastoral Ideal in America* (New York: Oxford University Press, 1964); Roderick Nash, *Wilderness and the American Mind* (New Haven and London: Yale University Press, 2001); Max Oelshlaeger, *The Idea of Wilderness: From Prehistory to the Age of Ecology* (New Haven and London: Yale University Press, 1991).
[17] NTS, File e4, "Environment and Landscape, 1985-6," Trevor Croft, Head of Policy Research, NTS, to Richard Ferguson, NSA Project Advisory Group Secretary, CCS, "Loch Rannoch and Glen Lyon National Scenic Area: A Strategy for Landscape Conservation and Development," 10 December 1986. Support for this

assertion is drawn from the individual attitudes towards remoteness reflected in this correspondence.
[18] Sheail, *Environmental History*, pp. 147-8.
[19] NTS, File e4, "Environment and Landscape, 1973-74," Richard Livingstone, Organising Secretary, APRS, to WDC Lyddon, Scottish Development Department, Edinburgh, 7 January 1974; APRS, MCM, 6 March 1973.
[20] The increasingly well-established Friends of the Earth applied to join the ccs-orchestrated "Forum on the Environment" in 1978, but found its initial advances blocked by reticent traditional bodies fearing its militancy and potential lack of objectivity. NTS, File e4, "Environment and Landscape, 1976-78"; "CCS: Forum on the Environment," "Note of Second Meeting of 1978," 16 May 1978, p. 2; NTS, File e4, "Environment and Landscape, 1976-78,"; "CCS: Forum on the Environment," "Note of Third Meeting of 1978," 7 September 1978, pp. 4-5. See also Sheail, *Environmental History*, pp. 147-8.
[21] Renewables Obligation (Scotland) Order 2002, Schedule 1, Articles 2(1) and 6(2).
[22] *Ibid.*
[23] NTS, "Annual General Meeting, National Trust for Scotland, Glasgow, 23 April 1983," transcript of speech delivered by Jock Elliott.
[24] CCS, *Draft Conceptual Review of Wild Land in Scotland* (1992), p. 23.

Chapter Three

Scenes from a Desert Wasteland

John Scanlan

Of the many natural features of desert regions that allow them to be distinguished one from another—from flora and fauna to animal life and geology—none, taken either individually or in combination, can reveal the nature of the desert-as-wasteland. Such a notion is inextricably bound up with the human encounter and perception of the desert as an unremittingly hostile place. Thus, just as notions of "waste" and "wasteland" are, more generally, meaningless beyond some human standard of judgement or value, so, too, the place of the desert in the human imagination tells us as much about ourselves as it does about the desert as a natural environment. For what "desert" has always referred to, in varying degrees, is the *absence* of the human—that is to say, it points to the unpopulated, the abandoned, and the hostile.

One might then go along with the self-declared "desert freak" Reyner Banham, who wrote in his *Scenes in America Deserta,* that "the ultimate definition of a desert may yet prove to be concerned with the number and type of people present, and what they think they are doing there."[1] But one might go further and say that in its relation to the human, and as a peculiar kind of wasteland, the desert is—or can quickly become—a place where what Martin Heidegger terms "authentic Being" encounters a temporal dislocation. This condition of being *out of joint*, or *out of time*, manifests itself in forgetting.

As Harald Weinrich has noted, the metaphors employed to reveal the nature of forgetting seem to invert the metaphors of remembering.[2] Where the ancient arts of memory famously rested, as Frances Yates wrote, on a "technique of impressing 'places' and 'images' on memory," forgetting is more readily assimilated to places devoid of the markers that might support such a mnemotechnics, or "art of memory."[3] What this means is that without places and details to remember—without stability of impression, and temporal duration—memory founders. When that happens we slip into a kind of reversal. So, perhaps it is no surprise that Weinrich can write that "the metaphorics of forgetting occupies primarily the desert wastes." Because, in the desert, everything so easily—like the sand—may soon be "gone with the wind."[4]

Through a consideration of the human relationship to the desert—in this instance, the Mojave desert of Nevada and California—it is possible to direct our attention to some important temporal dimensions of "waste" in contemporary life. Given the importance of the human to an understanding of the desert's very essence, it is arguable that the Mojave, on closer examination, presents a unique instance of the contemporary wasteland. During the twentieth century, for instance, it became home not only to a number of sites of what sociologist Thorstein Veblen would have called "conspicuous waste" (the cities of Las Vegas and Reno, to name but two), but more broadly to liminal, or existential, encounters through the pursuit of subjective oblivion and disappearance. It is such a forgetting of past and future that, for Heidegger, represented the "oblivion of Being."[5]

The other notable characteristic of the deserts in the Californian Nevada region is that they have, for decades, been the location of highly secretive government and military activity.[6] It is worth pausing here just to consider how secrecy and forgetting conjure up the spectre of waste. Harald Weinrich remarks on some of the interesting philosophical and literary associations of forgetting and, more broadly, how it figures in western thinking about the human condition. *Forgetting* is not merely defined in terms of the losing or banishing of something from memory or consciousness. It refers, more interestingly, to a kind of concealing, a covering up that might be countered or revealed in all of its dimensions by *truth.* And truth, Weinrich suggests, has since the time of the ancient Greeks, correspondingly been seen to involve a kind of memory work. *Alethia,* the ancient Greek word for truth, for instance, negates the term for forgetting, *lethe.* That is a word which also names the mythical river of oblivion; where "the hard contours of the remembrance of reality are dissolved and, so to speak, *liquidated.*"[7]

The activity of forgetting has thus always been regarded as something that defines us. Heidegger thought that there might be "an invisible cloud of forgetting … the oblivion of Being" hanging "over the whole sphere of the Earth and its humanity."[8] Perhaps we can say—with Heidegger—that there are "clouds" of forgetting in this desert. At its most serious, forgetting is evident in the fact that in the Mojave one can find a toxic nuclear threat of barely comprehensible dimensions. Aside from the obvious nature of its threat, what its existence reveals is the frailty of the work of human memory which, in consequence, may well be rendered obsolete.

Here, waste and oblivion—the unwanted and the forgotten—merge and become as one to reveal that not only are we are our wastes and our wastelands, but that they only exist as such because of our negative relation to them.

Oblivion

Deserts are known to excite mirages. In their vast openness, those wild, exuberant and ultimately unreal illusions induced by dehydration and heat exhaustion symbolize the disconnection of such an environment from routine life. Indeed, David Jasper notes, "all desert literature," which might provide some of the most telling insights into how we perceive abandoned places, "emphasizes the mirage—the image conjured by the place that is not there, yet is part of the place."[9] The kind of reversal such experience effects is illustrated in the most stereotypical of these illusions: life-giving oases await the one who has wandered endlessly without food or water, lush Polynesian greenery sprouts in the distance of an otherwise "dead" landscape and, as in those dreams that vanish on awakening, hidden treasure is thought to be buried under some lone desert palm tree.

In fact, while these mirages could stand as a description of what lies in wait at the "dangerous illusion" (to use Jasper's words) that is Las Vegas, visions of the unreal inform a history of encounters with desert spaces that for all their fantastical elements nevertheless animate the silent menace of desertification.[10] As Francesco Orlando notes, such uninhabited places are more likely to be "strewn with false signals, surprises, and dangers" than with anything necessary to human survival.[11] It is a fact of their emptiness, and one that undoubtedly explains why they remain largely abandoned or uninhabited. What we learn from looking at the foundations of western thinking is that desertification and barrenness were always implicitly in opposition to the kind of qualities that constituted the human. Thus, the kind of metaphors that were deployed to reveal the very vitality of human life tended to look towards nature and plant life or, in general, qualities that desertification would seem to negate.[12] This was an association especially marked in the writings of some of the ancients. The Greek poet Pindar, for instance, declared that "human excellence" could only be understood to flourish "like a vine tree, fed by the green dew, raised up … to the liquid sky."[13] It was a relation of the human to natural abundance that Clarence Glacken, in his *Traces on the Rhodian Shore,* reveals using an extended passage from Philo's *On The Creation,* of which the following is a snippet:

> Poets quite rightly are in the habit of calling earth “All-Mother,” and “Fruit-bearer” and “Pandora” or “Give-all” inasmuch as she is the originating cause of existence and continuance in existence to all animals and plants alike.[14]

Beyond its simple hostility to human life, though, the desert-as-wasteland is also defined by its *vastness*.[15] The desert region of Nevada-California—my focus here—loomed large in the imagination of westward migrants as “the mythical ‘Great Basin’,” according to Reyner Banham. The myth rested partly on the difficulty of simply grasping the immensity of its scale. This ended up making it “the most feared barrier to the westward expansion of the United States,” which stood in the way of the Promised Land of California: “the older the map one consults, the larger the Basin figures.”[16]

If the desert can be seen to exist beyond a human scale of comprehension then it is something equalled in its essential strangeness—and for the desert traveller its mysteries and secrets therefore often present themselves as anomalies. Not long after I embarked upon a trip through the Mojave desert from Los Angeles to Las Vegas in spring of 2000 I found myself two-thirds of the way to my destination, on Interstate 15, looking at a sign that seemed be confirmation that even the language of naming in such a place is exhausted by the emptiness of the environment. Perhaps no one but the most determined scientific explorer with knowledge of its sparse plant life and desert critters would be able to draw on more than a limited range of referents in settling on suitable place names. Instead the word “dead” has been commonly applied to name areas of the desert. But such names don’t seem as unusual as the most peculiar and unexpected road sign I observed. It directed drivers to a place that seemed to have been named by some lonely prospector who might have imagined that the ends of the earth had been reached, and decided that this place should be named something other than Dead Man Flats or Devil’s Plain. “Zzyzx Rd” the sign read. A directional arrow pointed to a right turn off the highway a few hundred yards ahead. Looking off to the distance along the imagined sweep of this road, there seemed to be no hint of a human presence, let alone any evidence of the strangely-named Zzyzx. There was only emptiness as far as the eye could see.

Writing now the sight brings to mind an observation made by Daniel Heller-Roazen in his study of language and forgetting, *Echolalias*. When a language seems to have run its course, he notes, “the name we are accustomed to [giving] such an end is the one we use in reference to organic being: death.”[17] Could Zzyzx have been one of the many ghost towns that were dotted throughout the Mojave; one of those places founded during earlier westward migrations and expeditions to locate minerals or other valuable natural resources? It wouldn't have been unusual because the "failed cities" of this desert, which happen to "outnumber those that are still going concerns," as the writer David Thomson suggests, can still be identified by their abandoned and ruined buildings.[18]

I later discovered that Zzyzx was not really one of these ghost towns, and that the road off I-15, which today leads to a University of Southern California desert outpost, had once arrived at the Zzyzx Mineral Springs and Health Spa. Or, as the authors of a travel companion titled *Weird California* note, this strange road delivered one to "a Utopia spelled Z, Z, Y, Z, X."[19] Zzyzx was, it seems, an oasis also once known as Soda Springs. Despite its invisibility from where I photographed the road sign in question, it was a destination only a few miles along a bumpy sand track off the Interstate, the details of which can now only be found in off-beat travel guides to the region—or notably in Reyner Banham's long out-of-print *Scenes in America Deserta,* which describes its location on the eastern side of the Great Soda Lake as "one of the greatest resort sites in the whole of America Deserta."[20] As Zzyzx, the settlement had been founded in the 1940s at a time when the desert was still largely unoccupied, by one of those California-based health freaks, a radio preacher who "touted the medicinal properties of the mud from Soda Springs."[21]

The identity of Zzyzx—its origin as the meeting place for a kind of corporeal spirituality induced by this magic mud—is a reminder that the settlement lies within the California area of the Mojave. The early history of that state—since its settlement by the westward migration that eventually overcame the terrifying prospect of the Great Basin in the nineteenth century—was, as Carey McWilliams has noted, shaped by a motley assortment of health cranks and religious non-conformists.[22] At some time in the development of the unique Californian identity, with its mixture of Eastern mysticism and hard-nosed individualism, spirituality and health were conflated, producing in the Golden State a new "religion" of the body beautiful.[23] One of California's chief attractions to migrants was the supposedly health-giving virtues of the land and climate. This usually referred to is proximity to the ocean and its abundant fruit orchards, but in the desert—and this was probably what made the site of Zzyzx uniquely attractive—the stripping away of the demands of routine life mark it already as a quasi-spiritual zone where "intoxications of sight" and "exaltations of perception" are common.[24]

Indeed, religion, of various sorts, has an intimate connection to the desert precisely because the reversal of human comforts one will experience in such an environment elevates the spiritual above the material. To survive in the desert was a test of faith in God for the early desert ascetics. It is no coincidence then that the world's monotheistic religions are seen to emerge from the desert, and have further constructed narratives about the role of the desert in the Creation. "Deserts," William Fox writes, "are where the Prime Mover was practicing before making the rest of the world."[25] Unsurprisingly, early studies of Christianity in particular reveal the desert to have been a perfect location for the kind of withdrawal and denial of bodily needs demanded by the monastic ideal—because it was the place where God was most likely to be encountered.[26]

But today the Mojave Desert has become an interesting region for reasons unrelated—in any direct way—to the desert of myth. Its peculiar geology, phenology and vegetation, for example, explain the existence of desert research centres, like the outpost of the University of Southern California that now occupies the site of the former utopia of Zzyzx. More recently the impact of human population growth and the activities of the United States government and military on the environmental sustainability of the desert have informed a number of interesting considerations of its future. What they have in common is that they all implicate the human, now, in the survival of what was once, by definition, a place unknown or abandoned.[27] Yet equally, the Mojave's uniqueness is determined by the existence of its many "desert playgrounds" and their proximity to the major metropolitan areas of California—in particular to Los Angeles. This human geography has ensured the desert's emergence as another kind of waste space: a destination for activities and entertainments that had either been designed for (or were engaged in) as a means to overturn the norms and expectations of everyday life. As we will see, the California-Nevada desert, site of oblivion in so many respects, and a metaphor for human reversal, provides a glimpse of a world turned inside out.

My own journey was to Las Vegas, the foremost of these desert playgrounds and gambling capital of the western world. The city's allure may now be global, but it was once projected primarily towards weekend revellers from Los Angeles, which was the centre of what Sally Denton and Roger Morris describe as Vegas's "Californian hinterland."[28] In decades past, before cheap air travel opened it up to the world, attractions at the desert resort were often advertised on huge billboards along Hollywood's Sunset Strip, beside other such signs for entertainments more local, giving the impression to the unwise that Vegas was located in a nearby locale rather than 195 miles away. Putting myself in the position of a weekend reveler set on the other Strip in Las Vegas, I put thoughts of the distance between the two cities to one side as I, like countless other travellers, set out from Santa Monica, the terminus of the old Route 66 by the Pacific

Ocean. From there it wasn't difficult to negotiate towards the on-ramp of Interstate 10, the San Bernardino Freeway, which leads directly Interstate 15, and links the two cities as the twin poles of American excess. Other parts of California—if one turns northwards—are within closer reach of desert destinations such as Reno, Lake Tahoe and Black Rock City, the latter of which is home to the annual Burning Man Festival and the newest of these desert playgrounds. Advancing beyond the Zzyzx Road exit the next significant point of divergence on my route was headed to Death Valley which, even in 1982, in the view of Reyner Banham, was approaching a kind of Disneyfication of over-regulation and prescribed visitor attractions.[29] According to other observers, the status of the Death Valley National Park amidst the surrounding no-go regions of the desert—which are correspondingly cast as *de facto* wastelands—reveals something of the politically charged history of this place. Valerie Kuletz argues that the labelling of pockets of the desert as either, "wasteland or national treasure demonstrates the arbitrary nature" of the way such descriptions are applied, but also contrastingly how they can "be precise and strategically used for political ends."[30] Yet Death Valley's location in the vast openness of the desert and its extremity of temperature has been enough to lure others in search of physical transcendence. It provided such a backdrop to a celebrated hippy "love-in" sequence in Michelangelo Antonioni's 1970 film, *Zabriske Point,* which thus identified the desert with the overturning of the social and sexual mores of the day.

These locations and the goings-on that characterize their pull on visitors from far and wide bear out Joseph Masco's observation that the "modern American desert is a fantasy playground … a place where curious things seen possible."[31] In terms of the desert reversal of human comforts, it is a place of liminality; which is to say that we experience in our encounters with it the suspension of social norms and institutions.[32] As Victor Turner has argued, liminality always involves the ritualized stripping away of the normal because it delivers us into a condition of indeterminacy, "betwixt and between the positions assigned and arrayed by law, custom [and] convention."[33] Liminality is also "frequently likened to death," Turner notes; and to other points of reversal or transition—invisibility, darkness, and the wilderness. Such a void presents a perfect location for the disappearance of all manner of unconventional, illegal, and outlaw activities—as well as the people who practice them. It was in Death Valley, in the 1970s, under a starry night sky and with the discordant sound of Karlheinz Stockhausen blaring, that philosopher Michel Foucault, accompanied by two Californian companions, dropped acid for the first time, an event that was said to have changed his life thereafter. The experience locates him within a tradition of transformative and spiritual desert encounters.[34]

The lure of desert invisibility being what it is, Death Valley earlier provided, at a remote box canyon known as Barker Ranch, an apparently suitable opening place into which the on-the-run Charles Manson and his notorious "family" temporarily vanished, escaping the reach of the Los Angeles Police Department as they sought the perpetrators of the notorious Tate-LaBianca murders of 1969. Much like an evil counterpart of the desert mystics of religious lore—who went into the desert in search of some kind of revelation from above—Manson, as Erik Davis notes, believed himself to be within reach of "a hole that he claimed would lead" below—"to an underground paradise."[35]

In choosing the desert as the most fitting base to launch a series of destructive forays into Los Angeles, Manson—consciously or otherwise—made a symbolic choice that followed in a tradition of outsiders and outlaws; people, as David Thomson notes, drawn to a life "apart from all semblance of order," and one that was most readily found, especially for those fleeing the confines of normality, in the "empty, arid and inhospitable" regions of the modern American desert.[36]

In recent years many in search of encounters in the liminal spaces of the desert have found themselves in Black Rock City, some 500 miles north of Las Vegas, for the annual Burning Man Festival. This gathering, which concludes with the immolation of a giant human-like figure, is a mix of carnival, sacrificial rite and—in practice—returns to a tradition of thinking about the desert, and uninhabited places more generally, as connected to the sacred. As Mircea Eliade has written, the sacrilization of uncultivated zones through such rituals has been a common way of negotiating and claiming places that otherwise would "still persist in the undifferentiated, formless modality of pre-Creation."[37] Again, one might refer back to the idea that the desert provided space for the Almighty to practice his moves before setting about the Creation. In the case of the Burning Man Festival, as Erik Davis suggests, the undifferentiated—waste stuff and wilderness—come together to point the way towards some extra-human encounter with the mysteries of existence. The festival thus "ties together the pursuit of intense experience" with "the alchemy of trash." Accordingly, the event generates a "bohemian embrace of sacred sensuality" drawing its participants closer to "nature unadorned."[38]

This could be seen as an instance of waste as the modern sublime, yet the so-called "alchemy of trash" refers to something more prosaic; and that is the common festival practice of reusing and refashioning discarded and scavenged objects and materials to produce a "flea market of parallel worlds."[39] This is recycling and reuse to transcend the prerogatives of the capitalist economy as they have come to inform contemporary life. Burning Man's particular allure, though, lies in the fact that it offers the desert experience as an aspect of the real; an unmediated encounter with one's own being, here in the middle of nowhere, stripped of all vanities and pretensions.

And, as it happens, "Welcome to Nowhere" was the headline that would have greeted some of the first Burning Man visitors in 1993 had they picked up a copy of the *Black Rock Gazette*. "Here on this sere and desolate plain," the newspaper began, "our community springs from the desert floor like a strange and beautiful mushroom":

> Its name is whatever you name it. Its wealth is whatever you bring. Next week it will be gone, but next week might as well be never. You are here now.[40]

You are here now. That may be the leitmotif of the desert experience that reverses the temporality of routine life. In the case of Burning Man, the peculiar sense of time and place it generates casts it, the *Black Rock Gazette* claimed, as "a post-modern Brigadoon in an ancient wasteland." Like Brigadoon its existence seems to be conjured out of nothing only for it to vanish soon after, and before its mysteries become clear. What we should not lose sight of here is the importance of *the moment,* and of the necessity of grasping the moment. This had always been a motive force in the history and the culture of the west, particularly in California. And, of course, it was largely they who allowed the desert as a place for such carnivals of forgetting to flourish.

Infinity

One might speculate that the way in which the desert experience easily permits a visitor to transcend the vanities of human agency results directly from the seemingly infinite dimensions of the environment. In the Mojave, as Rebecca Solnit notes, "it is geological time and geological scale that dominate."[41] In a place so vast, beguiling and deadly the human becomes less significant—and quickly *insignificant* should one be lost or abandoned—and this means that we tend to "read the desert as a dead or barren landscape."[42]

The temporal disconnection that the desert effects is underlined, in the case of the Mojave, by the consequences of its use for weapons testing, which includes the production and disposal of nuclear waste. The Nevada Test Site (NTS) is just one of several such locations established in the United States during the Cold War by what was then called the Atomic Energy Commission. The vast area of the NTS, as Joseph Masco writes, represents the desert as a "technoscientific wasteland." It is a playground of different dimensions to those already touched upon, one "where many of the most dangerous projects of an industrial, militarized society are located."[43] The NTS, which for a long time could not be found on maps of the desert, actually lies within the larger Nellis Air Force Base, a site that itself is home to a wide range of military training activities and scientific experimentation. It was in and around here, too, in the mysterious Area 51, that UFO sightings became commonplace, also during the Cold War. That there are several distinct sites within this area of the desert indicates something of the size of the space given over to military-scientific experiments—it is a region roughly the size of Switzerland. Valerie Kuletz notes that "military ranges in Nevada alone," which are dispersed in a variety of locations "amount to four million acres."[44] That is in Nevada. In California there are additional sites given over to similar activities. Here, in both sides of the Mojave, however, deep within what Masco terms this "dream space for a spectacular idea of progress" the government of the United States built a model town that was named Apple II. It was designed to be a simulacrum of the most recognizable, but inanimate, features of life under normal conditions. It included "a fire station, a school, a radio station, a library," and so on. But in the testing of weapons it was seen as necessary to try and assess the effects of nuclear warfare on humans, rather than just the structures they lived in. Thus:

> a dozen homes were built and furnished with everyday items (televisions, refrigerators, furniture, carpets, and linens). They were then stocked with food, populated with white-skinned mannequins, and neatly incinerated.[45]

Photographs taken during the experiments show mannequins pre- and post-test: a female figure surrounded by three "children"; a family at rest in their sitting room; another family sheltering in the basement. Later they can be seen lying amidst the wreckage of blown-out buildings.

As a frontier for experiments in future living, the desert always revealed its hostility to the human in its trash and leftovers. A century before Apple II nature was the oppressor. Some of the earliest migrants to the west gave up the dream of the Promised Land when they discovered that crossing the Great Basin was simply a frontier too far. It is a fact that was revealed in the journal of one 1850 migrant who, as Richard White has written, recorded passing some "two thousand abandoned wagons," all "in a single 40-mile stretch of the Nevada desert."[46] Today, the wagons are no more, but the desert nonetheless reveals traces of its forgotten futures not only in the many ghost towns abandoned to deadly time, but also in what Gerard DeGroot notes to be its "weird collection of atomic artifacts." Domestic settings aside, DeGroot records, there was, "amidst the rubble … a bank vault." Its purpose was much like that of the mannequins of the Apple II houses; the American banks wanted "to find out how to protect their money in the event of a nuclear war," so they delivered a standard vault, complete with its valuables, which was then subjected to the ferocity of the blast and its thermal effects.[47]

But the perception of these Mojave test sites as wastelands is not merely evident through encounters with the trash of nuclear experiments. It is rather an awareness that is raised by knowledge of the strategies required to come to terms with the by-products of the military and scientific testing, the most profound of which cannot be seen littering the ground. And that, precisely, is the problem. In this case, the everyday equation often applied to our logic of waste disposal—*out-of-sight, out-of-mind*—speaks of a more alarming dilemma. Namely, material that will remain dangerously toxic for periods estimated on such a timescale that it escapes the scope and comprehension of any model devised by the parameters of the study of human history invites a new kind of attentiveness that is at odds with how the polluting potential of waste has hitherto been managed. It involves a renewed commitment to understanding the interplay between place, memory and oblivion. Clearly, the absence of a meaningful temporal horizon for simply grasping the idea of the toxic *longevity* of nuclear waste means that the gravest danger lies in the fact that once it has been subjected to deep ecological burial, its existence and threat may—over a period of tens of thousands of years, at a minimum—be forgotten about. The desert seems to add complexity to the equation because it is such a void, and one in which any kind of surface marker or memorial might simply be lost over time in such its inhuman landscape.

The specific practical danger is that a toxic dump might be accidentally rediscovered and disturbed at some unknown time in the future. It is a possibility that has invigorated attempts to find some kind of durable monument to the remembrance of what lies buried here deep below the ground—some of these designs for suitable monuments are reproduced in Peter van Wyck's profound and disturbing *Signs of Danger*. With the Waste Isolation Pilot Plant (WIPP) site in the New Mexico desert, for instance, the concern has been not only with creating an over ground "sign of danger" that will persist in geological time, but one that might be read by a future people who could—fifty thousand years hence—have no knowledge of our languages, nor access to the symbolism that today might (like the common danger sign of a skull and crossbones) indicate that a place contains grave dangers. The sci-fi dimensions of the scenario are equalled in the fact that contingencies have even been made for the possibility that alien visitors might one day in the distant future move around on such sites. The threat to humanity from the (apparently safe) deep burial of toxic materials being disturbed is acute enough that the outer space aliens, too, would have to be warned, lest they inadvertently release the danger.

What this means is that nuclear waste, as distinct from all other kinds of waste, "is really matter *without* a place."[48] Its unique danger moves it beyond the simpler spatial nuisance of much other waste material, because it is:

> a kind of waste that resists its own containment. A kind of waste that operates in a radically different temporality: it is material whose toxicity requires a different conception of history and time.[49]

Some of this waste—most deadly of all, plutonium—will remain radioactive and dangerous for periods estimated to range from one-quarter to three-quarter of a million years. This geological lifespan reveals a connection between acts and consequences that is almost the opposite of the "here and now" of the Mojave's other desert playgrounds. Where their appeal is a more personal and subjective kind of oblivion, the forgetting implied here has global historical consequences. Because nuclear waste extends into a time almost beyond comprehension it represents, actually, a disconnection from life: a void of meaning and signification, evident in the impossibility of designing "signs of danger" that might last into a future that might as well be calculated as the infinite. Thus, the future that is invoked in the Mojave will remain in the shadow of a cloud of forgetting that will darken the lives of hundreds, perhaps thousands, of generations of human descendants, spread well beyond a Great Basin that once seemed so endless itself.

Even attempts to calculate with some degree of precision the conditions under which these waste materials may be safely buried turn out to look like guesses, producing what Peter van Wyck terms "lively fictions" about the ability of the best science to control the situation.[50] In Nevada's Yucca Mountain region—at a location mysteriously named "Jackass"—the US Department of Energy (successor of the Atomic Energy Commission) has been developing its principal deep geological burial repository for commercial nuclear waste, an initiative that is charged with guaranteeing the safety of buried materials for a minimum of ten thousands years.[51] This, too, is also the length of time that a "permanent" marker above ground will be required to persist in order to continue to identify the site as dangerous.[52]

The period of ten thousand years was arrived at, rather interestingly, because it was thought that any structure capable of lasting for such a time will *probably* last longer. Yet this appeal to such a loose probability does not get over the fact that this, as Joseph Masco notes, is "only a fraction of the lifespan the most dangerous nuclear materials."[53] A realization that was made even more worrying by an admission Masco elicited from the Yucca Project's tunnel engineer that he wouldn't guarantee the lifespan of the tunnel for any period longer than a mere *one hundred* years. After that, he said, it would probably be wise for the authorities to find another site to rebury the waste.[54] Thus, what we are dealing with here, van Wyck confirms, is a kind of temporality "that approximates pure future."[55] If humanity, as the philosophers have always said, is defined by its finitude, nuclear waste—as the infinite—is thus the *humanly-created* inhuman.

Nonetheless, just as the larger Great Basin (within which the Mojave lies) was once thought by early migrants to itself be the infinite, so too the government and military, especially during the Cold War, conceived of it as an outer space-like frontier. Like the void of the heavens, it was a place that seemed capable of absorbing anything humanity might throw at it. In fact, military communiqués circulating in us government circles before the site was finally chosen described it as a land essentially forgotten by the rest of the United States; it wasn't "good for anything but gunnery practice," they said. In fact, one "could bomb it into oblivion and never notice the difference."[56]

This draws me back to Harald Weinrich's revealing analysis of the connection between forgetting and "covering up." The relative invisibility of what Valerie Kuletz calls the "secret science cities" of the Great Basin—declared off-limits to the public—deflect attention from the consequences of their activities, and conceal the significance of nuclear waste more generally.[57] Attention was driven further into the depths of oblivion when weapons testing went underground in 1963, thereby seeming to vanish from even the limited public consciousness of the earlier testing that was visible in the skies many miles away. Eventually, as Rebecca Solnit writes, "Americans forgot that bomb after bomb was exploded here."[58] In this they have been repeatedly assisted by the United States government who, "with a remarkable talent for euphemism" as Gerry DeGroot notes, now describe the Nevada Test Site as "an environmental research park."[59] It is the kind of re-branding that embraces forgetting through the manoeuvre of what Norman Klein calls "simultaneous distraction." What Klein means is that memory is always under threat when the external points of knowledge that support it are effaced, or disappear altogether. "Forgetting," he notes, "is a twin" born in the "instant when one memory defoliates another."[60]

Today, signs around the perimeter of the NTS declare its use for "the testing of devices for defense and for peaceful uses of nuclear explosives." Such language further lives up to the "environmental research park" tag by elliptically alluding to the land's prior use by "prehistoric, aboriginal, cultures," as if what was now—and had been for decades—taking place formed part of some natural continuity and was an example of humanity's affinity with the land. But—unsurprisingly—these signs make no mention of the most important fact of the site's existence, which is its waste legacy. To look below the obfuscations of the official language surrounding the activities taking place in this part of the world, Mike Davis suggests, is to find a "landscape of incomprehensible devastation, sown with live ammo and unexploded warheads."[61] The truth, William Fox argues, is that the Nevada Test Site, the location for the storage of decades worth of radioactive waste, now "contains more than fourteen hundred toxic areas that need cleaning up" and some of these are allegedly already "leaking radiation" into groundwater "that may slowly move towards Las Vegas."[62] The danger has, however, already touched other areas of the desert, particularly, Davis writes, a "downwind population" in Utah already "exposed to the fallout equivalent of perhaps fifty Hiroshimas" who are now "being eaten away by cumulative cancers, neurological disorders, and genetic defects."[63]

Expenditure

But oblivion is most commonly encountered as a differentiated aspect of the everyday that is innocuously sold as leisure, some distance from the NTS in the city of Las Vegas. It was once characterised as "just a place in the desert to dump the leavings of a lifetime."[64] That seems to be a view that shared the assessment of the military that the desert is a landscape of forgetting. Probably the most important fact of Las Vegas's pull is its location in the desert, which provides the contrast of scarcity and abundance that makes it the "oasis-to-end-all-oases," Reyner Banham suggests.[65] Thus the means one employs to get through the desert to Las Vegas form an important part of how the city itself works its spell, as a host of writers have attested.[66] Because on the road, "once you get sixty miles out of L.A. and into the high desert," William Fox suggests, "your senses are loosened from reality."[67] Indeed, the interstate—from a driving point of view—often looks like an endless line extending ahead on the surface of some distant planet, effectively sucks one into a different real; one whose disorientating effects were undoubtedly more intensely felt in the days before vehicles had air conditioning.[68]

My own sense of accelerating into this liminal zone was accentuated by the sudden awareness of gigantic roadside billboards punctuating the endless desert plain. I was within one hour of the state line, which meant the signs had a temporal function, counting down to the transcendence of routine life. One sign, displaying the whirring barrel and numbers of a slot machine wondered if I, like countless others before, was "Ready to Spin?" The existence of such signs underlines the fact that the Nevada state line, wherever one finds it, is very much intended to represent the point beyond which ecstasy and oblivion are to be found. No one should then be surprised that the first structure encountered at the Nevada State line on I-15—keep in mind this is the middle of nowhere—is a lone casino (and hotel complex), Whiskey Pete's. Equally, visitors who arrive by air are often surprised to find that the airport passenger spaces look very much like the gaming floor of a casino, with slot machines all over the place.

In Jean Baudrillard's *America,* which reflects on just such travels along this road, the affinity between gambling and the desert is seen to be at once "mysterious," but no less illuminating in its mystery for the way it forces a consideration of the connection between "the sterility of wide open spaces" and the practice of gambling. For Baudrillard, it is the element of the non-human that seals the relationship: the desert as a "sublime natural phenomenon" is matched by "the abject cultural phenomenon" of Vegas.[69] In other words, the void of the desert is matched in its excess by the absence of limits in Las Vegas; both in terms of the way the city has plundered the wealth of the patrons who fill its casinos, and through its voracious use of the region's finite water supplies. Wilderness and excess thus share in their reach for the infinite and unsurpassable, but promise only death—something that has perhaps always been known, as the prevalence of place names like Dead Mountain, Dead Horse Point and Deadman's Canyon reveal, or even as the representation of decay and death in art that has gambling as its subject confirms.[70]

Unlike Baudrillard, preoccupied by desert metaphysics, Mike Davis is more concerned with the anomaly of this "strange amalgam of boomtown, world's fair, and highway robbery" in an environment—in both social and natural terms—that would seem unable to support it.[71] It conceals its anomalies, though, beneath the spectacle that sustains its pull. Under darkness and from a distance it looks less like the empty nowhere of its desert environment than a "gigantic funfair fallen from outer space … [or] the fragment of a still incandescent comet that has just smashed into the ground," Bruce Bégout notes. "With its thousands of fitful garish glimmerings, it illuminates the celestial vault, which puts on a pallid show by comparison."[72]

We can see that there is no balance in the relationship of the human to the natural—the city to the desert—in this pocket of the Mojave. Today's mega casinos, for instance, offer the merest of clues to the disconnection implied in the way that they depart from the simpler western stylings of old, which at least seemed to have one foot in the reality of the old west. Today's casinos more commonly present fantastical illusions that reinforce the mirage-like quality of the Vegas experience, featuring simulations of Polynesian abundance driven by what Davis terms an unsustainable "hydro-fetishism"—not to mention the postmodern copies of places like Manhattan, Paris, and Venice:

> No other city in the American West seems to be as driven by occult forces or as unresponsive to social or natural constraints. Like Los Angeles, Las Vegas seems headed for some sort of eschatological crackup.[73]

Davis refers to the city as "Oz," in reference to Dorothy's fantastical dream in *The Wizard of Oz*—like a mirage, or indeed the "Brigadoon" of Burning Man, it seems to be a place that tomorrow may easily vanish. In Bégout's apt and more reductive terminology, Vegas is "Zeropolis," a void of all values and, we read, a wasteland beyond compare in contemporary historical-cultural terms.[74]

The fact that the total gambling experience as it was designed in the twentieth century is found in a desert may be no more than historical coincidence, but in the success with which Las Vegas conjures up a world of reversed values, one might suggest that it is a necessary fact of its existence that it flourishes deep within in an inherently inhospitable zone that had until a few decades before the Mafia discovered it, been no more than a remote stop for prospectors on the Las Vegas and Tonopah Railroad.[75] Just as the surrounding barrenness of the desert brings into relief the abundance and gaudy excess of the attractions on The Strip, it is perhaps the seeming impossibility of such a fantasy world in the desolate territory of the Mojave that makes any transformation of personal fortunes—like those desert visions of palm trees or the intimations of buried treasure—seem possible in this city.[76] Its on-tap excess delivers a world of limitless and possibly life-changing futures, a place where the past can be easily and painlessly detached. Thus, as the sign suggested, better that one is "ready to spin."

It is useful at this point to consider the uniqueness of gambling in this place. Unlike the aristocratic gambling experience of old Europe, of the

French Riviera—the inspiration for Dostoevsky's *The Gambler,* to cite one example—it does not take place in what we would today call a socially exclusive environment.[77] Rather, Las Vegas does not try to obscure or limit the real appeal that gambling holds: on the contrary, it is the city's accessibility to every segment of society (and the common perception of it as a grown-up's playground) that sees it manifest gambling's ideal of the destruction of normal space-time relations.[78] Old casinos were proximate to the everyday world but exclusive; Las Vegas is remote from routine life, but open to all. The difference allows it to create more effective means towards oblivion.

Beyond the gambling, we see much the same elsewhere. The city's relationship to the real is perhaps best summed up by the impermanence of appearances at almost every level of a visitor's experience. These range from the manufactured natural environment to the seemingly temporary front presented by the glitzy facades of its main streets, which are regularly torn down to be replaced by newer and more alluring distractions. For Mike Davis, "extravagant demolitions," such as the staged obliteration of old style casinos now deemed out of date, "have become Las Vegas's version of civic festivals."[79] This sense of instability—which, not without coincidence, is the oxygen of the gambler's experience—exists in an environment which, at a level below outright demolition, sees "every structure, sign, and view" offer "metaphoric testimony to the insecurity that mobility implies."[80] Again, it is the familiar hint that it might all be gone tomorrow, so grab some of it while you can. This is a city unique, Hal Rothman suggests, in the fact that it lays no "claim to a past" and, indeed, "purposefully markets the destruction of its history."[81] Even what we would term the "natural" features are temporary stand-ins, with trees uprooted and moved if necessary.[82]

Undoubtedly, though, much of Las Vegas's allure lies in this constant surface reordering, which is arguably rapid and unsentimental enough to render the city devoid of the kind of place-holders that would define place in an anthropological sense.[83] The perception of it consisting of a series of movable stage sets is heightened also by contrasting its architecture and spaces with those one might find elsewhere.[84] The difference? Here, there is no single or notable architect associated with its most famous structures and public spaces—which is why there are, ultimately, no durable landmarks.[85] But the speedy removal of solid structures and the erasure of normal social conventions do not end with architecture and gambling (or sex, or any of the other pleasures on offer). In fact, Las Vegas presents itself as a panacea for all the disappointments of the world its visitors have left behind and has for decades been a destination for those set on a way to forget "the hardships and pain of life."[86] Instant marriage and quick divorce co-exist with the more pervasive dream that any life, however miserable and unfulfilling, might be transformed at speed under cover of its glowing lights.

Before it attained its current status as the fasted growing city in the western United States, it had long been known as precisely a nowhere; which is to say, more often a destination than a place chosen to live. And, as we have seen, it is duration and settlement that the desert has always prevented. Sally Denton and Roger Morris, in their account of the development of modern Las Vegas, observe that it was unique in luring people "in search of what they and their world might yet be" with "a self-consciousness and intensity" that validated the city's mythical power of transformation. This, of course, it promised to deliver via "entertainment, money, sex, escape, deliverance, another chance, a last chance," and so on.[87] Such means are an accurate reflection of what Martin Heidegger terms "inauthentic Being"; which is to say a Being that has vanished in forgetfulness.[88] To exist in a time and place of forgetting is evident in the appeal of "another life for a few hours, days, forever."[89]

Perhaps the most revealing evidence of how deeply the dream of desert transformation penetrated the popular imagination is found in the example of Elvis Presley. Largely eclipsed in the 1960s by a new generation of musicians and performers, Presley was given one more chance in Las Vegas, but in the end it consumed him just as it does most of those who stay too long. Unsurprisingly for one stuck in such a fabled dreamworld, Presley ended up as an exaggerated and white-suited cabaret ghost of the original or "authentic" Elvis. It is a caricature reinforced through Hollywood films such as *Honeymoon in Vegas,* which featured a scene with a ten-man troupe of parachuting white-suited Elvis clones known as "The Flying Elvi," who are seen to descend from the skies above the city. That they were not invented for the purposes of the film tells its own story. Such scenes, nonetheless, have ensured that the singer is now forever associated with tacky kitsch. Given Presley's seismic contribution to the popular culture of the twentieth century, this is a fate not without tragedy.

His last years were perhaps also tied up with the gambling debts his manager had been accumulating for years since the two had arrived in the city in the late 1960s. It was a situation that has been said to have seen Elvis transformed one more time into a kind of deluxe gambling chip; a star turn to be traded in exchange for the credit that would keep Colonel Tom Parker, the ex-circus carny who oversaw Presley's career, at the tables where he lost millions.[90]

Yet Elvis's association with the city finds its truest expression in a song he recorded over a decade before he ended up trapped by the tackiness and kitsch of the place. "Viva Las Vegas," an unmatched performance that inadvertently encapsulates the essence of a life given over to oblivion, identifies the city at once with a life of excess in its celebration of round-the-clock liberation, but also in figuring gambling as a kind of restlessness that intensifies and accelerates the movement of a body—the supposedly autonomous subject—into the kind of *non-human* transformation that would be wrought in its ecstatic playgrounds:

> Bright light city, gonna set my soul
> Gonna set my soul on fire
> Gotta whole lotta money that's ready to burn
> So get those stakes up higher

The song also implies that an excess of desire and sexual energy underlies the drive towards the destruction of resources which, in the world of utility, would not be spent.[91] For, as Baudrillard has observed, gambling is "a crazed activity on the fringes of exchange" that makes a mockery of society and economy. It is, indeed, "a desert form," which is to say that it touches on the "inhuman."[92] For in a Las Vegas money is stripped of all its daily, taken-for-granted associations and "metamorphosed into pure circulation, into pure fascination, into an absolute passion."[93] The reward for *getting the stakes up higher,* therefore, lies in the sense of the limitless possibilities of this destination where temporal constraints do not apply, and in which—as Presley sings—a "strong heart and a nerve of steel" are required:

> Viva Las Vegas, with your neon flashing
> And the one armed bandits crashing
> All those hopes down the drain
> Viva Las Vegas, turning day into nigh time,
> turning night into daytime
> If you've seen it once, you'll never be the same again[94]

When Presley sings those lines "turning day into night time, turning night into daytime," he invokes the reversal that Las Vegas had sought to formalize since the city's unlikely appearance in the desert. Like some kind of shaman, driven by the juddering and relentless rhythm of the song (and one has to hear the music as it relates to the sentiments in order to understand this), his performance seems to will a rupture in time and space that can, for good or ill, fuse new realities.

But what the song ultimately directs us to is the experience of the gaming room, in which, one might say, the world is turned inside out using means both subtle and obvious. Here, for instance, the regulated time of modernity is made obsolete. The absence of clocks, along with the lack of all natural light—and thus, the division of lived experience into day and night—is a renunciation of "normal" space-time conditions. Because the gaming room is so out of joint with time, it is, really, a virtual world that exists in a mythical causality. People, objects and other phenomena appear from—and disappear into—nowhere, according to a logic that has little to do with reason. It is the logic of chance.

Consider the fate of money here in Las Vegas. There are, on the one hand, the huge sums of dirty money that the city was arguably established in order to launder or make "disappear" like it does nowhere else (and hence the Mafia's deep involvement in the city since its earliest days).[95] On the other hand, one can't overlook the destination of money lost in the gaming rooms, which was revealed in a detailed sequence of Martin Scorsese's celebrated film *Casino,* which takes us, so to speak, behind the illusion that conceals the depressing reality and secret of this desert mirage. As if pulling back the curtain to reveal the identity of the Wizard of Oz, Scorsese's set-up draws viewer into the comings and goings of the counting room, which is necessarily hidden from the players in the gaming rooms. The workings of this backstage counting room are coldly described by the voice of the casino's mob-appointed manager, Ace Rothstein (played by Robert DeNiro). After revealing the detailed mechanisms of the economics of gambling, as the viewer observes men wheel around large containers of cash, he declares that "this is the end result of it all. The bright lights and broads and booze; it's all been arranged just to get your money."[96]

Like the Nevada Test Site, Las Vegas was run for decades in the belief that it could conceal its real workings under the cloak of oblivion—for the casinos that referred to the patrons, rapt in the moment, who would temporarily lose sense of their own being. For those who ran the place, however, its magical disappearing money, sucked into the void of the desert, represented the "perfect setup." As Hal Rothman notes, it meant nothing less than "unearned, unaccounted for and untraceable money, forever."[97] Or so it seemed. What cannot be denied, however, is that to achieve this kind of forgetting it took a wasteland setting—a desert—and the kind of disconnection that it could effect from time, memory and reality.

NOTES

[1] Reyner Peter Banham, *Scenes in America Deserta* (London: Thames and Hudson, 1982), p. 205.
[2] Harald Weinrich, *Lethe: The Art and Critique of Forgetting,* trans. Steven Rendall (Ithaca, NY: Cornell University Press, 2004), p. 4
[3] Frances A. Yates, *The Art of Memory* (1966; Chicago, IL: University of Chicago Press, 1974), p. xi
[4] Weinrich, p.4
[5] Martin Heidegger, *Nietzsche, Volume 1: The Will to Power as Art,* trans. David Farrell Krell (London and Henley: Routledge and Kegan Paul, 1981), p. 194
[6] Thorstein Veblen, *The Theory of the Leisure Class* (New York, Dover, 1994); "the principle of conspicuous waste requires an obviously futile expenditure," p. 108
[7] Weinrich, pp. 4-6
[8] Martin Heidegger, *Parmenides* (Bloomington, IN: Indiana University Press, 1998), p. 28.
[9] David Jasper, *Sacred Desert: Religion, Literature, Art and Culture* (Oxford: Blackwell, 2004), p. 109.
[10] Quotation from Jasper, *Ibid.,* p. 7
[11] Francesco Orlando, *Obsolete Objects in the Literary Imagination: Ruins, Relics, Rarities, Rubbish, Uninhabited Places, and Hidden Treasures*, trans. Gabriel Pihas, Daniel Seidel and Alessandra Grego (New Haven and London: Yale University Press, 2006), p. 141.
[12] Richard B. Onians, *The Origin of European Thought about the Body, the Soul, the Mind, the World, Time and Fate* (Cambridge: Cambridge University Press, 1988), p. 221.
[13] Quoted in Martha C. Nussbaum, *The Fragility of Goodness: Luck and Ethics in Greek Tragedy and Philosophy* (Cambridge: Cambridge University Press, 1986), p. 2.
[14] Clarence J. Glacken, *Traces on the Rhodian Shore: Nature and Culture in Western Thought from Ancient Times to the End of the Eighteenth Century* (Berkeley, CA: University of California Press, 1967), p. 14.
[15] The contemporary uses of the word "waste" can be traced to its origin in the word "vast"; as in the *vast* of the oceans, uninhabited places, and so on. See John Scanlan, *On Garbage* (London: Reaktion, 2004), ch.1, but particularly pp. 22-33, for a discussion of the meanings and etymology of the word "waste."
[16] Banham, *Scenes in America Deserta,* p. 192.
[17] Daniel Heller-Roazen, *Echolalias: On the Forgetting of Language* (New York: Zone Books, 2005), p. 53.
[18] David Thomson, *In Nevada: The Land, the People, God and Chance* (London: Little, Brown, 1999), p. 23.
[19] See Greg Bishop, Joe Oesterle and Mike Marinacci, *Weird California* (New York: Sterling Publishing, 2006).

[20] Banham observes that the sign strikes most who see it as a "typographical error," *Scenes in America Deserta,* p. 37.
[21] The founder was a "quack" by the name of Curtis Howe Springer, who established Zzyzx in the 1940s. The resort was eventually closed by Federal authorities in 1974, which reclaimed ownership of the land. It is now home to an outpost of California State University's Desert Studies Center. See L. Wilcox, *Desert Dancing* (Walpole, MA: Hunter, 2000), pp. 70-71.
[22] See Kevin Starr, "Carey McWilliams's California: The Light and the Dark," in *Reading California: Art, Image and Identity, 1900-2000* (Berkeley, CA: University of California Press, 2001), p. 18.
[23] See, for instance, Erik Davis, *The Visionary State: A Journey Through California's Spiritual Landscape* (San Francisco, CA: Chronicle Books, 2006); Kevin Starr, *Coast of Dreams: California: A History of Contemporary California* (London: Allen Lane The Penguin Press, 2005), pp. 10-13.
[24] Banham, *Scenes in America Deserta,* p. 63.
[25] William L. Fox, *The Void, the Grid & the Sign: Traversing the Great Basin* (Reno, NV: University of Nevada Press, 2005), p. 2.
[26] The desert, Philip Sheldrake notes, "is both a paradise, where people may live in harmony with wild animals and at the same time a place of trial where ascetics encounter the inner and outer demons." See Sheldrake, *Spaces for the Sacred: Place Memory and Identity* (Baltimore, MD: The Johns Hopkins University Press, 2001), p. 91; also, Douglas Burton-Christie, *The Word in the Desert: Scripture and the Quest for Holiness in Early Christian Monasticism* (Oxford University Press, 1993), pp. 3-5.
[27] See, for instance, Mike Davis, *Dead Cities and Other Tales* (New York: The New Press, 2002), pp. 21-106; Rebecca Solnit, *Savage Dreams: A Journey into the Landscape Wars of the American West* (Berkeley, CA: University of California Press, 2000); Valerie L. Kuletz, *The Tainted Desert: Environmental Ruin in the American West* (London: Routledge, 1998); Fox, *The Void, the Grid & the Sign.*
[28] Sally Denton and Roger Morris, *The Money and the Power: The Making of Las Vegas and Its Hold on America, 1947-2000* (London: Pimlico, 2002), p. 98.
[29] Banham, *Scenes in America Deserta.*
[30] Kuletz, *The Tainted Desert,* p. 66.
[31] Joseph Masco, "A Notebook on Desert Modernism: From the Nevada Test Site to Liberace's Two-Hundred Pound Suit," in *Histories of the Future,* ed. Daniel Rosenberg and Susan Harding (Durham, NC: Duke University Press, 2005), p. 23.
[32] In the case of the Mojave, however, this effect will depend on the objectives of the desert traveller. While it is still true that the desert is so vast as to contain destinations remote enough from more "civilized" society, there is, as Reyner Banham points out, plenty of human activity in the place. See Banham, *Scenes in America Deserta,* p. 194.
[33] Victor W. Turner, *The Ritual Process: Structure and Anti-Structure* (New York: Aldine De Gruyter, 1995), p. 95.

[34] For Foucault the experience was said to have clarified his views of his own sexuality. The events are retold in James Miller, *The Passion of Michel Foucault* (London: Harper Collins, 1993), pp. 245-54.
[35] Davis, *The Visionary State*, p. 184; on the desert mystics / desert fathers, see Jasper, *Sacred Desert,* pp. 25-41.
[36] Thomson, *In Nevada,* p. 33.
[37] Mircea Eliade, *The Myth of the Eternal Return, or Cosmos and History,* translated from the French by Willard R. Trask (Princeton, NJ: Princeton University Press, 1971), pp. 9. For a consideration of waste and the sacred within the context of contemporary literature and culture, see William G. Little, *The Waste Fix: Seizures of the Sacred from Upton Sinclair to The Sopranos* (New York and London: Routledge, 2002)
[38] Davis, *The Visionary Landscape,* p. 245.
[39] *Ibid.,* p. 246.
[40] Quoted in Thomson, *In Nevada,* p. 44.
[41] Solnit, *Savage Dreams*, p. 8.
[42] *Ibid.,* p. 8.
[43] Masco, "A Notebook on Desert Modernism," p. 23.
[44] Kuletz, *The Tainted Desert*, p. 70. On the California side of the desert, one can find Edwards Air Force Base, and try to find the highly secret, long off-the-map, China Lake, which Kuletz describes as "one of the largest weapons research and testing facilities in the world' where scientists and technicians "work hidden behind mountains and in desert territories that are strictly guarded" (pp. xiv-xv).
[45] Masco, "A Notebook on Desert Modernism," pp. 28-29.
[46] Richard White, "Trashing the Trails," in *Trails: Towards a New Western History,* eds. Patricia Limerick, Clyde A. Milner ii and Charles E. Rankin (Lawrence, KS: University Press of Kansas), p. 26.
[47] Gerard J. DeGroot, *The Bomb: A Life* (Cambridge, MA: Harvard University Press, 2005), p. 279.
[48] Peter C. van Wyck, *Signs of Danger: Waste, Trauma, and Nuclear Threat* (Minneapolis, MN: University of Minnesota Press, 2004), p. 5.
[49] *Ibid.*
[50] *Ibid.,* p. 23.
[51] Writing in 1998, Valerie Kuletz noted the extent of the United States' nuclear waste production, by pointing to the fact that Yucca was to be the destination for the spent radioactive fuel rods of twenty American states. See Kuletz, *The Tainted Desert,* p. 93.
[52] And as van Wyck notes, this can only be a model for a viable future when one takes into account that some nuclear by-products "such as Uranian-238 … take 4.5 million years to decay only partially." See van Wyck, *Signs of Danger,* pp. 17-18.
[53] Masco, "A Notebook on Desert Modernism," p. 33.
[54] *Ibid.*, p. 36.
[55] *Ibid.,* pp. 5-6.
[56] Quoted in Kuletz, *Tainted Desert,* p. 67.
[57] On the science cities of the desert, see Kuletz, *Tainted Desert,* pp. 38-120.

[58] Solnit, *Savage Dreams,* p. 6.
[59] DeGroot, *The Bomb,* p. 340.
[60] On "simultaneous distraction', see Norman M. Klein, *The History of Forgetting: Los Angeles and the Erasure of Memory* (London: Verso, 1997), p. 13.
[61] Davis, *Dead Cities,* p. 37.
[62] Fox, *The Void, the Grid & the Sign*, p. 39.
[63] Davis, *Dead Cities,* p. 43.
[64] John Gregory Dunne, from his 1974 novel, *Vegas.* Quoted in Neil Campbell, *The Cultures of the New American West* (Edinburgh: Edinburgh University Press, 2000), p. 154.
[65] Banham, *Scenes in America Deserta,* p. 41.
[66] For example, Jean Baudrillard, *America*, trans. Chris Turner (London: Verso, 1988); William L. Fox, *Desert of Desire: Las Vegas and the Culture of the Spectacle* (Reno, NV: University of Las Vegas Press, 2005); Banham, *Scenes in America Deserta.*
[67] Fox, *Desert of Desire*, p. 3.
[68] *Ibid.,* p. 3.
[69] Baudrillard, *America*, p. 67.
[70] For examples, see Arthur Flowers and Anthony Curtis, *The Art of Gambling through the Ages* (Las Vegas, NV: Huntington Press, 2000). On the association between gambling and death, see Sigmund Freud, 'Dostoevsky and Parricide,' in *The Standard Edition of the Complete Psychological Works of Sigmund Freud*, Vol. 5, ed. J. Strachey (London, 1953); Arlene Kramer Richards and Arnold David Richards, 'Gambling, Death, and Violence: Hollywood Looks at Las Vegas,' in *The Psychoanalytic Review*, vol. 84 (5) (October 1997), pp. 769-788.
[71] Davis, *Dead Cities,* p. 86. On the other hand I can't resist recording for posterity at this point the fleeting observation of the uk-based Australian television presenter, Clive James, who, on a visit to Las Vegas, attributed a deal more cunning to the mobsters who established the city; specifically in Bugsy Seigel's reasoning that there was no need to rob people if you could provide them the means to rob themselves.
[72] Bruce Bégout, *Zeropolis: The Experience of Las Vegas* (London: Reaktion, 2003), p. 17.
[73] Davis, *Dead Cities,* p. 87.
[74] Bégout, *Zeropolis*, p. 22.
[75] The railroad arrived at Las Vegas Rancho, as it was, in 1904-05. See David F. Myrick, *Railroads of Nevada and Eastern California: the Southern Roads* (Reno, NV: University of Las Vegas Press), p. 458: "In 1912 Las Vegas, Nevada was a typical frontier town comprised of tents, false-front buildings and more substantial structures. Horses were the prime motive power and dirt streets prevailed."
[76] On Lansky see Denton and Morris, *The Money and the Power*, pp. 27-29. Mike Davis discusses the impact of the city's major commercial interests belonging to the unincorporated township called "Paradise," rather than within the city of Las Vegas proper which, he suggests, creates a "diabolical political geography con-

ceived to separate tax resources from regional social needs." Davis, *Dead Cities*, p. 93.

[77] On high society or, aristocratic, gambling, see Thomas M. Kavanagh, *Dice, Cards, Wheels: A Different History of French Culture* (Philadelphia, PA: University of Pennsylvania Press, 2005).

[78] On the space-time of gambling experience see, for instance, my essays "Combustion: An Essay on the Value of Gambling," in *Gambling: Who Wins? Who Loses?,* ed. Gerda Reith (New York, 2003), pp. 348-354, and "Gambling in Mythical Temporality: Ontological Excess and Virtual Reality," in *International Political Anthropology,* No. 2 (2009), pp. 179-196.

[79] Davis, *Dead Cities,* p. 86.

[80] Peter Goin, "Discordant Infrastructure," in *The Grit Beneath the Glitter: Tales from the Real Las Vegas,* ed. Hal Rothman and Mike Davis (Berkeley, CA: University of California Press, 1992), p. 59.

[81] Hal Rothman, *Neon Metropolis: How Las Vegas Started the Twenty-First Century* (London and New York: Routledge, 2003), p. xxviii.

[82] *Ibid.*, p. 59.

[83] Sara Chapin, however, argues that the city now plays a "game of effective history," assimilated into the tourist economy, and evident in attractions such as the Yesco company's Neon Museum. One might suggest that these dismantled, yet iconic, signs of old attractions, scattered indiscriminately across this "boneyard" attain some kind of poetic reversal that tugs at the seemingly insatiable contemporary craving for nostalgia. For Chapin, this "mythologised collective memory … plays with the image of Las Vegas as a place of constant change." See Sarah Chapin, "Heterotopia Deserta: Las Vegas and Other Spaces," in *Intersections: Architectural Histories and Critical Theories,* eds. Iain Borden and Jane Rendall (London: Routledge, 2000), pp. 203-220; quotations, p. 215.

[84] Famously celebrated in Robert Venturi, Denise Scott Brown and Steven Izenour, *Learning from Las Vegas: the Forgotten Symbols of Architectural Form* (1972; Cambridge, ma: mit Press, 1977)

[85] F. Anderton and J. Chase, *Las Vegas: A Guide to Recent Architecture* (London: Ellipsis, 1997).

[86] The quotation is from Mario Puzo, *Inside Las Vegas* (New York: Grosset & Dunlap, 1976).

[87] Denton and Morris, *The Money and the Power,* p. 6.

[88] Heidegger, *Nietzsche, Volume i,* p. 194.

[89] Denton and Morris, *The Money and the Power,* p. 6.

[90] Peter Guralnick, *Careless Love: The Unmaking of Elvis Presley* (London: Abacus, 1999), pp. 447-48.

[91] In this sense, one might equally consider the relevance of Georges Bataille's *The Accursed Share, Volume 1* (New York: Zone, 1989) to such an analysis.

[92] Baudrillard, *America,* p. 128.

[93] Baudrillard, *Revenge of the Crystal* (London: Pluto Press, 1999), p. 184.

[94] Doc Pomus and Mort Shuman, "Viva Las Vegas" [song], recorded by Elvis Presley (rca Records, 1963). In Aristotle, the term *to automaton* is usually translat-

ed as "chance." For an exploration of this connection, see Scanlan, "Gambling in Mythical Temporality."

[95] Denton and Morris, *The Money and the Power,* pp. 3-7.

[96] Martin Scorsese (dir.) *Casino*, from a screenplay by Nicolas Pileggi, 1995. I am quoting from the film dialogue.

[97] Hal Rothman, "Colony, Capital and Casino: Money in the Real Las Vegas," in *The Grit Beneath the Glitter,* ed. Rothman and Davis, pp. 307-08.

Chapter Four

Exhaust: On Aerial Rejectamenta

Steven Connor

The history of air pollution and efforts to combat it has been well-served. There are histories of the growth of different polluting influences, from industrialism to the automobile. There are also studies of the growth of environmentalist pressure groups and legal frameworks for the control of aerial effluents like smoke and exhaust fumes. The cultural phenomenology, or interior life, of these developments has been attended to in much less detail. My aim in this talk is to follow out some of the changes in sensibility brought about by the idea and experience of air pollution. How, I will wonder, has the *idea of air* changed as a result of the increased awareness of the sensitivity and the fragility of the atmosphere?

Most creatures deal with their waste products in two ways. First of all, they discharge, or put their wastes away from them. More importantly, they themselves move away from their wastes. The nomad does not merely put waste aside—he puts it behind, in the past, which it then becomes. As human beings settle, cultivating rather than grazing the resources, two alternative movements are set up. On the one hand, there must be a concentration, an influx of resources—fields must be irrigated, workers brought together, food stocks built up. There must be synchronization, not just of human affairs with natural, but also of human beings with themselves, in regular work patterns. At the same time, waste products that might previously have been left behind, must be mobilized. As human beings become sedentary, their goods and wastes must be got on the move. Nomadic peoples are themselves fluent, fluctuant; sedentary peoples must create and manage flow, the itineraries of input and output, around them. Hence the importance of rivers, by which many early settlements are formed: rivers bring resources in and take them away, creating a movement through and around spaces of human habitation. But rivers may gradually prove inadequate to cope with the volumes of waste that cities produce, as happened during the years of London's expansion during the nineteenth century. The creation of a sewage system created the need for a new form of map, which had not previously existed, or been necessary—a detailed map of gradients which would ensure that sewage could flow out of the city, at the right speeds and in the right directions.

Previously, human temporality been formed by itinerary and circulation. Space was temporal (the hills in summer, the plains in winter). The establishment of settlements and cities spatialized time. Time had to be managed, in a series of flows inwards and outwards.

Waste of Space

This active concern with expulsion means that sedentary human beings are forced into more and more intimate relation to their wastes, in the very intensity of their efforts to drive them away. The fortunes of the word "waste" itself may help to dramatize this. "Waste" derives from Latin "vastus," which means empty, wild, uninhabited, and perhaps uninhabitable. It referred to wilderness and desert and, as its afterlife in the word "vast" suggests, its principal component is the idea of immeasurability or formlessness. The "vast" is that which human beings cannot encompass or inhabit, in fact or imagination. Already in Latin, the word "vastare" suggested the possibility that human beings could themselves lay waste, could "devastate." Gradually, "waste land" came to mean, not mere unoccupied vastness, but land that had been exhausted, used up, rendered unfit for human habitation.

For a mobile people, waste is the past. It no longer exists. A turd is exactly equivalent to an uttered word—no sooner has it risen to the lips than it is irretrievably gone. Once settlements and civilizations arise, and records begin to be kept, once peoples begin to keep time, litter and literature together ensure and require a continuing relation to one's past. Perhaps this is the opposite of the situation endured by Benjamin's Angel of History, blown irresistibly backwards into the future by the gale of time that blows from Paradise, watching the wreckage of history accumulating at his feet. Modernity must always confront the fear that the propelling gale will flag, or worse still, change its course, blowing backwards from the future, and stranding it in the midst of its own asphyxiating waste.

One may suppose that animals do not suffer from agoraphobia, from a sense of the pressing, exposing, terrifying emptiness of the world. To be sure, most animals have a strong instinct for the particular niches or environments in which they exist and on which they depend: mice, spiders, frogs, all depend on darkness and cover, and very few animals, not even the birds of the air, are able to make a living in the open. Animals do not have a conception of the world, do not have a way of including what excludes them. Human beings, by contrast, whose architecture and culture seem agoraphobic, designed to keep vastness at bay, nevertheless have a relation to vastness, have a conception of the inconceivable, the infinite. They can make of the vast a niche.

The hugeness of the mountain or the plain, stretching out beyond the eye, or the vastness of the ocean, have traditionally stood as synecdoches of the illimitable. But, more than earth or sea, both of which it has long been possible for human beings to traverse, it is the element of air that has stood for the vast. For most of human history, the air has been on the one hand intimately present to human beings and, in the form of the weather, actively influential in human affairs while yet being unknowable and unencompassable. This is because the air is more than a mere element. The air is a quality, a dimension, rather than a mere substance. Where the other elements occupy space, air has habitually been thought of as space itself, the space within which the other elements are disposed. As such, more than the earth or the sea, the air has been seen as the inexhaustible, a pure gratuity.

Luce Irigaray has attempted to reinstate this tradition of thinking in her anamnesis of air. For all its talk of opening and unconcealing, she urges, the philosophy of Heidegger, and the male philosophy he exemplifies, effects a claustration of the pure openness of the air, a hardening and appropriation of what otherwise might be apprehensible as pure excess:

> Is not air the whole of our habitation as mortals? Is there a dwelling more vast, more spacious, or even more generally peaceful than that of air? Can man live elsewhere than in air? Neither in earth, nor in fire, nor in water is any habitation possible for him. No other element can for him take the place of place. No other element carries with it—or lets itself be passed through by—light and shadow, voice and silence. No other element is to this extent opening itself . . . Were there not air—and always more of it than its consumption by living beings requires, and always more than that air now surrounding them, in which everything appears to them—were there not an unlimited and always irritatingly excessive resource of air, the open expanse would not take place.[1]

The air is vast, in both senses, then: it is immense, and it is waste, never fully inhabitable by man. The air is a nothing, an out of sight and out of mind. Air does not present itself—"Air never takes place in the mode of an 'entry into presence' … in air [the philosopher] does not come up against a being or a thing."[2] This enables air to be thought of as a universal solvent, the origin and destination of all things. The air thus can become an edgeless, bottomless receptacle for waste.

Time is of the essence of waste management. Aeroplanes and trains used to dribble their wastes unconsciously as they went along, in real time, like birds or cattle. All one had to remember was not to flush the toilet in a station, which was equivalent to fouling one's own nest. But to excrete in transit or at speed was to return to the mode of the nomad, who risks no contamination from his wastes, because he leaves them behind him, in the sand, on the road, in the past. One used to be able to look down through the bottom of the lavatory pan and see the railway track whizzing along underneath. The ideal of shitting into air is to be found enacted early on in the location of privies in castles, which tended to be on the outside corners, in order to allow the wastes to drop straight into the moat below. Early writers on privies praised the ideal location of the privy between the earth and sky, and recommended it for that reason as an appropriate place for contemplation and prayer.

Nowadays, train and aeroplane toilets do not simply trail human waste behind them in clouds of glory. Rather, they gather and conserve wastes in sealed chemical cisterns for later disposal. The wastes are thereby taken out of time, just as the journey itself is an intermission in time. And yet the manner in which these wastes are removed intimates the survival of an earlier belief in the air's infinite powers to consume and dissolve. There seems to be no water involved in the vacuum-driven flushing of a toilet on a plane. Rather, the excremental products are gulped by a sudden, violent spasm of air (strong enough to seize a wallet, we are assured), as though just for a moment a vent had been open to the freezing, howling, healing waste outside. We may know rationally that the human waste will form part of the aircraft's cargo, rendering it a giant pooper-scooper, which must be sluiced out on arrival at its destination. We have all of us, after all, heard the stories of windshields being shattered by strange blue lumps of ice falling out of the sky underneath flight paths, but our fantasy is fed that our feeble stream has been plucked up by the jet stream.

The air has been the guarantee of the non-historical temporality of the earth—a temporality founded on fluctuations and recurrences without direction or purpose. The air has time only in one of the two senses in which the French word "temps" can be used—weather rather than history. This sense of the ahistorical temporality of the elements provided the limitless resource on which human accumulation through history has been built. Now, that relation seems to be failing. As Michel Serres has said, not only has history entered nature, but nature has entered history. In a special application of that principle, the element of air, the bearer of the weather that has always embodied pure fluctuation and cyclicity as against the irreversible linearities of history, has crossed over: now, it seems, the air does have a history, and one in which we are implicated.

In fact, Irigaray's female dream of infinitude is the metaphysics of air, not Heidegger's alleged encrusting reduction of air. The metaphysics of air makes air bear the burden of the illimitable, the inexhaustible, Confronted with the illimitable and the inexhaustible, man will always see it as a toilet. Like the ocean, the atmosphere has functioned for centuries as a vast oubliette, into which noxious fumes and vapours could be discharged, in the certainty that they would dissipate. The problem, not to say the disaster, is not the "male" finitizing of the air, of which Irigaray accuses Heidegger and others, but the infinitizing of the air, the belief in the air as a horn of plenty, a bottomless fund of vastness, emptiness, openness, which can never be overdrawn. The air is not so much a way of having your cake and eating it, as of making your waste and never having to see it again.

The loss of belief in the powers of air to vaporize and dissipate waste is part of that finitizing of the air which has taken place from the late eighteenth century onwards - the loss of belief in the air as the abode of the endless. Air has come down to earth. It has become a quantity rather than a dimension. Rather than extending beyond sight, knowledge and belief, merging into the celestial ether of the upper air, the air has come to be seen as no more than a thin and increasingly patchy rind of breathable gas clinging wispily to the surface of the planet.

The word exhaust derives from *exhaurire*, to draw out or draw off. This was its primary meaning up until the beginning of the seventeenth century. Exhausting was the word used for medicinal drawing of blood, for example. The idea of giving or being drawn off could suggest abundance, as in a sermon preached by Lancelot Andrewes on Easter Day 1606, which refers to Christ's blood "the vertue and vigor whereof, doth still continue as a fountaine in exhaust, never dry; but flowing still as fresh, as the very first day, His side was first opened."[3] The exhaustion of Christ's blood is the sign that "*His goings out are everlasting.*"[4] Only by interpretation did exhaustion come to be extended from the idea of drawing off to that of emptying out, for example in Robert Boyle's cylinder, exhausted of air. In a later sermon preached by Nathaniel Heywood, we are told that:

> 'tis not enough to have Christ, unless you use him … so we must be always drawing, and drinking, and deriving good from Christ, as the branch makes the best of the root, and draws from it, as though it would exhaust all its life and vertue.[5]

Van Helmont saw in the universal and ceaseless process of exhaustion the secret of death itself:

> unless the blood, yea the very sanies or purulent effluxions from Wounds and Ulcers, the Urine, and that subtle effluvium, which by insensible transpiration evaporates through the pores of the skin, did continually exhaust, and carry with them some part of the vital spirit; and unless these had also some participation of vitality, and conspiracy with the whole body, after their remove from the whole concretum: Undoubtedly the life of man could not be so short. For indeed this is the cause of our intestine calamity, and that principle of death we carry about us, ambuscadoed in the very principles of life.[6]

Thus the object of exhausting came gradually to refer, not to that which was drawn off, but rather the receptacle from which it was drawn off—in Boyle's case, for example the glass which was exhausted by his air-pump, or the chamber in the famous experiment at Magdeburg.[7]

It was not until the end of the nineteenth century that the nominal form of the word "exhaust" started to refer, not to the action of drawing out or emptying, but to the product of the process. Despite the chimneys that sprouted up across Britain during the nineteenth century, it was not until the coming of the automobile that the gases vented from exhaust pipes came regularly to be known as "exhaust."

Exhaust therefore becomes a kind of diffused negativity. Rather than drawing out the air, or the life-principle it embodied, the air itself was rendered exhausted.

The Ambivalence of Air

After nearly 70 years of debate and attempted legislation, from the first Act to try to mitigate the effects of urban smoke in 1821, through to the Alkali Acts of the 1870s, which attempted to regulate sulphurous and other less visible emissions, the 1890s saw a succession of intense fogs and smogs, probably caused by a mixture of environmental conditions and domestic and industrial emissions.

The late nineteenth century cultivated this ambivalence towards exhaust and aerial waste with a curious intensity and versatility. As Peter Brimblecombe has indicated, the smokes and fogs of London were seen as much as an enchantment as a curse. Even Dickens, who took fog as the image of a benighted condition of obfuscation and paralysis in public life, praised the fogs of London. Sara Jeannette Duncan, a Canadian novelist who visited London, even found something to admire in its smell:

> There was the smell to begin with … always more pronounced in the heart of the City, than in Kensington for instance. It was no special odour or collection of odours that could be distinguished—it was a rather abstract smell—and yet it gave a kind of solidity and nutriment to the air, and made you feel as if your lungs digested it. There was comfort and support and satisfaction in that smell.[8]

Fog is also bound up with a growing suspicion of transparency and attraction towards obscurity and blur among artists and writers of early modernism. In my essay "Haze," I have suggested that a kind of nebular modernism arose during the late nineteenth century, a modernism focussing on the rich ambivalence of mist, fog and haze, rather than the burning clarity and transparency.[9]

The ambivalence towards air pollution is registered in a number of apocalyptic fictions having as their subjects the devastation of the air. One of the earliest to appear was Robert Barr's story "The Doom of London" (1892).

The story is told by an old man, reflecting back on the 1890s in which Barr wrote his story from the mid twentieth century. Both gas and electricity have been rendered obsolete by the discovery of "vibratory ether', which provides all the necessary power (the 1890s was the high point of the ether theory in nineteenth-century popular understanding, the ether being conceived as a kind of a higher, purer, more powerful air). The narrator tells the story of a lethal fog that gathers one day after London, formed from a combination of water vapour rising from the ground and smoke belched out from domestic chimneys; the latter pressing down upon the former, prevented both the light and the heat of the sun from getting through. The narrator is approached by an American trying to persuade him to adopt a portable oxygen-dispenser. He declines him, but is made a gift of the dispenser anyway. The fog comes, getting thicker and thicker:

> We were, although we did not know it, under an air-proof canopy, and were slowly but surely exhausting the life-giving oxygen around us, and replacing it by poisonous carbonic acid gas.[10]

Once the oxygen in the air begins to fail, Londoners expire in their millions. Only the narrator survives, in the "oasis of oxygen" furnished by his acclimatizing machine. He eventually escapes London on a train, sharing the machine with its driver, though by now all the passengers on the train are dead. The fate of London is regarded by the inhabitants of the mid twentieth century as a purgation.

Barr's story appears to be elaborated by Arthur Conan Doyle's story of urban apocalypse, *The Poison Belt*, is not the build-up of industrial and domestic wastes that represents the danger, but a belt of poison gas in space through which the earth itself is passing. The story, narrated by a young journalist called Malone, describes the last day and night of life on earth that he spends in the company of Professor Geoffrey Challenger, and the other members of the band of adventurers who first appeared in Doyle's story *The Lost World.* The very possibility of the story rests upon a central incoherence. The belt of instellar poison is said to consist of a disturbance in the ether, the almost infinitely tenuous substance extended throughout all space, and, in the undulatory forms of heat, light, electricity and so on, also passing through all matter. The four adventurers survive the passage through the poison belt when the rest of the planet is exterminated because they remember to bring with them to Geoffrey Challenger's house in Sussex cylinders of oxygen. One of the band, the sceptical Summerlee, objects to the narrative *donné* which Doyle has devised to keep them alive and narratable:

> But what can oxygen effect in the face of a poisoning of the ether? There is not a greater difference in quality between a brick-bat and a gas than there is between oxygen and ether. They are different planes of matter. They cannot impinge upon one another.[11]

The other peculiarity of this poisonous ether is that it seems to behave much more like a flood of water than a cosmic gas. Those on high ground are affected much later than those in the plans, and women, children and inferior peoples succumb to the effects of the poison much more readily than sturdy, stiff-backed Englishmen like Challenger and his indomitable party. The first signs of the effects of the cosmic poison are felt among "the indigenous races of Sumatra," who seem to respond more quickly to the cosmic conditions than "the more complex peoples of Europe." Although death for all seems certain in the end—painless but inevitable—death for young and old, for weak and strong, for rich and poor, without hope or possibility of escape. There is a differential topography of annihilation, partly because the preliminary effects of the poison involve disturbances of the rational faculties. The telephone in Challenger's house brings news of the slow spread from south to north:

> The great shadow was creeping up from the south like a rising tide of death. Egypt had gone through its delirium and was now comatose. Spain and Portugal, after a wild frenzy in which the Clericals and the Anarchists had fought most desperately, were now fallen silent. No cable messages

> were received any longer from South America. In North America the southern states, after some terrible racial rioting, had succumbed to the poison. North of Maryland the effect was not yet marked, and in Canada it was hardly perceptible. Belgium, Holland, and Denmark had each in turn been affected.[12]

This is interesting despite, or in fact because of its implausibility. "Ether," which for so many was imagined as a kind of idealization of the air—air refined and attenuated to the ultimate degree, and thus spiritualized—here becomes an annihilating, asphyxiating agent. The poison belt of malignant ether, into which the earth passes purely by chance, provides an image of the air both desublimated and moralized. The uncertainty as to whether ether can be thought of as material or immaterial carries this ambivalence towards the air. Doyle is attempting to hold together the two dimensions of the air—the air as infinite space, maximal communicability, and the air as finitely subject to the laws of spatial distribution. Ether represents the infinite space that has been slowed and thickened into place.

In fact, it turns out that the great extermination is merely temporary—that the gas acts as an anaesthetic inducing catalepsy, rather than actually killing its victims. This means that the end of the world is followed by a great awakening, which is spiritual as well as physical, and has longer-term effects, in the turn away from a culture of excess and consumption:

> Surely we are agreed that the more sober and restrained pleasures of the present are deeper as well as wiser than the noisy, foolish hustle which passed so often for enjoyment in the days of old—days so recent and yet already so inconceivable. Those empty lives which were wasted in aimless visiting and being visited, in the worry of great and unnecessary households, in the arranging and eating of elaborate and tedious meals, have now found rest and health in the reading, the music, the gentle family communion which comes from a simpler and saner division of their time. With greater health and greater pleasure they are richer than before, even after they have paid those increased contributions to the common fund which have so raised the standard of life in these islands.[13]

Thus, the poison turns out to be sanative and restorative. Doyle fantasies a world of proportion and limit, brought about by the accident of a cosmic encounter.

As we have come to draw the air into our sphere, even to come ourselves to occupy the mid-air in which our mass communications and movements take place, we have indeed, as Irigaray charges, forgotten the dream of the infinitude of the air. But that dream is fatally implicated in the history of man's efforts to put waste definitively put of sight and out of mind in the all-encompassing and all-consuming invisibility of the air. The philosophical discussion of waste has, since Bataille, been tempted by the pull of the unaccountable, of pure gratuity or excess.[14] The compromising of the idea of air that I have briefly followed through here suggests that we should suspect and resist the return of infinitude, the modernist yearning to escape place, limit and economy. By contrast, this narrowing, this claustration, this accounting for and economizing on air may be the only way in which we can temper our dream of the air as a vastness that we can never lay waste.

Notes

[1] Luce Irigaray, *The Forgetting of Air in Martin Heidegger*, trans. translated by Mary Beth Mader (Austin, TX: University of Texas Press, 1999), pp. 8, 40.
[2] Irigaray, *The Forgetting of Air in Martin Heidegger*, p. 9.
[3] Lancelot Andrewes, xcvi, *Sermons by the Right Honorable and Reverend Father in God, Lancelot Andrewes, late Lord Bishop of Winchester* (London: George Miller, for Richard Badger, 1629), p. 584.
[4] *Ibid.*, p. 584.
[5] Nathaniel Heywood, *Christ Displayed as the Choicest Gift, and Best Master...* (London: Thomas Parkhurst, 1679), p. 118.
[6] Johan Baptist Van Helmont, *A Ternary of Paradoxes: The Magnetick Cure of Wounds, Nativity of Tartar in Wine, Image of God in Man,* trans. Walter Charleton (London: James Flesher for William Lee, 1650), p. 15.
[7] Robert Boyle, *Experiments, Notes, &c. About the Mechanical Origine or Production of Divers Particular Qualities ...* (London: E. Flesher for R. Davis, 1676), p. 29; Robert Boyle, *New Experiments Physico-mechanical, Touching the Air* (London: Miles Flesher for Richard Davis, 1682), p. 17.
[8] Sara Jeanette Duncan, quoted in Peter Brimblecombe, *The Big Smoke: A History of Air Pollution in London Since Medieval Times* (London: Methuen, 1987), pp. 85-86.
[9] Steven Connor, "Haze: On Nebular Modernism," a paper given at *Modernism and Beyond: Interdisciplinary Seminar in Art Theory and Literary Theory*, Trinity College, Oxford, 12 May 2006 (available: www.stevenconnor.com).
[10] Robert Barr, "The Doom of London." (1892) Accessed online at: www.online-literature.com/robert-barr/face-and-the-mask/7/ (Aug. 2009).
[11] Arthur Conan Doyle, *The Poison Belt: Being an Account of Another Amazing Adventure of Professor Challenger* (Lincoln, ne: University of Nebraska Press, 2001), pp. 35-36.
[12] Conan Doyle, *The Poison Belt*, p. 39.
[13] Conan Doyle, *The Poison Belt*, pp. 82-83.
[14] On Georges Bataille and waste, see Steven Connor, *Theory and Cultural Value* (Oxford: Blackwell, 1992).

Chapter Five

Debating Waste, Nature, and Justice: Spencer, Huxley, Morris

Piers J. Hale

Thomas Huxley, the eminent nineteenth-century scientist, once remarked, “It was a shrewd man of the world who, in discussing sewage problems, remarked that dirt is riches in the wrong place!”[1] As Huxley’s comment suggests, exactly what is and what is not waste must necessarily be a discussion that is alive to the contingencies of context—not only of place, but of time. In this essay I suggest that we can profitably look to past debates about the nature and meaning of waste to help us through our own discussion of the issues at stake, to see how others have deployed and debated the term, and the ends to which they did so. What becomes clear from such an analysis is that waste has always been at least an implicitly pejorative term that has always had significant political consequences. Indeed, a look to the past shows that “waste” has frequently not only had environmental implications, but, and often as a result, has also had significant implications for social justice as well.

As the historian Richard White has pointed out, a development of our own industrial and urban history is that waste has come to be thought of largely as a by-product of those most necessary of human activities: production and consumption.[2] However, ever since Thomas R. Malthus wrote his famous essay on the principles of population, commentators have stressed that labour and consumption are also the most necessary part of our existence. Meeting even our basic needs requires the transformation and consumption of nature, and as Malthus pointed out, population will always tend to grow at a higher rate than our ability to provide for it.[3] Indeed, it is in the context of the inherent disequilibrium of the Malthusian ratio of population relative to resources that waste has taken on its current association with entropy in a closed system. Of course, such a neat definition draws our attention towards consideration of standards of efficiency and away from the many other diverse and rich meanings that waste has had throughout its history. While efficiency has certainly dominated discussion of waste since the early-twentieth century, this interpretation has not been either timeless, or without its critics.[4] The juncture of production and consumption has also been the site of the expression of waste in terms of both nature (in terms of both human nature and the natural environment) and social justice. Indeed, if the history of environmental thought tells us anything, it is that these concepts need to be taken seriously, and taken together.[5] Given these concerns, one particularly fruitful field of past debate to which we might turn is that of nineteenth-century political economy and its discontents. Especially worthy of note, given the resonance of their concerns with our own, were the views and opinions of those involved in "early green politics," and who also found their home in the various branches of the British socialist movement.[6] Amongst them, and of particular importance, was the artist, craftsman, and socialist, William Morris.

Morris remains so relevant to this debate because despite the historically antipathetic relationship between environmentalists and socialists (and it might be added non-socialist representatives of labour), he is at once recognized as an important or even founding figure in the respective histories of both environmentalism and socialism. Not only have present-day environmentalists lauded Morris's conception of the human relationship with nature, but people who are deeply concerned with labour and work—socialists amongst them—have also articulated a deep appreciation of Morris's unique and passionate views on the subject.[7] A look back at Morris highlights not only his very real awareness of the diverse implications of the transition from rural to urban living that accompanied the industrial revolution, but also his very real awareness of waste in terms of both social justice—and what we would now call environmental justice.[8]

Urbanization and industrialization not only created the problem of what to do with the unprecedented amount of waste that was produced in such a close-knit and confined area, but also—and important for thinking of waste in the much broader context that I wish to highlight here—raised awareness of the human cost of these processes. Social surveys by Mayhew, Rowntree and others told the tale of the often tragic consequences of industrialization, exposing the squalor and destitution of families crammed into unsanitary slum housing, of the unemployed starving and impoverished, and pushed towards a life of crime or vice as a result.[9] Variously called "the residuum," the "submerged sixth," or similarly dismissive and pejorative term, these urban-dwellers not only lived in the midst of their own waste, but were viewed by many social commentators, from a number of perspectives, *as* waste. At the extremes of the spectrum they were either viewed as the inevitable, if unfortunate, by-product of a successful and thriving economy, or alternatively, as the immoral waste of humanity that resulted from a system that placed the profits of a few above the very lives of the many. Morris was amongst those taking the latter view, and he railed against what he perceived to be the wanton waste of life that went on in the name of commercialism and profit. Morris was by no means alone in this belief, and his views on the subject were influential far beyond the confines of the socialist movement. Robert Blatchford, the editor of and contributor to the popular *Clarion* newspaper also expressed his own deeply moral concerns about industrialization in terms of the waste of human life that resulted from the capitalist focus upon profit rather than people. In his widely read series of letters advocating socialism, which were published together in 1894 as *Merrie England*, he argued:

> It is not only the wrong of this it is the *waste*. Look through the slums ... and see what childhood, girlhood, womanhood and manhood have there become. Think what a waste of beauty, of virtue, of strength, and of all the power and goodness that go to make a nation great is being consumed there by ignorance and injustice … Cast your eyes, then, my good friend, over the Registrar General's returns and imagine if you can how many gentle nurses, good mothers, sweet singers, brave soldiers, and clever artists, inventors and thinkers are swallowed up every year in that ocean of crime and sorrow, which is known to the official mind as "the high death rate of the wage earning classes.[10]

Blatchford openly acknowledged that, in addition to what he witnessed with his own eyes, it had been Morris that had brought him to recognize capitalism as the true cause of this waste of humanity.[11] Morris, however, had been deeply influenced on this point by the art critic and social reformer John Ruskin.[12] Ruskin was no socialist, but his views on labour and art were deeply imbued with moral significance. Ruskin had also lambasted the immoral human consequences of a system based upon unfettered laissez-faire and the pursuit of profit without responsibility:

> The great cry that rises from all our manufacturing cities, louder than the furnace blast, is all in very deed for this—that we manufacture everything there except men; we blanch cotton, and strengthen steel and refine sugar, and shape pottery; but to brighten, to strengthen, to refine or to form a single living spirit never enters into our estimate of advantages.[13]

Before going on to consider Morris's views on waste it is necessary to put them in context, and in doing so to recognize that nineteenth-century languages of waste, like our own, need to be understood not only in relation to the development of hotly contested discourses of political economy, but as we shall see, also in relation to the equally disputed field of natural history. What is ultimately at stake here at this intersection between the political and the natural is what we might call "the naturalization of politics," the attempt to present certain political ideas as "natural," and thus, to a great extent, to imbue them with a hegemonic status that all but puts them beyond question. From this view of life the human cost of laissez-faire could be explained, and thus justified, as merely the inevitable outcome of natural processes. In the context of debate about nineteenth-century political economy and how it relates to waste, there was perhaps no one as influential as the Radical journalist and philosopher, Herbert Spencer.

Spencer, as well as an avid advocate of laissez-faire economics and natural rights also coined the phrase "the survival of the fittest" that Darwin later adopted as a less anthropomorphic description of the mechanism of "natural selection" at the heart of his theory of evolution. Significantly, Spencer based his early Radical claims to bourgeois natural rights upon the occurrence of waste in natural physiological systems and processes and indeed was responsible for popularizing the metaphor of society as a "social organism."[14] Beginning from a description of the depletion of physical energy and the "waste" of the human muscles when put to work, Spencer constructed an argument for natural rights upon the existence of basic physiological needs. This was an element of his thought that persisted. In *The Great Political Superstition* of 1884 he wrote:

> Let us, then, transfer this question of natural rights from the court of politics to the court of science—the science of life. The reader need feel no alarm: the simplest and most obvious facts will suffice … Animal life involves waste; waste must be met by repair; repair implies nutrition. Again, nutrition presupposes obtainment of food; [and] of locomotion; and that these powers may achieve their ends, there must be freedom to move about … Those who hold that life is valuable, hold, by implication, that men ought not to be prevented from carrying on life-sustaining activities. In other words, if it is said to be "right" that they should carry them on, then, by permutation, we get the assertion that they "have a right" to carry them on.[15]

Having naturalized individual rights of action and movement through the concept of organic waste, repair and development, Spencer transferred similar organic assumptions back onto society. Just as in the operation of the muscles there was a certain amount of natural and inevitable wastage, so, too, in the operation of society's muscles, its industrial productive forces, there would also be inevitable natural wastage. In the case of industrial Britain, however, the wastage was not just the excess material in the production of goods, but was most pointedly realized in the poverty and destitution of the over-worked, under-worked and unemployed and in the pitiful condition of the men, women and children who had been laid low by the demands of unfettered capitalism.

Because natural wastage was tied to Spencer's conception of the proper functioning of the organism, whether biological or social, as a unified confederation of once-autonomous parts, he argued against what he perceived to be the misplaced and mistaken humanitarian sentiments of those "New Liberals" who advocated state intervention. Such authoritarian interventions, he believed, would disrupt the natural and democratic interactions between these parts, which could only derail their progressive and evolutionary development "from homogeneity to heterogeneity."[16]

Also, inspired by Adam Smith's theory of the division of labour, it did not require much of Spencer's analogy for him to claim that state interventions into the natural workings of the capitalist economy were similarly dangerous to progressive social development. Such a move, he believed, could only weaken Britain's position *vis-à-vis* international competition, whilst at the same time swelling the ranks of the unfit.[17] "Animal life involves waste," Spencer concluded, and the throwing-off of the natural waste elements of society would allow society to continue its long-standing progressive evolution.

Locating a defense of free agency in a claim to a natural right to life and, at the same time, in the natural progressive processes of evolutionary development, Spencer justified the basic tenets of laissez-faire economics, a move that found significant support throughout the 1840s and '50s. By the 1870s and '80s, however, and in light of the tragic plight of a great many of the urban poor made clear by Mayhew, Rowntree and others, an increasing number of people began to feel moral qualms about the waste of humanity that resulted from laissez-faire economics. Although Spencer continued to advocate for economic purism, Liberalism capitulated to humanism. Spencer's colleague and formerly close associate, the talented comparative-anatomist and evolutionist Thomas Henry Huxley, who had famously advanced Darwinism as a "veritable Whitworth Gun in the armoury of liberalism," was one such liberal reformer.[18] Huxley had—like Spencer—read the Darwinian competition evident in a nature that he perceived to be very much "Red in Tooth and Claw" as an endorsement of capitalist society. In "The Struggle for Existence in Human Society," written in 1888, Huxley had drawn analogies from nature to society very much to this effect:

> from the point of view of the moralist the animal world is on about the same level as a gladiator's show. The creatures are … set to fight—whereby the strongest, the swiftest and the cunningest live to fight another day. The spectator has no need to turn his thumbs down, as no quarter is given.[19]

Drawing analogous conclusions from natural history and onto society, Huxley continued:

> There is only one reason why our goods should be preferred to those of our rivals—our customers must find them better at the price. That means that we must use more knowledge, skill, and industry in producing them, without a proportionate increase in the cost of production; and, as the price of labour constitutes a large element in that cost, the rate of wages must be restricted within certain limits. Cheapness, then … is essential to our success as competitors in the markets of the world.[20]

Central to Huxley's conception of labour here, however, is not only the notion of cheapness, but also of efficiency, a concept that went on to become a major part of early twentieth-century discourse about labour, political economy and competition.[21] In contrast to Spencer, however, and much to his chagrin, Huxley had come to see the barbarism that resulted from unrestricted competition as ample justification for state intervention and the institution of a measured social reform.[22] Indeed, as articulated most forcefully in his 1871 paper "Administrative Nihilism," Huxley had also been instrumental in the rejection by the scientific community of Spencer's conception of "compound individuality" that was at the heart of his laissez-faire conception of the social organism.[23]

Huxley was acutely aware of the Malthusian economics that had led Darwin to theorize natural selection and fully appreciated the ferocity of the competition for scarce resources that followed. Huxley was sensitive to social issues, and had championed social and educational reform through the Workingmen's College in Red Lion Square. His awareness of the plight of the working poor made him recoil from Spencer's conclusion that struggle in nature confirmed that unfettered competition was not only natural, but necessary for the continued progress of human social evolution.[24] Huxley's convictions regarding the necessity of ethical intervention into the working out of natural processes is nowhere more emphatic and poignantly expressed than in his 1894 "Prologomena" to *Evolution and Ethics.* Writing just one year before his death, Huxley argued that although the competition between men was indeed thoroughly natural, so too was the human impulse to cooperate in order to resist it. Asserting a different conception of social evolution to that outlined by Spencer, he concluded that whilst the struggle for existence might be a fact of nature, there were no grounds for taking it as a model for human ethics. In the March of 1894, he wrote:

> There are two very different questions which people fail to discriminate. One is whether evolution accounts for morality, the other whether the principle of evolution in general can be adopted as an ethical principle. The first, of course, I advocate, and have constantly insisted upon. The second I deny, and reject all so-called evolutional ethics based upon it.[25]

Thus, Huxley argued that ethical interventions to alleviate the suffering that resulted from the unhindered operation of natural law was not only the basis for a humane society, but was the sentiment that had allowed humanity to evolve from its less social forbears. It was this ethical sentiment of "mutual sympathy" that distinguished humanity from the animal world of nature and that separated civilized man not only from his primitive forbears, but from the barbarous instincts that, he contended, lie within us all. He argued that:

> since society differs from nature in having a definite moral object … it comes about that the course shaped by the ethical man—the member of society or the citizen—necessarily runs counter to that which the non-ethical man—the primitive savage, or man as a member of the animal kingdom—tends to adopt. The latter fights out the struggle for existence to the bitter end like any other animal; the former devotes his best energies to the object of setting limits to the struggle.[26]

Significantly, though, this ethical concern was what raised humanity above the amorality of nature and natural law, and thus reinforced rather than challenged the belief that the waste that resulted from the capitalist organization of labour was the result of natural progressive tendencies.

Many socialists denied this view of humanity and of nature, and William Morris was arguably foremost among them.[27] In contrast to Spencer's articulation of natural wastage of strength as a basis for a natural right to freedom of action and free enterprise, Morris utilized the same concept of natural waste as an argument against overwork, against bad conditions, and for adequate periods of rest and recuperation in the working day, adequate nourishment, as well as for work that was stimulating to the mind as well as the body. "This restoration of the waste of life, ought to be looked on as a kind of sacrament," he argued. "It is Nature herself who leads us to desire this pleasure."[28]

Further to this, in "How I Became a Socialist," written in 1894 for *Justice*, the paper of the Social Democratic Federation, Morris had also made clear that Huxley's vision of Man's place in nature was not akin to his own. Indeed, it was largely his refusal of Huxley's portrayal of the same that had led him to give up on his earlier Radical sympathies and move towards socialism.[29] To Morris's mind, Huxley had quite erroneously portrayed humanity as naturally individualistic, stuck in an ongoing struggle for existence against his fellow men as well as against a nature that was antagonistic to human interests. In contrast Morris believed that humans were cooperative and social by nature, and had the potential to live in harmony with their environment. Indeed, he believed that this had been the case prior to the advent of industrialization and the subjection of humanity to capitalism, and thus, as he portrayed in his utopian novel *News from Nowhere* (1891), he believed they might do so again in the future.

Central to Morris's discontent was his denial that the "foolish Malthusian craze" of liberal political economy was based in nature.[30] Rather than believing that the existence of scarcity and poverty in the world was the result of any immutable natural law, as was implicit in the liberal appeal to Malthus, Morris argued that it was through the unjust and wasteful organization of labour in society that people were forced to live degraded and degrading lives, compelled to struggle to make even a mere subsistence wage. Importantly, Morris clearly conceived that waste under capitalism was a matter of justice rather than merely one of efficiency—or, to be more accurate, a matter of grave injustice that was perpetrated in the name of efficiency and natural law. Specifically Morris argued that commercialism wasted labour, humanity, and nature on a number of levels. First, he argued that the fact that only a few people provided the labour that supplied society meant that a great deal of potentially productive labour was left idle. Second, he believed that the vast majority of the labour that was undertaken was spent in the production of useless and shoddy goods. Third, he argued that the pursuit of such work wasted natural resources. Further to this conception of wasting nature, however, and one of the reasons why Morris has had such strong appeal among present-day environmentalists, is that he also argued that capitalism did not just waste nature in terms of irresponsibly depleting limited natural resources, but that the heedless pursuit of profit also wasted both the aesthetic and—on a deeper level—the intrinsic value of the natural world. This had implications for nature, of course, but also, Morris argued, for the possibility of a just and happy future for humankind.

The Waste of Labour

Morris argued that far from being the natural state of things, scarcity resulted from the fact that capitalism was wantonly wasteful of labour. Capitalist society, he observed, was organized so that only a small proportion of the people who were capable of working actually did so. Morris was convinced that if everyone laboured, for even but a few hours a day, then there would be sufficient for all. The fact was, however, that only one class of people laboured to support the whole of society, and "could be said to do nothing else than work" as a result. At the same time there was another class of people, the aristocracy and landed-gentry, who refused to work while they consumed by far the majority, and the best, of what was produced. In between these two classes was a third class of people, the bourgeois middle class. Morris argued that although by all appearances they worked hard, they actually produced nothing useful, and thus were ultimately parasitic upon the labour of those they employed: "As to the middle class," he wrote:

> including the trading, manufacturing, and professional people of our society, they do, as a rule, seem to work quite hard enough, and so at first sight might be thought to help the community, and not burden it. But by far the greater part of them, though they work, do not produce, and even when they do produce, as in the case of those engaged (wastefully indeed) in the distribution of goods, or doctors, or (genuine) artists and literary men, they consume out of all proportion to their due share.[31]

Included among those that were idle or unproductive, were those who Morris classed as "merely parasites of property."[32] Among the ranks of the superfluous and unproductive were the soldiery, domestic servants, clerks, shop-assistants, and others, "who are engaged in the service of the private war for wealth."[33] Indeed, in *News from Nowhere* Morris portrayed children tending the shops and market stalls that the Guest in his utopian future encounters along his way, because he believed shop-keeping, although a pleasant-enough amusement, was "such very easy work that anybody can do it."[34] One of the greatest injustices of capitalism, however, was that in addition to those who lived comfortably enough upon the work of others, there was also a large class of men and women who were forcibly kept from working who would willingly do so. These poor and destitute, bereft of the means to make their own living, were kept from the factories to ensure the continued competition for work that kept wages low.[35]

The workings of this system had results that Morris was not alone in perceiving as profoundly immoral. The idle accumulated ever more wealth, while those that laboured became ever more degraded and impoverished. Although Spencer might attempt to naturalize this growing inequality by reference to the inevitable wastage of a natural economy, Morris could only see in it the perpetration of the most outrageous injustice and "waste most grievous."[36]

Exacerbating this waste of potential labour was the fact that, even of the work that was done, the majority of it was wasted. A great proportion of those who were employed were either forced into the production of harmful and adulterated goods intended for consumption by themselves and their fellow labourers, or else engaged in making useless frippery to appeal to the degraded and unmanly tastes of the rich. "I tell you I feel dazed at the thought of the immensity of work which is undergone for the making of useless things," Morris lamented:

> I am talking of the merely useless things; but there are other matters not merely useless, but actively destructive and poisonous which command a good price in the market; for instance, adulterated food and drink. Vast is the number of slaves whom competitive Commerce employs in turning out infamies such as these. But quite apart from them there is an enormous mass of labour which is just merely wasted; many thousands of men and women make *nothing* with terrible and inhuman toil which deadens the soul and shortens mere animal life itself.
>
> Next there is the mass of people employed in making all those articles of folly and luxury, the demand for which is the outcome of the existence of the rich non-producing classes; things which people leading a manly and uncorrupted life would not ask for or dream of. These things, whoever may gainsay me, I will for ever refuse to call wealth: they are not wealth, but waste."[37]

Foremost among those whose work was not only wasted but harmful were those engaged in both the advertising industry and those responsible for adulterating the goods they ordered made. "The waste is further increased," Morris argued:

> by the necessity of these organisers of the commercial war for playing [sic] on the ignorance and gullibility of the consumers of these two processes … It would be hard to say how much ingenuity and painstaking have been wasted on these incidents in the war of commerce.[38]

The waste of the labour-power involved in these processes was one thing, but much more significant, was that it was also wasteful of the very lives, not only of those who were forced to produce them, but those who consumed them as well.

The Waste of Humanity

Morris believed that labour was not only a necessary occupation, but that it was also the most human of occupations. Through work humankind expressed its natural creative impulses and aspirations. Thus, although labour was certainly undertaken to meet the needs of subsistence, this was not its only, or, indeed, even its most significant function. Rather, through labour Morris believed that humanity might finally realize its most true and fulfilled expression. Indeed, although many have viewed Morris as anti-industrial in his aspirations, it is clear that he saw great potential in the evident productive capacity of the mechanization of labour. However, instead of this being employed to make vast profits for the few, as was the case under capitalism, he believed it might otherwise be used to minimize the drudgery of many necessary tasks, and thus set the workers free to labour in an unhurried, thoughtful, and creative manner. Under capitalism, this opportunity was not only denied, but the very practice of labour was twisted and distorted so that instead of being the most fulfilling enterprise and a joy to the labourer, it had become the most degrading and dehumanizing of occupations.[39] Morris was only too aware of the stark contrast between his hopes for labour and the common experience of his contemporaries, a point he made clearly in *News from Nowhere*. One of Morris's utopians, named Dick, upon hearing tales of the nineteenth-century past, remarks: "Fancy people not liking to work!—It's too ridiculous." At this William Guest, the visitor to this utopian future, remarks in a bitter aside to the reader, "I laughed with him for company's sake, but from the teeth outward only; for I saw nothing funny in people not liking to work, as you may well imagine."[40]

The results for humanity of the imposition of the capitalist organisation of labour were not inconsequential. Morris argued that capitalism not only denied the satisfaction of these most basic of human instincts, but, and as a result, it degraded the body, the mind, and the spirit of the worker. Indeed, it is clear from Morris's writings that he believed that the physical degeneration that resulted from such a system would have hereditary implications. Like many of his contemporaries, Morris believed the way in which one lived one's life would have hereditary implications.[41] This was the case not only for the workers, who were forced to expend their lives in the service of capital, but also for the middle classes, whom Morris imagined would gradually become more and more ineffectual and demoralised as the generations passed.[42] In Nowhere, Dick recounts the tale of the middle-class women who in the past had tended the shops and stores as a full-time occupation. As a result of their inactivity, they became "as little like young women as might be: they had hands like bunches of skewers, and wretched little arms like sticks; and waists like hour-glasses, and thin lips and peaked noses and pale cheeks."[43] Where a lack of labour had been the cause of this degeneration, just as forced and degrading labour was the cause of the degeneration of the workers, so unalienated socialist labour was the remedy. "Indeed," Dick reflected,

> I believe that at one time they [the middle class] were actually *compelled* to do some work, because they, especially the women, got so ugly and produced such ugly children if the disease [of idleness] were not treated sharply, that the neighbours couldn't stand it.[44]

As Hammond, the old man who guides the Guest through much of his journey through Nowhere adds, "how to take the sting out of heredity, has for long been one of the most constant cares of the thoughtful men amongst us."[45]

In addition to being physically harmful, Morris believed that capitalist labour was also mentally and spiritually draining. If correctly organised, however he believed that work could potentially be a deeply aesthetic occupation through which humanity might take the materials provided by nature and create things of even greater beauty, as they imparted their creative spirit into the artefact they crafted.[46] It was in this sense that Morris believed that labour could become art. In denying this aesthetic aspect of productive life capitalism spiritually demoralised all who came into contact with it. "There is," wrote Morris, "a great deal of sham work in the world, hurtful to the buyer, more hurtful to the seller, [but] if only he knew it most hurtful to the maker." As a result of this alienation Morris contended that the workers were also divorced from every other aspect of human life. "You will surely see what a hideous nightmare that profit market is," wrote Morris,

> It keeps us sweating and terrified for our livelihood, unable to read a book, or look at a picture or have pleasant fields to walk in, or to lie in the sun, or to share in the knowledge of our time, to have in short either animal or intellectual pleasure and for what? That we may go on living the same slavish life til we die, in order to provide for a rich man.[47]

Waste of Nature

Thus far what is compelling about Morris's conception of waste is that it was intrinsically linked to his ideas about social justice. Implicit to this Morris also clearly recognized that capitalism was wasteful of the earth's natural resources. As I have shown above, he was keenly aware that it would be better if a great many of the things that capitalism produced had never been made, and he famously urged people to "have nothing in your house that you do not know to be useful or believe to be beautiful." Capitalism was not only a waste and degradation to the producer and consumer, but thus also a most unnecessary use of natural resources. In contrast, Morris believed that the goods that were produced under socialistic unalienated working conditions would be much more valued than the shoddy and disposable produce churned out by the capitalist division of labour. Morris anticipated that quality hand-crafted goods might make the basis of a new aesthetic of labour, in which the product would not only be beautiful, but would stand as physical testament to, and embodiment of, the human creative spirit—a creativity that was squandered under capitalism. Goods made to such a benchmark of quality, with human skill and passion, would not be disposable, but rather would be treasured as only true art is:

> A work of Art, be it ever so humble, is long-lived; we never tire of it; as long as a scrap of it holds together it is valuable and instructive to each new generation … for from the first there was a soul in [it], the thought of man, which will be visible in [it] so long as the body exists in which [it] was implanted.

In such a manner resources might be conserved rather than wasted. It was evident to Morris however, that until such a time as society might be transformed, "Civilization therefore wastes its own resources, and will do so as long and the present system lasts."[48]

Going beyond many of his socialist colleagues in this respect, Morris also clearly recognized that capitalism was wasteful of nature at a much deeper level. Not only was it wasteful of raw materials, but it was also wasteful of the naturally occurring beauty of the landscape, and this was tragic not only for its impact upon humanity, but also for nature's own sake. Morris was only too aware of the inroads that industry was having into previously rural and beautiful areas. In "Under an Elm Tree," a short prose essay that he wrote for his own socialist journal *Commonweal* in 1889, he had made exactly this point. Reflecting upon the likely fate of the rural areas of the Thames valley, Morris lamented:

> The beauty of the landscape will be exploited … in quiet places like this it will vanish year by year (as indeed it is now doing) under the attacks of the most groveling commercialism.[49]

No one, Morris argued, should "be allowed to cut down, for mere profit, trees whose loss would spoil a landscape":

> neither on any pretext should people be allowed to darken the daylight with smoke, to befoul rivers, or to degrade any spot of earth with squalid litter and brutal wasteful disorder.[50]

Labour was thus to be measured against its impact upon the beauty of the natural world, as well as upon humanity. "Everything made by man's hands," Morris wrote, "has a form, which must be either beautiful or ugly: beautiful if it is in accord with Nature, and helps her; ugly if it is discordant with Nature, and thwarts her."[51] Indeed, Morris hypothesized that it might be better to do without the benefits of any labour that did not meet these criteria.[52] Unlike his socialist contemporaries, Morris's imagined future ideal was based upon conservation strategies in the present. On this basis, he argued for human responsibility for the future beauty of the earth. "We who are part of civilization are responsible to posterity for what may befall the fairness of the earth in our own days," he wrote. "Each one of us … must keep watch and ward … lest we deliver to our sons a lesser treasure than our fathers left to us."[53]

The practice of labour such as that undertaken by Morris in the production of his furniture, wallpaper designs and the Kelmscott Press was an intrinsic part of Morris's strategy for social change. The production of aesthetically beautiful objects would serve to expose the shoddy nature of the work that capitalism forced upon people and would act to remind the workers of what they had lost.[54] "We at least have remembered …," Morris declared in an address to fellow craftmen, "that it is possible to be happy, that labour may be a pleasure; and that the essence of pleasure abides in labour."[55] Indeed, objects produced by such labour would embody the very life and labour of the craftsman who had created them. Morris argued that because "the ethics of the artist bid him put as much of himself as he can in every piece of goods he makes," he would thus instil the very "soul" and "thought of man" into his product.[56] Such an object would be "long lived, [people would] never tire of it; as long as a scrap of it holds together it [would be] valuable and instructive to each new generation."[57] The existence of the products of socially produced labour would connect the labour of the present and the future with that which had gone on before capitalism and thus create an awareness of the historical ties that link humanity. "Memory and imagination help him as he works," Morris wrote, "not only his own thoughts, but the thoughts of men of past ages guide his hands, and as a part of the human race he creates."[58] In addition, the very existence of these objects would help to transform both humanity and society. Morris asserting that "it helps the healthiness of both body and soul to live among beautiful things."[59]

Conclusion

Although today we may tend to think of waste in a much more limited way than Morris did, it is clear that if we focus upon waste as merely the outcome of inefficiency, as seems to be the case in much of the current environmental literature, then the focus of our concerns will tend to fall upon such issues as the relative merits of carbon-credits versus the taxation of emissions, with only incidental attention paid to conditions of social and environmental justice. Thus a reconsideration of the ways in which waste was deployed in the context of what Peter C. Gould has termed "early green" responses to industrialization and urbanization is important for a number of reasons.[60] First, it reminds us that the concept of waste is perhaps much more contingent and contested than is evident from the parameters of how we currently tend to think about it. "Waste," it seems, is a term that has always been bound up with ideological assumptions and agendas. Second, it reminds us that when we talk about waste in terms of pollution and the by-products of industrial processes, what is also at stake, although often unspoken, are the working conditions, and working practices of those employed. Further to this, however, these past discussions of waste demand that we take our consideration beyond even this. Collectively, we must also seriously consider the implications of our present practices of production and consumption for what it means to be human, to ask what kind of lives we want to lead, and to think beyond short-term profitability to imagine seriously—as Morris did—what kind of world we most desire to live in. To Morris's mind:

> This would at the least be a society which would try to perform those functions of seeing that everyone did his fare share of work and no more, and that no labour was wasted.[61]

Notes

[1] Thomas Henry Huxley, "Prologomena" to "Evolution and Ethics" [1894], reprinted in James Paradis and George Williams, *Evolution and Ethics: T.H. Huxley's Evolution and Ethics, with New Essays on Its Victorian and Sociobiological Context* (Princeton: Princeton University Press, 1989), p. 39. [The pagination for Huxley's "Evolution and Ethics" refers to the facsimile reprint of the 1894 edition included in this volume, in which p. 1 corresponds to p. 59 of the 1989 reprint].

[2] Richard White, "Are You an Environmentalist Or Do You Work For A Living? Work and Nature," in *Uncommon Ground: Rethinking the Human Place in Nature,* ed. William Cronon (New York and London: W.W. Norton & Co, 1996).
[3] Thomas Robert Malthus, *An Essay on the Principles of Population* (London: J. Johnson, 1798).
[4] G. R. Searle, *The Quest for National Efficiency: A Study in British Politics and Political Thought, 1899-1914* (London: Ashfield, 1990).
[5] Piers J. Hale, "For Eco-Socialism: Re-Reading the Works of William Morris, Edward Carpenter and Robert Blatchford on Labour, Nature and Embodiment." Unpublished PhD Thesis, University of Lancaster, 2003.
[6] Peter C. Gould, *Early Green Politics: Back to Nature, Back to the Land and Socialism in Britain, 1880-1890* (Harvester Press: Brighton, 1988).
[7] Piers J. Hale, "For Eco-Socialism."
[8] Piers J. Hale, "Labor and the Human Relationship with Nature: The Naturalization of Politics in the Work of Thomas Henry Huxley, Herbert George Wells, and William Morris," *Journal of the History of Biology*, Vol. 36 (2003), pp. 249–284.
[9] Gareth Stedman Jones, *Outcast London: A Study in the Relationship between Classes in Victorian Society* (Oxford: Clarendon, 1971).
[10] Robert Blatchford, *Merrie England* (London: Clarion Press, 1894), p. 54. Emphasis in the original.
[11] Laurence Thompson, *Robert Blatchford: Portrait of an Englishman* (London: Victor Gollancz, 1951), p. 23.
[12] Fiona McCarthy, *William Morris: A Life for Our Time* (New York: Alfred A. Knopf, 1995), pp. 69-73.
[13] J. Ruskin, "The Nature of Gothic," in *The Victorian Prophets: A Reader from Carlyle to Wells,* ed. P. Keating (London: Penguin, 1981), p. 163.
[14] As James Elwick has pointed out, Spencer first made mention of the "social organism" in a series of letters written to the Nonconformist in 1843. These were later reprinted as *The Proper Sphere of Government*. Spencer's famous paper "The Social Organism" was published in 1860. See James Elwick, "Herbert Spencer and the Disunity of the Social Organism," in *History of Science*, Vol. XLI (2003), pp. 35-72, 46, 58.
[15] Herbert Spencer, "The Great Political Superstition," in *Political Writings,* ed. John Offer (1884; Cambridge: Cambridge University Press, 1994), pp. 140-169.
[16] Elwick, "Herbert Spencer and the Disunity of the Social Organism," p. 39.
[17] Herbert Spencer, "On the Proper Sphere of Government," in *Man and the State and Other Essays* (Indianapolis: Liberty, 1982)
[18] Thomas Henry Huxley, "On the Origin of Species," *Westminster Review* (London, 1860).
[19] Thomas Huxley, "The Struggle for Existence in Human Society," in *The Collected Essays of T.H. Huxley*, Vol. IX (1888; London: Appleton & Co., 1902)
[20] *Ibid.*, p. 213
[21] Searle, *The Quest for National Efficiency.*
[22] Herbert Spencer, "The New Toryism" in *Man and the State and Other Essays* (Indianapolis, IN: Liberty, 1982).
[23] Elwick, "Herbert Spencer and the Disunity of the Social Organism," p. 60.

[24] Adrian Desmond, *Huxley, From Devil's Disciple to Evolution's High Priest* (Reading, MA: Helix, 1997) pp. 573-4.
[25] Huxley, quoted in Paradis, *Evolution and Ethics*, pp. 46-7.
[26] Huxley, "The Struggle for Existence in Human Society," p. 331.
[27] Gould, *Early Green Politics*, p. 15.
[28] William Morris, "The Arts and Crafts of Today," in *Collected Works of William Morris*, Vol. XXII, pp. 358-9.
[29] William Morris, "How I Became a Socialist," in *News from Nowhere and Other Writings* (1894; London: Penguin, 1993), pp. 379-83, 382.
[30] William Morris and Henry Hyndman, *A Summary of the Principles of Socialism* (London: The Modern Press, 1884), p. 46.
[31] William Morris, "Useful Work versus Useless Toil," *Collected Works of William Morris*, vol. 23, p. 101.
[32] *Ibid.*, p. 102.
[33] *Ibid.*, p. 102.
[34] Morris, *News from Nowhere*, p. 75.
[35] Morris, "Useful Work versus Useless Toil," p. 117.
[36] Morris, "True and False Society," in *Collected Works of William Morris*, ed. May Morris (London: Longmans, Green & Co., 1910-15), Vol. XXIII, p. 219.
[37] Morris, "Useful Work versus Useless Toil," p. 103.
[38] William Morris, "True and False Society," p. 227.
[39] Morris, *News from Nowhere*, p. 292.
[40] *Ibid.*, p. 76.
[41] Piers J. Hale, "William Morris, Human Nature and the Biology of Utopia," in *William Morris in the Twenty-First Century,* eds. Phillippa Bennett and Rosie Miles (Bern: Peter Lang), pp. 107-128. See also Peter J. Bowler, *Evolution: The History of an Idea* (Berkeley, CA: University of California Press, 1989).
[42] Indeed, H. G. Wells, a former science student of Huxley's and an early admirer of Morris, had explored exactly this scenario in his first attempt at science fiction writing, *Time Machine*, published in 1885.
[43] Morris, *News from Nowhere*, p. 76.
[44] *Ibid.*, p. 75. Emphasis in the original.
[45] *Ibid.*, p. 95.
[46] *Ibid.*, p. 160. The role of aesthetics in Morris' politics is also the central concern of B. J. MacDonald's *William Morris and the Aesthetic Constitution of Politics* (Maryland, 1999).
[47] William Morris, *On Art and Socialism, Essays and Lectures* (London: John Lehmann, 1947), p. 222.
[48] Morris, "Useful Work versus Useless Toil," p. 104.
[49] William Morris, "Under an Elm Tree," in *News from Nowhere and Other Writings* (1889; London: Penguin, 1993), pp. 361-364.
[50] William Morris, "Art and Socialism," in *Collected Works of William Morris*, Vol. XXIII, p. 210.
[51] William Morris, "The Lesser Arts," in *Collected Works of William Morris*, ed. May Morris (London: Longmans, Green & Co., 1910-15), Vol. XXII, p. 4.
[52] Morris, "Useful Work versus Useless Toil," p. 119.

[53] William Morris, "Art and the Beauty of the Earth," in *Collected Works of William Morris*, Vol. XXII, pp.119-21.
[54] John Mackenzie ties Morris's sense of loss and idolisation of an imagined Golden Age of medieval art to a broader reaction to the apparent decline of art and morals as a result of British industrial manufacturing processes. See, John M. Mackenzie, *Orientalism: History, Theory and the Arts* (Manchester: Manchester University Press, 1995), p. 120.
[55] Morris, *On Art and Socialism*, p. 243.
[56] *Ibid.*, p. 100.
[57] Morris, "Art and Socialism," p. 196.
[58] Morris, "Useful Work versus Useless Toil," p. 100.
[59] Morris, "Art of the People," in *Collected Works of William Morris*, Vol. XXII, p. 38.
[60] Gould, *Early Green Politics.*
[61] Morris, "True and False Society," p. 233.

Chapter Six

H. G. Wells's Taxonomy of Waste

William Kupinse

Popular conception remembers H. G. Wells as the author of science-fiction novels and short stories—what in Wells's day would have been called scientific romances—such as *The Time Machine*, *War of the Worlds*, and *The Invisible Man*. To a scholar of late-nineteenth- and early-twentieth-century fiction, Wells will be recalled as the author of dozens of novels, several of which—such as *Tono-Bungay* and *Kipps*—were justly praised in their time for their aesthetic merit, despite Wells's sometimes ambivalent reception by the literary establishment. To the historian, Wells is also known for his early involvement in the Fabian Society, his tireless advocacy of a world state, and his two unsuccessful campaigns as Labour Party candidate for Parliament.[1] This essay introduces another perspective on Wells, though a view that encompasses all of the above personae; Wells should be understood as one of modernity's pre-eminent theorizers of waste. More specifically, I argue that for Wells, waste serves as a term that mediates between human subjects and society, between science and literature, between the often slippery categories of nature and culture. Famously asserting that "human history becomes more and more a race between education and catastrophe," Wells understood that educating the public about modernity's unsustainable practices was crucial to preventing such catastrophe.[2] Diagnosing the myriad and interdependent forms of waste within the society of late-nineteenth- and early-twentieth-century England, and attempting, however futilely, to move that society toward what we would now term sustainability—in all its economic, social, and environmental aspects—motivated both his public life and his writings in several genres.

Such a perspective was peculiarly enabled by Wells's early training as a biologist. Rescued from his parents' plans to apprentice him in the drapery trade by qualifying for a student assistantship at Midhurst Grammar School, Sussex, the young Wells went on to win a scholarship to the Normal School of Science in South Kensington, London, where for a year he studied biology with Darwin's famous popularizer, Thomas Henry Huxley, and later taught biology himself. Indeed, two years before the publication of *The Time Machine* would make H. G. Wells a household name, W. B. Clive's University College Correspondence College Press published Wells's first book: *Text-Book of Biology.*[3]

What strikes the modern reader most immediately about the two-volume *Text-Book* is its unusual organization and its holistic approach. Proceeding in order of reverse complexity, volume one of the *Text-Book* covers the animal kingdom, while volume two treats plants and a few invertebrates. Even within the first volume, this schema of reverse complexity prevails, beginning with the morphology of the rabbit, and then moving on to consider that of the frog, dog-fish, and amphioxis, a small marine filterfeeder.[4] A fold-out illustration[5] of a rabbit's alimentary canal is almost the first thing one learns about the subject, an odd choice when one considers other possibilities—the rabbit's skeleton, musculature, or nervous system—not to mention the possibility of beginning with a line-drawing dissection of a simpler organism.[6] The explanatory copy also conveys an oddly anthropomorphic tone in places; when Wells writes that the rabbit "is a pure vegetarian, and a remarkably moderate drinker," we may imagine for a moment that we have stumbled across the description of a New Age socialist from one of the later novels. In describing the rabbit's digestion, Wells concludes, in more properly biological fashion, "Some, but only a portion, of the vegetable matter it eats, leaves its body comparatively unchanged, in little pellets."[7] Even in glossing this static rendering, Wells gestures toward the greater ecology of organism and environment.

But when it comes to biology, and in particular Darwinian natural selection, rabbits signify more broadly for Wells than a textbook dissection example. In his essay "Human Evolution, an Artificial Process," he contrasts the relatively modest fecundity of humans with the fertility of rabbits, which produce offspring earlier, more frequently, and in litters of greater number than do people.[8] In a thought experiment, Wells compares a human mother and her offspring to those of a rabbit matriarch over the course of a century:

> In the hundred years the [rabbit] family had therefore been picked over two hundred times by the hand of Natural Selection. Two hundred times had there been a rejection of unsuitable variations and a premium upon adaptive novelties. In the case of the human being this had occurred only five times. By virtue of its rapid breeding alone, then, it is evident that a rabbit must be, to express the thing in a numerical image, at least forty times more amenable to the process of Natural Selection than a type of human

> being without the slightest sexual restraint, and at least fifty times more than the civilized human being.[9]

We see the rhetoric of waste at work here, with Nature's hand "pick[ing] over" two hundred rabbits and "reject[ing] unsuitable variations." Within the realm of early understandings of population biology, the rejected rabbits represent the sweepings of natural history's dustbin, the set of unfit individuals living prior to the emergence of their better adapted successors, who will in turn have the opportunity of passing their genes on to the next generation.

Waste is inherently a strange term, however. Originally signifying uninhabited, uncultivated, and therefore empty land, waste has also come to mean that which is not profitably used, and thus it also indicates "a profusion or lavish abundance of something."[10] Like *cleave*, which can mean both to split apart and to cling to, waste is one of those odd words that invokes both itself and its opposite. But in the discourse of nineteenth-century biology, waste is a more problematic concept still, in that it is an articulable yet inoperative concept, since every wasted thing—every non-adaptive mutation, every dead-end strategy, in short, every unfit organism—represents a necessary if unacknowledged step in the evolution of a species.

The famous last paragraph of *Origin of Species* makes this clear, when Darwin, rhapsodizing about an "entangled bank" full of plants, birds, insects, and worms, considers the complexity of interdependency and predation, and concludes: "Thus, from the war of nature, from famine and death, the most exalted object which we are capable of conceiving, namely, the production of the higher animals, directly follows."[11] Famine and death may seem a waste from the human perspective, but they are nature's motive force.

Wells will echo Darwin's sentiment nearly exactly at the end of *The War of the Worlds.* At an early point in the novel, Wells's unnamed narrator remarks casually that "Microorganisms, which cause so much disease and pain on earth, have either never appeared on Mars or Martian sanitary science eliminated them ages ago."[12] This turns out to be a significant bit of foreshadowing, since what finally saves humankind from the technologically advanced invaders is not human ingenuity, but the Martians' susceptibility to terrestrial disease. As the narrator observes a crater filled with the rotting corpses of dead Martians, his tone enters a register that would be more at home in a nineteenth-century biology textbook:

> These germs of disease have taken their toll on humanity since the beginning of things—taken toll of our prehistoric human ancestors since life began here. But by virtue of this natural selection of our kind, we have devel-

> oped resisting power; to no germs do we succumb without a struggle, and to many … our living frames are altogether immune. But there are no bacteria in Mars, and directly these invaders arrived, directly they drank and fed, our microscopic allies began to work their overthrow. Already when I watched them they were irrevocably doomed, dying and rotting even as they went to and fro. It was inevitable. By the toll of a billion deaths man has bought his birthright of the earth … For neither do men live nor die in vain.[13]

As with Darwin's entangled bank, none of the "billion" human deaths has been in vain, for each death has contributed to the increasing fitness of the species. In this scientific romance, whose *deus ex machina* arrives in microbial form, the moral may well be to beware too vigorous a campaign of sanitary reform. For an age in which sanitary reform was synonymous with social progress—the Victorian era comprehended everything from its colonial undertakings to its treatment of the poor through the rhetoric of hygienic science—Wells's critique of modernity's germophobia is surprisingly transgressive.

Social critique notwithstanding, Wells's argument, in "Human Evolution, an Artificial Process," assumes that the human debt to natural selection that he would invoke in *War of the Worlds* has already been paid and thus that biological evolution has effectively stopped for humans. Instead, what humankind presently faces is the challenge of cultural evolution, a different thing altogether. Shifting to the example of primates, our closest animal cousins, Wells goes on to assert:

> That in civilized man we have (1) an inherited factor, the natural man, who is the product of natural selection, the culminating ape, and a type of animal more obstinately unchangeable than any other living creature; and (2) an acquired factor, the artificial man, the highly plastic creature of tradition, suggestion, and reasoned thought. In the artificial man we have all that makes the comforts and securities of civilization a possibility. That factor and civilization have developed, and will develop together.
>
> The artificial factor of mankind—and that is the one reality of civilization—grows, therefore, through the agency of eccentric and innovating people, playwrights, novelists, preachers, poets, journalists, and political reasoners and speakers …[14]

Far from being that which makes nothing happen, poetry is portrayed by Wells with unexpected Shellyean sincerity; it and all related literary, creative, and political arts turn out to be crucial in the evolution of human society, even as they must contend with the legacy of humankind's biological history, its vestiges of rabbits, apes, and all the other creatures inhabiting nature's entangled bank. This natural history can never entirely be effaced, particularly for a thinker such as Wells. Thus human-animal tension arises frequently in Wells's fiction, most obviously in texts such as *The Island of Dr. Moreau*, in which the scientist of the book's title conducts horrifying vivisection on a variety of animals in an effort to shape them into human form.

But human-animal tension also takes subtler forms in Wells's fiction, with rabbits serving several times as the emblem of this anxiety via metaphoric renderings of theriomorphism—that is, animal-human hybridity.[15] The social satire *Tono-Bungay*, for example, features a character named Rabbits, a butler who spends much of his time following the goings-on of the English nobility, and who professes fear of "colonials," whom he describes as "queer people … [who] make me nervous. They have an eye on you. They watch you—as you wait."[16] Here Rabbits' fascination with the upper classes and distrust of individuals with indeterminate status within the social hierarchy suggest an unease rooted in naturalism and social Darwinism. In *Mr. Britling Sees It Through*, a First World War novel published in 1916 and written when the Allied victory was far from assured, the novel's writer-protagonist paces the grounds of his Essex estate and muses:

> The rabbit is nimble, lives keenly, is prevented from degenerating into a diseased crawling eater of herbs by the incessant ferret. Without the ferret of war, what would life become? … War is murder truly, but is not Peace decay?[17]

Wells's use of the term "degenerating" is significant, as it situates Mr. Britling's metaphor within late-nineteenth- and early twentieth-century debates that we now understand as pseudoscience, but which were taken quite seriously at the time.[18] Occurring under the related rubrics of eugenics and degeneration, in England these debates were informed to a large extent by concerns over military readiness.[19] As Donald J. Childs notes in his introduction to *Modernism and Eugenics*, by 1871 statistical data on the vigour of British army recruits had raised fears of widespread physical degeneration, fears which by the century's turn early British losses in the Boer Wars had magnified.[20] Whether as class-obsessed servant or symbol of nationalist apprehension, Wells's two exemplary rabbits carry with them suggestions of anxiety inherent in the human reflection on biological processes of natural selection—an anxiety motivated in part by the apprehension that an individual human might be consigned to history's dustbin in much the same way as the unfit rabbits of Wells's *Fortnightly Review* essay.

But if rabbits emerge as emblems of anxiety in Wells's fiction, monkeys, apes, and other primates form the occasion for greater concern, and one more threatening to established notions of the human subject. This concern is not exclusive to Wells, of course; humans have a tendency to view our evolutionary relatives as uncanny "others," and the reminder of our animal affiliation has elicited disquietude historically and continues to do so.[21] As geneticist David Page observes in a recent news article reporting that evolutionary biologists had discovered that early humans and chimpanzees had continued to inter-breed after an initial evolutionary split: "We'd like to have a more Victorian view of the genome."[22] But in the late-Victorian, Edwardian, and Georgian eras in which Wells wrote, the view of monkeys and humans as biological cousins could provoke still greater discomfort, even for a stalwart Darwinian. What is interesting about Wells's work is that simian theriomorphism, which by the early twentieth century had accrued a long history of complicity in forms of racist representation,[23] appears in his fiction to have been wholly transposed to the realm of economic class, seemingly unaware of its racist legacy. ("Race" in Wells's writing most often indicates collectively the "human race"). In *Tono-Bungay*, narrator George Ponderevo is sent as a boy to live with the poor yet fertile family the Frapps, sharing a bed with a cousin whom he describes as "a singularly thin and abject, stunted creature … whose chief liveliness was to pretend to be a monkey, and who I am now convinced had some secret disease that drained his vitality away."[24] As with Mr. Britling's account of the nimble rabbit, eugenic discourses of degeneration inform the description of the Frapp cousin. In this latter example, the boy's "abject" physique and his monkey-playing are juxtaposed with a framework of retrospective interpretation that suggests their connection to the malady sapping his "vitality"—another term borrowed from the eugenics.

While both Britling's rabbit and the Frapp cousin serve as emblems of degeneration-inspired anxiety, a significant distinction separates these two examples. Britling's rabbit exists solely within the realm of biological natural selection, and thus can be prevented from degeneration by the "incessant ferret"—what Malthusian thought would have termed a "positive check"—*positive* in that it operates actively—to rabbit population and what Wells's era would have understood as the natural world's form of "negative eugenics." The frail cousin in *Tono-Bungay*, however, might presumably be supported by Victorian England's social safety net of workhouses and church-based charity—as sparse and fraying as that net may have been (we learn that several of his siblings have died)—and thus the negative eugenic force of starvation has been ameliorated, if not eliminated. Narrator George Ponderevo's description of his cousin's family voices derision at both the Frapps' religiosity and their fate—"God sent them many children, most of whom died, and so, by their coming and going, gave a double exercise in the virtues of submission,"[25] he muses—but George's attitude is hardly one endorsed by the novel, which tends to hold its narrator-protagonist at ironic distance, particularly at this early stage in his development. Allowing active checks to what were presumed to be eugenically inferior populations—in other words, consigning living humans to death, even by negligence—would have been out of the question for Wells the socialist; fortunately, despite the philosophy of Social Darwinism popularized during Wells's era by Herbert Spencer, most societies have come to deem abhorrent the prospect of allowing the disadvantaged to starve, the political careers of Ronald Reagan and Margaret Thatcher notwithstanding.

A less dire instance of metaphoric simian theriomorphism occurs in Wells's 1911 political novel, *The New Machiavelli*. Here the narrator-protagonist Richard Remington recalls the evening "cruising" of himself and other lower middle-class youngsters, all sporting the hard-earned accoutrements of class and adulthood—"collars and ties, chiffon hats, smart lace collars, walking sticks, sunshades or cigarettes"—and notes that critics deride these strolls as "Monkey's Parades."[26] How these "Monkey's Parades" are to be interpreted would depend on the disciplinary frame of reference employed. A biological reference frame—what today would fall under the heading of evolutionary psychology—might anticipate biologist Richard Dawkins's understanding of such risky displays (risky in that they waste the limited capital of young men inhabiting a precarious economic position) as human versions of the peacock's tail.[27] A cultural interpretative reference frame would perhaps employ a more anthropological reading and understand the money spent on this posturing as a form of "glorious expenditure," to borrow a phrase from George Bataille, an expenditure akin to that of potlatch ceremonies in which the squandering of ready money in the service of elevating status is precisely the point.[28]

But what these simian comparisons reveal for Wells is that while strict Darwinian discourse refuses the category of waste—as the truism that nothing is wasted in nature suggests—waste signifies resonantly, perhaps even foundationally, when biology is considered within the domain of human culture. The degenerative cousin in *Tono-Bungay* signifies as abject only by the considerations of class and urban geography that delineate the borders of Chatham slum within the larger socio-economic geography of Edwardian England. The "Monkey's Parades" in *The New Machiavelli* represent the lower middle-class youngsters' attempts at fashion, and are clearly wasteful of their meagre economic means; it is both ironic and significant that these parades signify a mating ritual within a novel in which the adult narrator's promising political career is laid waste when his sexual impropriety is revealed publicly. In each example, the narrative commentary suggests that these biologically-rooted forms of waste assume their meaning only when imported into the social realm. In the animal world, such individuals and such displays by individuals are natural; in the human world they are wasteful. We may understand these theriomorphic renderings as sites of anxiety for Wells, who, despite his distance from the Social Darwinism of Spencer, nevertheless struggled to come to terms with a society that could paradoxically perpetuate and punish degenerative behaviour, given the waste that both activities entailed.

Wells's response to society's paradoxical attitude toward waste of all forms was to imagine a world in which waste would be progressively and humanely eliminated, both as a feature of social mismanagement and as a marker of social opprobrium. As a first step toward realizing this admittedly utopian world, Wells's project in his fiction, as in all his output as a public intellectual, was the systematic analysis of the function that waste plays in the cultural realm when viewed against the backdrop of biological necessity; of equal importance for Wells was transmitting that understanding to society at large. Although, as noted earlier, Wells's first training was as a biologist, his first significant employment was as a teacher, and this was a role that he maintained throughout his life, long past his time in the classroom or dissecting hall.

Wells's project of bringing waste to public consciousness was an undertaking of such sustained and methodical nature that perhaps the best way to understand its significant contours is to employ the systematizing technique that appears at the end of Wells's *Text-book of Biology*: the Linnaean system of biological taxonomy, which categorizes organisms according to the categories of Kingdom, Phylum, Class, Order, Genus, Species, with the last two terms forming the Linnaean system of binomial classification.[29] In Wells's day, categorization was based upon factors including the morphology (form), physiology (function), and behaviour of the organism; taxonomy today is heavily influenced by the tools of genetic analysis that have become available in recent years. This essay's speculative taxonomy is consistent with Wells's analytic methodology as evidenced in his prose, which often applies a scientific lens to examination of problems of human culture; in *Tono-Bungay*, George Ponderevo claims that his method of understanding England is one of "social comparative anatomy."[30] Wells's fictional methodology employs biological similes and metaphors to explicate cultural wastage and atavism; the English "are like that little shell the Lingula, that is found in the oldest rocks and lives to-day: it fitted its easy conditions, and it has never modified since," says Mr. Britling.[31] Constructing a visual schema of Wells's taxonomy of waste further highlights the rigour and systemic quality of his thought on the subject, and it allows for interconnections among various categories of waste to be more clearly apprehended. It is these linkages—situated in chartable regions, manifest in economic processes, materially affecting the lives of identifiable subjects, embodied (and addressable, believed Wells) in specific texts—that gives value to this imaginative project of articulating Wells's systematic thought on modernity's waste. In a passage in his credo *First and Last Things* in which both his abundant sympathies and his modernizing biases are in full view, Wells makes clear his conviction that his perception of the various and interpenetrating manifestations of waste informs his life's work:

> I see human life as avoidable waste and curable confusion. I see peasants living in wretched hits knee-deep in manure, mere parasites on their own pigs and cows [...]; I see the grimy millions who slave for industrial production; I see some who are extravagant and yet contemptible creatures of luxury, and some leading lives of shame and indignity; tens of thousands of wealthy people wasting lives in vulgar and unsatisfying trivialities [...]; Their disorder of effort, the spectacle of futility fills me with a passionate desire to end waste, to create order, to develop understanding ... All these people reflect and are part of the waste and discontent of my life, and this coordinating of the species in a common general end, and the effort of my personal salvation are the social and individual aspect of essentially the same desire ...[32]

Given this expression of urgency and interconnection, I offer the following taxonomic outline as a heuristic, mindful of Wells's own caveat—borrowed from the "constant refrain" of a former professor—that members of all humanly organized categories, from biological specimens to mineral specimens, "pass into one another by insensible gradations."[33] In other words, taxonomies of all kinds are blunt instruments. They are also useful instruments, as long as we recognize their provisionality and their limits.

CLASSIFICATORY SUMMARY

Wells's Taxonomy of Waste

a. *Wastelands*
a. i. Domestic
a. ii. Colonial

b. *Unsustainable Economic Practices*
b. i. The Stock Market
b. i. 1. "Corners"
b. i. 2. "Cover Gambling"
b. ii. Invented Consumer Desire
b. iii. Disposable Commodities/Serial Acquisitions

c. *Abject Individuals*
c. i. Members of Economically Dispossessed Classes
c. ii. Women
c. iii. Colonial Subjects and Racial "Others"

d. *Wasted Cultural Productions*
d. i. Advertising
d. ii. "Morally Instructive" Writing
d. ii. 1. Sentimental Fiction
d. ii. 2. Religious Literature
d. iv. Elite Cultural Productions that Substitute Aestheticism for Social Relevance

To illustrate this taxonomy, I drawn upon examples taken from two of the novels I have discussed earlier, which I suggest are among Wells's most successful: *Tono-Bungay*, his 1909 satire of a patent medicine empire by the same name, and *The New Machiavelli*, his 1911 story of a rising Member of Parliament who shifts political parties and who is disgraced by public revelation of an affair.

The category of wastelands divides into two sub-classifications: domestic and colonial zones of abjection. Richard Remington's account of the suburban wasteland of Bromstead in *The New Machiavelli* offers a representative example of the former classification. Surveying a once pastoral region in his fictional boyhood town, Remington finds a suburban wasteland in which "broken glass, tin cans, ashes, and paper abounded." He continues, "Cheap glass, cheap tin, abundant fuel, and a free untaxed Press had rushed upon a world quite unprepared to dispose of these blessings when the fullness of enjoyment was past." Challenging the notions that this concomitant industrialization, commercialization, and suburban development represents progress, Remington indicts the age in which he lives: "No, the Victorian epoch was not the dawn of a new era; it was a hasty, trial experiment, a gigantic experiment of the most slovenly and wasteful kind."[34] This surprising attention to domestic issues of suburban sprawl, solid waste disposal, and resource extraction demonstrates a prescient environmental consciousness. But Wells also demonstrates a strong interest in the way in which colonial spaces are exploited for their labour and cheap raw materials. In *The New Machiavelli*, Richard Remington meditates upon England's management of India, the fabled jewel in the Empire's crown. What strikes him most about the imperial relationship is the effacement of any real understanding of this colonial engine of imperial wealth. "Most English people cannot even go to this land they possess; the authorities would prevent it," observes Remington.

> If Messrs. Perowne or Cook organised a cheap tour of Manchester operatives, it would be stopped. No one dare bring the average English voter face to face with the reality of India, or let the Indian native have a glimpse of the English voter.[35]

Wells's comparative explication of domestic and colonial wastelands offers two insights, one into the definition of the modern industrial nation and one into the definition of the modern imperial power. For Wells, the modern industrial nation is one that relegates certain domestic regions to the status of waste, defending this procedure as progress; the modern imperial power does the same thing with colonial geographic zones, but on a larger scale and with a greater degree of effacement. It is not difficult to see how these insights would motivate Wells's desire for a world state as a means of combating these modes of geographical abjection and the human inequality they entail.

Tono-Bungay offers perhaps the most scathing indictment of the unsustainable economic practices that Wells saw driving the boom of Edwardian England.[36] The title of the novel refers to its central product, a patent medicine chock full of deleterious and likely addictive ingredients, which forms the immense fortune of the huckster Edward Ponderevo—who earlier has lost money in the stock market via "corners," meaning, cornering the market and "cover gambling," or trading on margin.[37] To realize his patent medicine empire, Edward, however, requires the engineering talents of his scientist nephew, George. George reluctantly signs onto the project of marketing the patent medicine, terming it "mischievous trash."[38] What is especially interesting about *Tono-Bungay* is that it costs nearly nothing to make, and, in fact, becomes less expensive as demand increases, since the Ponderevos simply dilute the formula. Edward recognizes as much when he asserts that what the Ponderevos are selling is "confidence"; as a confidence man, Edward is so successful that his fortune has enabled him to purchase series of artworks, motor-cars, and houses.[39] By the end of the novel, his nephew George is far from confident, however; the Ponderevo empire has collapsed, Edward has died after being charged with forgery, and Edwardian England itself seems to be in decay. Reflecting on the economic and spiritual devastation that surrounds him, George, who is the supposed author of the novel we are reading, muses that "I have called it Tono-Bungay, but I had far better have called it Waste . . ." [40]

But while George Ponderevo experiences a sense of spiritual abjection at the end of *Tono-Bungay*, even as a failed capitalist he is not brought as low as the abject individuals who populate Wells's fiction. These characters span a wide range of social groups: the poor, such as the monkey-imitating Chatham cousin; women, such as the protagonist of Wells's suffragette novel *Ann Veronica*; and individuals, including colonial subjects, who are depicted as racial "others," which in Wells's fiction tended to signify people who were neither "white" nor European. The attitudes Wells's novels evince toward these abject individuals vary, but once the biases of suspect narrators are accounted for, the broader narrative perspective is one of empathy and the novelistic method one of attempted ideological demystification. Ann Veronica is presented as occupying a position rendered impossible by the warring dictates of affection and socially rigid notions of female propriety; the unnamed inhabitant of a fictional West African island whom George murders in *Tono-Bungay* is depicted as the victim of imperial greed. At times Wells can devolve nearly to melodrama in his portrayal of the social forces responsible for disenfranchising these abject others, as an example from *The New Machiavelli* demonstrates. In this scene, the protagonist Richard Remington visits the factory of a wealthy uncle, who has made his fortune through the manufacture of high-end bathroom fixtures, such as those shown in a period drawing of a model bathroom. This technology, with its focus on smooth, white surfaces and the effacement of waste and disease, in fact occasions the production of waste and disease, as the following exchange between Richard and his uncle reveals:

> We passed a sickly-looking girl with a sallow face, who dragged her limbs and peered at us dimly with painful eyes. She stood back, as partly blinded people will do, to allow us to pass, although there was plenty of room for us.
>
> I glanced back at her.
>
> "THAT'S ploombism," said my uncle casually.
>
> "What?" said I.
>
> "Ploombism. And the other day I saw a fool of a girl, and what d'you think? She'd got a basin that hadn't been fired, a cracked piece of biscuit it was, up on the shelf over her head, just all over glaze, killing glaze, man, and she was putting up her hand if you please, and eating her dinner out of it. Got her dinner in it!"[41]

Liberally employing lead as the glaze of his quality bathroom fixtures, Richard's uncle subjects his workers to the perils of lead-poisoning—"ploombism." His blame-the-victim strategy suggests just how forlorn a role she occupies in terms of her lack of economic agency and her exposure to environmental toxins. A domestic subaltern, her narrative is misrepresented by Richard's uncle, depriving her of speech. Richard's effort to recount the girl's story may offer the best hope of bringing her abject situation to public consciousness, though the mediation itself is problematic, as is the episode's reliance on what ecocritic Lawrence Buell has termed "toxic discourse."[42] Defining toxic discourse as "expressed anxiety arising from perceived threat of environmental hazard due to chemical modification by human agency," Buell observes that while it has the potential for building consensus among dissimilar social groups, toxic discourse also carries with it "the liability of discourse to become its own sanctuary"—that is, to substitute aesthetic representation for real-world representation.[43] Thus as a strategy for addressing the plight of abject individuals and geographic zones, toxic discourse proves a promising but problematic genre—promising if it enlarges the reader's environmental consciousness, problematic if it stagnates at the level of discourse alone and thus becomes a wasted cultural production. Moreover, Wells's use of melodrama to render the scene does not bode well for his ecocritical insight to extend beyond the level of pure discourse, since melodrama's reliance on stereotypical notions of villains and victims tends to elide the complex structural causes of toxicity and waste.

Wasted cultural productions form perhaps the most expansive category in Wells's taxonomy of waste, signifying those texts that reify dominant ideological constructions and thus perpetuate the role that waste—in the form of wastelands, unsustainable economic practices, and the abjection of marginalized individuals—plays in the constitution of modern society. They thus include advertisements of all kinds (*Tono-Bungay* is especially rife with these), as well as morally instructive literature, from the pious, quasi-socialist novels of Mrs. Humphrey Ward (whose anti-suffrage activism would not have endeared her to Wells) to sentimental fiction such as Susan Warner's *Wide, Wide World* (this despite the occasional strains of sentimentality and melodrama that punctuate Wells's own fiction). Religious periodicals receive particular censure for constructing ideologically compliant subjects who perpetuate the unsustainable social practices detailed above. In *The New Machiavelli,* Richard Remington comments on a journal called *The Home Churchman*, to which his mother subscribes:

> Every month there came to my mother a little magazine called, if I remember rightly, the HOME CHURCHMAN, with the combined authority of

> print and clerical commendation. It was the most evil thing that ever came into the house, a very devil, a thin little pamphlet with one woodcut illustration on the front page of each number; now the uninviting visage of some exponent of the real and only doctrine and attitudes, now some coral strand in act of welcoming the missionaries of God's mysterious preferences, now a new church in the Victorian Gothic. The vile rag it was![44]

Despite Richard's use of an epithet based upon the journal's inexpensive paper content ("rag" in "vile rag" is analogous to "pulp" in today's "pulp fiction"), it is important to note that Wells's critique of empty cultural productions does not neatly follow an Arnoldian high-versus-low culture division. This is, of course, a wise decision on Wells's part, since as a popular writer he is himself a purveyor of mass culture; and nearly all assessments of his work—from his famous literary feud with Henry James to the criticism that many of his novels are more journalism than art—have participated in this dichotomous framing. Indeed, Wells often professed ambivalence toward elite literary culture, criticizing its impulse to substitute aestheticism for social relevance.

Stepping outside this limited cultural division, we nonetheless find ourselves faced with the question of what Wells would envision as the counter to the wasteful cultural productions he identifies—or, to phrase the issue in different terms, how Wells imagines cultural productions could contribute to social and environmental sustainability.

I would like to suggest that *The New Machiavelli*, despite its occasional lapses into melodrama, offers a representative example of this impulse toward sustainability in Wells's fiction —that is, toward the exposure and reduction of waste. This impulse manifests itself in an unexpected fashion, however, in a novel that resists easy generic categorization. The fictional autobiography of failed politician Richard Remington, who, like his model Niccolo Machiavelli, is writing from exile at the age of forty-two, *The New Machiavelli* is at once a *Bildungsroman* of Remington's political education, a love story of his extramarital romance with the young political activist Isabelle Rivers, the narrative of his political conversion from the Liberal to the Conservative party, and a critique of a society whose narrow-minded moralism deprives it of its ablest political thinkers. But, in the tradition of its intertext, *The New Machiavelli* is foremost a political credo, and its narrator's desire to contribute toward what he calls the national project of "social construction" needs to be understood for its sincerity. At the moment of his transition from political Liberalism, which he finds to be primarily an oppositional ideology, and thus incapable of true constructionist perspective, Remington reflects,

> I see human progress, not as the spontaneous product of crowds of raw minds swayed by elementary needs, but as a natural but elaborate result of intricate human interdependencies, of human energy and curiosity liberated and acting at leisure, of human passions and motives, modified and redirected by literature and art ...[45]

Here we see Remington's conception of society as a "natural" ecological system, a set of "intricate human interdependencies" that must be further shaped by progressive cultural productions, by literature and art. Lurking just beneath the surface of this sentiment is Shelley's notion of poets as the world's unacknowledged legislators. What saves *The New Machiavelli* from didacticism is its subtlety and overdetermination—in short, its effective political realism. Rejecting such ideologies as Labour politics, Liberalism, and Socialism on the basis of his intimate knowledge of each, Richard Remington stands for the seat at Handich for the Conservatives, but he runs on a decidedly anti-Conservative platform. Asserting that he "would attempt to biologise Imperialism," Remington proposes establishing an "Endowment of Motherhood," so that the state, rather than individual men, would provide the financial support for England's next generation, thus liberating childbirth from matrimony.[46] It is worth noting that this concept of endowed motherhood represents a form of what would have been termed "positive eugenics"—that is, an effort to increase the reproduction of fit individuals—as opposed to the negative eugenics—which depends upon the death of unfit individuals—discussed earlier.

A favourite idea of Wells in his mode as public intellectual, the appearance of endowed motherhood as a plot element in *The New Machiavelli* did nothing to discourage the critical reception of the novel as a *roman a clef* satirizing leading public personalities of the day, chief among them Fabian leaders Beatrice and Sidney Webb, who appear in the novel as the dishevelled social reformers Oscar and Altiora Bailey. To be sure, the correspondences between the characters in *The New Machiavelli* and the figures in Wells's own life are striking. Isabelle Rivers, the young woman with whom Richard Remington engages in a career-sabotaging affair, is in both name and rendering a slightly concealed version of Amber Reeves, the Cambridge student with whom Wells had an extramarital relationship, thereby suffering the censure of the Webbs.[47]

But it is one thing to observe that a novel contains elements drawn from a writer's experience, and quite another thing to dismiss a novel as mere autobiography, as critics of *The New Machiavelli*—and, indeed many of Wells's novels—did; seeing the book less as art than a collection of "thinly veiled personalities and bitter attacks on leading political and social figures."[48] This critical trend of misapprehending the novel's genre puzzled no less careful a reader than Wells's friend Arnold Bennett, who, in defending *The New Machiavelli*, mused, "There are some critics of fiction who talk about autobiography in fiction in the tone of a doctor who has found arsenic in the stomach at a post-mortem inquiry."[49] The question of genre proves to be far more than a matter of nomenclature; as Heather Dubrow observes, generic assumptions both structure our strategies of interpreting texts and offer writers sets of conventions that can be deliberately transgressed in the service of artistic purpose.[50] "Art is selection and so is most autobiography," asserts Richard Remington, before going on to profess that his goal is capturing the "more tangled business [of] contemporary man in relation to the state and social usage."[51] By the same measure, we should understand that Wells's rendering of the novel depends upon a similar process of selection, as well as a related strategy of pressuring generic assumptions in the service of greater social comprehension.

Literary genre, like taxonomical classification, is nothing more than a heuristic, a provisional method of exposition to be used as long as it proves of service, and to be modified or discarded as necessary. The connections between genre and taxonomy run deeper still; deriving from the Latin *genus*, for kind, genre shares it root with the first term of Linnaean binomial taxonomy, and Dubrow notes that from the nineteenth century onward, "critics have energetically pursued parallels between [literary] genres and biological species."[52] Where critics fail is where they mistake generic conventions as absolutes, or see generic revision as failure. For Wells, any generically pure text would likely prove a wasteful cultural production, in that fulfilment of an established aesthetic form cannot hope to do more than reify the existing wasteful social order. Indeed, from the earliest stages of his career, Wells seems bent on transgressing generic conventions, even in so non-literary a work as his biology textbook, which he concludes by asserting that:

> In the book of nature there are written … the triumphs of survival, the tragedy of death and extinction, the tragic-comedy of degradation and inheritance, the gruesome lessons of parasitism, and the political satire of colonial organisms.[53]

Blurring the boundaries between biological fact and poetic figuration, Wells concludes by eliding the disciplinary boundaries between science and art: "Zoology is, indeed, a philosophy and a literature to those who can read its symbols."[54]

In his fiction, Wells sought to pressure existing generic conventions in the service of envisioning an art that would encourage a more sustainable relationship between social practice and biological necessity. That this effort is ultimately utopian in outlook does not diminish its urgency for Wells. While we should be suspicious of the unconsidered elision of autobiography and art, at the close of the scene in which *The New Machiavelli*'s Richard Remington assesses the wasteland of suburban Bromstead, we may not be wrong to imagine that we hear something of the voice of the developing ecocritic in Wells, reflecting on development in real-life town of Bromley. Meditating on the detritus of early twentieth-century consumer capitalism, Remington concludes that our only remedy lies in bringing this wastefulness to public consciousness. In his words, "Failure is not failure nor waste wasted if it sweeps away illusion and lights the road to a plan."[55]

NOTES

[1] Norman MacKenzie and Jeanne MacKenzie, *H. G. Wells* (New York: Simon and Schuster, 1973), pp. 333-4.
[2] H. G. Wells, *The Outline of History* (Garden City, NY: Garden City, 1930), p. 1100.
[3] H. G. Wells, *Text-Book of Biology*, 2 vols (London: W.B. Clive, 1893).
[4] The *Text-Book*'s organization is to some degree dictated by the fact that it follows the structure of the Intermediate Examination in Science at London University, for which it is intended to serve as a study guide.
[5] This illustration is taken from the second edition of the *Text-Book of Biology,* and was drawn by Wells's student Amy Catherine Robbins, who would become his second wife.
[6] For example, Henry James Clark's *Mind in Nature, or the Origin of Life and the Mode of Development of Animals* (New York: Appleton, 1865) begins with microscopic organisms such as the amoeba and difflugia.
[7] Wells, *Text-Book of Biology*, vol. 1, p. 4.
[8] H. G. Wells, "Human Evolution, an Artificial Process" *Fortnightly Review*, 1 October 1896, pp. 590-595.
[9] Wells, "Human Evolution, an Artificial Process," p. 591.
[10] "waste, *n*[1a,f]," *The Oxford English Dictionary*, 2nd edn (Oxford: Oxford University Press).
[11] Charles Darwin, *On the Origin of Species* (Cambridge, MA: Harvard University Press, 1964), pp. 489-90.
[12] H. G. Wells, *War of the Worlds* (Peterborough, ON: Broadview, 2003), pp. 146-7.
[13] *Ibid.*, pp. 181-82.
[14] *Ibid.*, p. 594.

[15] For a broad historical discussion of human-animal representation, see Steve Baker, *Picturing the Beast: Animals, Identity, and Representation* (Manchester: Manchester University Press, 1993).
[16] H. G. Wells, *Tono-Bungay* (New York: Oxford University Press, 1997), p. 25.
[17] H. G. Wells, *Mr. Britling Sees It Through* (New York: Macmillan, 1917), p. 290.
[18] Wells's pronouncements on degeneration in his nonfiction writings are varied. They include his infamous assertion that "whole masses of human population are, as a whole, inferior in their claim to the future, to other masses … " and his speculation that euthanasia will be practised upon the unfit who challenge society's "sufferance" of them by procreating. See *Anticipations of the Reaction of Mechanical and Scientific Progress upon Human Life and Thought* (New York and London: Harper & Bros., 1902), pp. 314, 324. Elsewhere, he insists that "though there are racial differences" between "Aryan" and "Negro," "Kelt" and "Teuton," "the artificial man differs far more profoundly between nation and nation"; in other words, cultural evolution—and the attending spectre of cultural degeneration—have largely superseded biological variation ("Human Evolution, an Artificial Process," endnote, p. 595). It is also instructive to compare Wells's prescriptions for countering biological degeneration—which included both positive and negative eugenic initiatives (positive initiatives being those intended to encourage the reproduction of "fit" individuals and negative initiatives those intended to discourage the reproduction of "unfit" individuals)—to the negative eugenic programs articulated by his modernist counterparts D. H. Lawrence and Virginia Woolf, who respectively, if offhandedly, endorsed genocide and euthanasia of degenerate individuals. See Donald J. Childs, *Modernism and Eugenics: Woolf, Eliot, Yeats, and the Culture of Degeneration* (Cambridge: Cambridge University Press, 2001), pp. 10, 23.
[19] An account of the history of discourse on degeneration is beyond the scope of this essay. Important primary texts in the field include first degeneration theorist Benedict Auguste Morel, *Traité des dégénérescences physiques, intellectuelles, et morales de l'espèce humaine et des causes qui produisent ces variétés maladives* (1857) founder of eugenics Francis Galton, *Hereditary Genius: An Inquiry into its Laws and Consequences* (1869) and *Natural Inheritance* (1889), Italian criminologist Cesare Lombroso, *The Man of Genius* (1889), and aesthetic reactionary Max Nordau's *Degeneration* (1892). For contemporary scholarship on degeneration, in addition to Child's *Modernism and Eugenics*, see William Greenslade, *Degeneration, Culture, and the Novel: 1880-1940* (Cambridge: Cambridge University Press, 1994), Richard A. Soloway, *Demography and Degeneration: Eugenics and the Declining Birthrate in Twentieth-century Britain* (Chapel Hill, NC: University of North Carolina Press, 1990), Daniel Pick, *Faces of Degeneration: A European Disorder, c.1848-c.1919* (Cambridge: Cambridge University Press, 1989), and Daniel J. Kevles, *In the Name of Eugenics: Genetics and the Uses of Human Heredity* (New York: Knopf, 1985).
[20] Childs, *Modernism and Eugenics*, p. 1.
[21] The biological order primate includes apes, monkeys, and lemurs; humans are counted along with apes as members of the superfamily *Hominoidea*. See John G.

Fleagle, *Primate Adaptation and Evolution,* 2nd edn (San Diego, CA: Academic Press, 1999), pp. 577-80, 235.

[22] Nicholas Wade, "Two Splits Between Human and Chimp Lines Suggested," *New York Times*, sec. a, May 18, 2006, p. 23.

[23] See, for example, Vincent Cheng's discussion of simian depictions of Irish and African-Americans on both sides of the Atlantic in journals such as *Punch* and *Harper's Weekly* in *Joyce, Race, and Empire* (Cambridge: Cambridge University Press, 1995), pp. 35, 38.

[24] Wells, *Tono-Bungay,* p. 50-51.

[25] *Ibid.,* p. 42.

[26] H. G. Wells, *The New Machiavelli*, 1911 (London: Bouverie House, 1926), p. 65.

[27] Richard Dawkins, *The Selfish Gene*, 1976 (Oxford: Oxford University Press, 2006), p. 164.

[28] See, for instance, Georges Bataille, *The Accursed Share,* Vol. 1 (New York: Zone, 1991).

[29] Elaborations have since been made to the Linnaean system; domain has been adduced as a category prior to kingdom, and subdivisions of various categories (e.g. superfamilies, infraorders) have been added.

[30] Wells, *Tono-Bungay,* p. 107.

[31] *Ibid.,* p. 32.

[32] H. G. Wells, *First and Last Things*, 1908 (New York and London: The Knickerbocker Press, 1908), pp. 130-131.

[33] *Ibid.*, p. 25.

[34] Wells, *The New Machiavelli,* pp. 47-48.

[35] *Ibid.*, p. 322.

[36] I discuss the economics of *Tono-Bungay* in "Wasted Value: The Serial Logic of H. G. Wells's *Tono-Bungay,*" *Novel: A Forum on Fiction* 33.1 (1999), pp. 51-72; also reprinted in *Modern Critical Views: H. G. Wells*, ed. Harold Bloom (Philadelphia, PA: Chelsea House, 2005).

[37] Wells, *Tono-Bungay,* p. 73.

[38] *Ibid.*, p. 147.

[39] Wells, *Tono-Bungay,* p. 238.

[40] *Ibid.*, p. 412.

[41] Wells, *The New Machiavelli,* p. 159.

[42] Lawrence Buell, *Writing for an Endangered World* (Cambridge, MA: Harvard University Press, 2001), p. 31. For a fuller discussion of the subject, see Buell's initial article, "Toxic Discourse," *Critical Inquiry* 24.3 (1998), pp. 639-665.

[43] *Ibid.*, p. 50.

[44] Wells, *The New Machiavelli,* p. 59.

[45] *Ibid.*, p. 304.

[46] *Ibid.*, p. 374.

[47] MacKenzie, *H.G. Wells*, pp. 233-269.

[48] Unsigned Obituary, "Mr. H. G. Wells," *Manchester Guardian*, 14 August, 1946, p. 6.

[49] Arnold Bennett, "The New Machiavelli," in William J Scheick, ed., *The Critical Response to H G Wells* (Westport, CT: Greenwood, 1995), pp. 95-97, 96.
[50] Heather Dubrow, *Genre* (London: Methuen, 1982)
[51] Wells, *The New Machiavelli,* p. 360.
[52] Dubrow, *Genre,* pp. 4, 7.
[53] Wells, *Text-Book of Biology,* p. 131.
[54] *Ibid.*
[55] Wells, *The New Machiavelli,* p. 48.

Chapter Seven

The Future as a Virus in the Midst of our Waste

Alberto Duman

Introduction

This paper is divided into two main parts: the first is an overview of some evidence of the centrality of the management of waste matter and hygienic practices in utopian projections of nineteenth century British urban planning, and the second directs our attention to two texts of the second half of the twentieth century in which the confrontation with waste in urban settings gives way to dystopian scenarios. I will then attempt to bring some of these concerns about the future, waste management, consumption and urban development into a specific contemporary context through the description of my own artistic project, called the "Lamby Way Time Capsule," as part of which the deliberate burial of a time capsule within the imminently closing landfill facilities at Lamby Way in Cardiff, will attempt to preserve some particular historical records in this most unlikely area of conservation. This time capsule is also the *virus* referred to in the title, in that its purpose is to function as a *reversed* viral presence, keeping alive a particular discourse of the landfill in the face of a development process that will see the site transformed from its waste origins into a leisure facility.

In the project the notion of waste is conceptualized as a harbinger of the future. It's main propositions are: (i) to reappraise the importance of the relationship between modernity and waste through cultural historical evidence from the nineteenth and twentieth centuries; and (ii) through such a reappraisal, to highlight the importance that such a relationship still plays both within any possible construction of an idea of the future, and any critical engagement with current socio-ecological and political-economical affairs.

One

Given the current insistence of the theme of waste within ecological arguments against contemporary aspects of modernity, global growth patterns and capitalism, it would appear at first that waste is an unforeseen consequence and the entropic other of modernity, one of the most resilient and ultimate obstacles towards the realization of its project.

However, such recent placements of the notion of waste as a fundamental limit to our ability to manage or order the world and the various attempts to reconceptualise such limits in the wishful propositions of sustainable cities and Zero Waste policies must not deflect our attention from the fact that the central position that waste occupies within the tropes of modernity has indeed a long traceable history. What I am arguing is that the temporary disappearances from public debates of such an omnipresent, implicit, even necessary human activity are moments of cultural amnesia partly caused by the very belief that waste is a problem to which a solution will ultimately be found, whether through management devices or technological innovation. In their latest incarnation, the attempts at the re-incorporation of waste within production cycles through the rhetoric of recycling and resource efficiency, appears as the ultimate form of confrontation between a drive towards continuous processes of expansion of western lifestyles into a global dimension and the realization that such desires are beginning to fall short of the time-space circumstances and conditions that produced those lifestyles in the first place.

Indeed, waste management, as well as hygienic and medical control, were from the very beginning a major focus of the project of modernity and one of its defining characters, a "sanitizing" project, guided by order and cleanliness as its main tenets. Issues of water quality, sewage discharge, waste collection and disposal were often the "executive powers" that greatly characterized some of the defining moments of modernity in action, particularly in urban development. The notion of sanitation, then, beside its connotations of cleanliness, waste management and public health, here must be intended as an organising principle, a quest for order through socio-economic processes in which "hygiene" is a process of separation, which today can create all kind of waste including—as Zygmunt Bauman points out—superfluous human lives who are finding themselves not needed or indeed desired, in the midst of the local-global processes of constant restructuring of land, economic and social relations.[1]

So here I will only sketch out a series of some points along a path—without actually being able to trace its course in full detail—of the preeminence of discourses on waste and their impact on western urban development, an impact that persists to this very day. The four sources that will provide the thread of my discourse are: Thomas Beddoes's *Hygëia: Or, Essays Moral and Medical on the Causes Affecting the Personal State of Our Middling and Affluent Classes* (1802); Benjamin Ward Richardson's *Hygeia: A City of Health* (1876); Vance Packard's *The Waste Makers* (1960); and Italo Calvino's *Invisible Cities* (1974). Through these texts I will try to make clear the path of a transition from a conception of waste through a "sanitary science"-based management, which lay at the centre of reformist utopias, to waste being depicted in the guises of pollution and resource depletion, as main ingredients in dystopian projections of the twentieth century.

The impression that hopefully will emerge is that whilst the utopias of the nineteenth century were mostly born out of reformist impulses of restraint, the dystopias described in the twentieth century appear to be caused largely by behavioural excesses, in terms of the consumption, resource use and waste generation "engineered" by marketing forces and the consumer's desire in a double-bind relationship. The matter that was the object of control in the first instance is that which spirals out of control in the latter; in this transition, waste shifts from being one of the main targets of progressive, reformist policies, to eventually becoming the defining characteristic of a society in which to waste, to consume, is to be progressive. And during this long period, gradually, the issue of waste receded from public debate, in the contentedness fostered by the "achievements" of plumbing, the water closet and public waste collection, which all had the capacity of removing the grimy evidence of our existence away from both our private spaces and the very public urban spaces in which the spectacle of these achievements could be displayed. In reality, waste, though pictured for some time at the margin of the great debates of society, nevertheless always remained at the centre of its organization: it was just made invisible, or to appear not worthy of notice, relegated from a public presence into a codified set of civic and industrial practices, as if the achievement of utopia was now simply an operational task.

Furthermore, before we come to the matter at hand, I would briefly like to raise two issues of context. The first of these is a matter of the time in which such shifts took place, and what ties it to our own moment: as the sanitary utopian projections of the nineteenth century were being constructed, the reality in which they took shape was the very first example of a global economic order bent on a search for planetary resources capable of supporting its overstretched metabolism. The second such issue of context more directly concerns our own moment, the context of this paper itself, and the contemporary discourses on waste in which it seeks to intervene. Our contemporary society, perhaps affronted with having to deal with life problems that in its utopian checklist of obstacles to human happiness appeared to have slipped off the edge of the balance sheet, seems to encounter great difficulties with the prospect of retracing its steps, with having to reconfigure issues of waste and hygiene once again placing them, as it needs to at the centre of discussions of urban development. It is perhaps a symptom of such denial that the neo-utopian discourses of Sustainability and Zero Waste, whilst taking into consideration all the charges leveled at the lack of ecological sense of most current cultural-economic practices, still appear to provide intellectual shelter, cultural comfort and a comfortable mode of repentance for the core of our economic structure that is the global capitalist system.

Two

The first text that I would like to discuss is Beddoes's aforementioned *Hygëia.*[2] I am bringing this text into focus for one main reason, clearly highlighted by Robert Mitchell in the introduction to the volume, and that is the important fact that *Hygëia* is premised on a belief that Britain's commercial ascendancy had resulted in new modes of disease.[3] It is this link between *wealth* and *pathology* that distinguishes Beddoes' contribution to my thesis here. Beddoes points to the historically contingent nature of desire, arguing that:

> In the social arrangements which have gradually formed themselves in Europe, wealth, the most general object of power, becomes the most general object of desire ... The multiplication of the roads to wealth unavoidably keeps pace with the multiplication of gratifications, and of contrivances for displaying accomplishments, which wealth can command.[4]

The term used by Beddoes to define such acts of desire is "the lust for things" and Mitchell contends that "the increasingly dominant belief that the satisfaction of desire was inextricably linked to purchases—was a drive at best indifferent to health, and more frequently, led directly to illness."[5] Very poignant here is the delineation of one of the first instances of *critique of consumption*, or to be more precise a critique of the type of desire that the new modes of consumption—lust for things—and the promise of moral-aesthetic satisfaction that those desires would engender.

Roughly a century later, we might argue, this is the very same desire that the newly developed "sciences" of marketing and PR would be contracted to stimulate on behalf of both state and corporations in order to sustain an economic system based on consumption in the very name of democratic advancement and lifestyle enhancement. The pathological character which Beddoes projects onto consumerism (a consuming desire to possess and display possessions) and attaches to the condition of the affluent classes at the dawn of the nineteenth century would by then be enlarged through economic availability and by means of the persuasion aimed at all social group, in the establishment of a "mass culture" of consumption in which consumption, deliberate acts of obsolescence and therefore waste will be transformed in acts of emancipation within a modern ethical paradigm.

Three

I would now like to switch our attention to another text, also inspired in its title by the mythological figure of Hygeia, who was the representative of cleanliness, and one of the six daughters of Asclepius, the demigod of Medicine and Healing in ancient Greek mythology. This text is Benjamin Ward Richardson's *Hygeia: A City of Health.* This work is a detailed and empirically evidenced account of what a city could be under the direction of a Sanitary Science, delivered as a lecture by Richardson in 1875. Richardson dedicates this address to Edwin Chadwick, his mentor, tutor and the most eminent reformist in the field of public hygiene in nineteenth century Britain, and whose 1842 *Report on The Sanitary Conditions of the Labouring Population of Great Britain* argued that disease was directly related to living conditions and that there was a desperate need for public health reform.

Richardson's text is unflinching and passionate in its convictions, stating from the very outset his ambitions: "Our object, to make present existence better and happier."[6] A sanitised city is, for Richardson, a community of dwellings, hospitals, schools and workshops, for which anything that is designed is done so with two main preoccupations: to avoid diseases, and to prolong life expectations. This amounts, in short, to a community which begins to resemble a factory; one that rather than producing goods, it is meant to produce a healthy and happy workforce itself. His city of hygiene is a "future now" project, it presents itself as simply a project to speed up the progress to an endpoint which is, for Richardson, already predestined as inevitable, turning wish into fate and fancy into destiny, justifying the project by seeming to render its outcome as unavoidable as it is desirable.

Despite the fact that most of the text is by now an anachronism, particularly in regards to Richardson's own emphasis on temperance, what remains interesting in *Hygeia* is the firm belief that a proper application of scientific norms, will undoubtedly bring us happiness because of cleanliness. In other words, the conditions of the present can be changed only if we are to embrace a specific future, a scientifically determined one, guided through technological expertise. This is the future city as a sanitised place, an enclave where all abjections are duly guided out of sight and out of mind. For Richardson:

> Utopia itself is but another word for time; and some day the masses, who now heed us not, or smile incredulously at our proceedings, will awake to our conceptions. Then our knowledge, like light rapidly conveyed from one torch to another, will bury us in its brightness.[7]

Enshrined in the capitalist myth of flow, and realised through the innovation of plumbing, Hygeia's utopia is founded on a total reliance on "distancing" the abject matter, on safely separating the "us" and the "it," which encapsulates a problem-solving strategy which still largely applies to urban development today.

It was not long before Richardson's publication of *Hygeia* that the events recalled as the London "Great Stink" of 1858 forced the House of Parliament to interrupt their sessions urging upon the legislative body a series of interventions which culminated in the decision to assign to Joseph Bazalgette the design and the construction of a sewage system for London, at the time the most densely populated city in the world. Prior to this final decision and the approval of its modalities, the Metropolitan Commission for Sewers commissioned a series of studies and contracted various consultants from UK and beyond to canvas for suggestions and different approaches. As a result of these exercises, many contrasting opinions emerged, one of which is particularly poignant for our discussion, as it exposed for the first time on a scientific basis the idea of waste as resource as well as putting forward the nowadays highly popular concept of "urban footprint."

Augustus Liebig was the foremost chemist of his days, thanks to his discovery of the part that the soil plays in plant's nutrient cycle. Because of this knowledge, he was asked to report to the Metropolitan commission in charge of deliberating the sewage construction. His proposal for the sewage system of London entailed a separation of the solid and liquid discharges in order to divert to the agricultural land around London human manure deployed as fertilizer, in an attempt to replete the already poor quality of the overused soil struggling to cope with the demands to feed such a densely populated city, therefore recognising waste matter as a potential resource. Such discourses are currently being reinvigorated through the potential of composting organic food waste, the spreading of sewage sludge as fertiliser for agricultural use or its compaction into combustible pellets for energy production.

However, the proposal was rejected and in 1862, when the seventh edition of his pioneering scientific work was published, he included a new introduction in which he declared the intensive "high farming" methods of British agriculture to be a "robbery system" as they necessitated the transportation over long distances of food and fibre from the country to the city with no provision for the recirculation of soil nutrients, a process that he called "the law of restitution."[8] This new introduction was omitted in the English translation of the same text. This principle was for Liebig at the core of a rational agricultural practice and it was utterly simple in its thesis: "the minerals taken from the earth had to be returned to the earth. The farmer had to restore to his land as much as he had taken from it."

It is easy for us today to recognise in Liebig's soil management philosophy, one of the basic principles of contemporary organic farming. It is also apparent that his critique of the way in which London managed its food subsistence was based on an early formulation of a concept of "urban footprint" nowadays one of the main tenets of an ecological critique of urban development.

FOUR

As the nineteenth-century utopian promises of a final sanitary solution waned with the evidence of the resilience and long-term polluting effect of the byproducts of the processes which had established a fully developed industrialized society, what instead started to emerge was the notion that there was a pernicious systemic correlation between consumption, affluence, marketing, design and the rising quantities of waste that could not simply be hidden from view. For Vance Packard and many around him, it became only too apparent that the prosperity and lifestyle attained by the US during the first half of the century had come at the expense of environmental degradation, resource depletion, diseases brought about by the heavy use of chemical substances in most production processes and a dependence on growth and consumption which institutionalised as "progressive" the concept of obsolescence.

The emerging problem of matching consumption rates with production output through deliberate obsolescence of goods, as described in Packard's *The Waste Makers*, is the basis for a city—Cornucopia—dreamed up by the marketing gurus of the 1950s, but long forecasted by novelists, marketing pioneers and, in one of the most explicit cases Edward Bernays a blood nephew of Sigmund Freud, considered the "father of public relations" and author in 1928 of *Propaganda.*[9] From this text we read: If we understand the mechanism and motives of the group mind, is it not possible to control and regiment the masses according to our will without their knowing about it? The recent practice of propaganda has proved that it is possible, at least up to a certain point and within certain limits.[10]

He called this scientific technique of opinion-molding the "engineering of consent." It is also important for us to briefly consider another one of these instances before venturing into an analysis of *The Waste Makers*; the text in question is "Selling Mrs. Consumer" published by Christine Frederick in 1929.[11] Part market analysis and part advocacy for a more intense profligacy of American women consumers, Frederick's text was a consumerist manifesto in the guise of home economics in which unashamed strands of essentialism—woman as "natural" efficient housekeeper—blended with an equally unambiguous call for unrestrained purchasing habits in the name of emancipation and identification with the success of America's progress. "[Our] triumph and rapidity of progress are based on progressive obsolescence":

> We have not been aware of it, either as consumers or producers, but we have an attitude that is quite different from the rest of the world ... It is the ambition of almost every American to practice progressive obsolescence as a ladder by which to climb to greater satisfaction through the purchase of more of the fascinating and thrilling range of goods and services being offered today. We obtain a sense of speed and progress and increase fullness of life as a result.[12]

Frederick elevated consumption to the level of a progressive doctrine, posing profligacy as a socially necessary, morally dutiful and fashionably advanced practice, the more so because it was posited as a suppression of an aristocratic tradition in which the purchase of objects was a sign of distinction, taste and status. She used as a device the polarization of the appeal of "antique," unfashionable objects as those resisting transition and opposing friction to the acceptance of the new, with the aura of the new, mass-produced, contemporary objects and their democratizing of taste. In a sweeping critique of the detractors of obsolescence, Frederick effectively ushers the age of "mass consumers" as a radical departure in social habits, but at the same time reinforces the social order in strongly maintaining the woman's place as the housekeeper, who: *no longer takes pride in the great square ebony piano of excellent tone her mother handed down to her, but on the contrary unsentimentally considers it a horror.*[13]

If Beddoes ascribed a pathological accent on consumption practices in the new climate of Great Britain as it edged into the nineteenth century,

Frederick reverses this and instead signifies that what was pathological was the notion of hanging on to objects, effectively transforming the "lust for things" into a praised new stage of evolution, where to waste is to be modern.

By equating modernity with practices whose only ultimate consequence is to produce waste, Frederick sealed the binomy of waste and modernity as inseparable. Waste, once the enemy of nineteenth-century reformist efforts, was now a steady partner in the creation and destruction of value that characterised the stages of modernity which were to follow and will almost certainly determine its future trajectory. In making it such a visible issue, Frederick and Bernays laid the ground for Vance Packard's more critically barbed emphasis on waste as he analyzed the essential practice of a consumerist, *progressive-obsolescence* led society such as the one he witnessed in the US of the late 1950s. The transformations of social ethic and etiquette that had occurred in the following decades, coupled with the surplus of production that the US faced in the aftermath of World War II, were now utterly visible and formed the basis of Packard's case against the prospect of a future model city such as the Cornucopia City described in his book *The Waste Makers*.

As it turned out, by the time his indignation was manifested it was already at a later stage of such transformations towards a society of waste; their trajectory was already too advanced to be practically opposed. His description of Cornucopia is filled with sardonic takes on the idea of obsolescence, such as papier-mâché houses that are trashed at the end of every season, or cars built in a plastic that deliberately shows fatigue after 4,000 miles and destroys itself. Born out of exuberance rather than necessity Cornucopia reveals its wealth through the capacity of its citizens of constantly exceeding their needs and of participating in an act of collective potlatch precipitated by the self-degrading nature of all they buy and consume.

Packard's moral tale and his argument that limited resources will hinder the possibilities for the construction of Cornucopia-like Utopia, bring him to declare that: "Prodigality is the spirit of the era. Historians, I suspect, may allude to this as the Throwaway Age," and later in the text to state:

> that pressures to expand production and consumption have forced Americans to create a hyperthyroid economy that can be sustained only by constant stimulation of the people and their leaders to be more prodigal with the nation's resources.[14]

But it is in his description of Cornucopia city centre that Packard displays his most prescient moment:

> The heart of Cornucopia City will be occupied by a titanic push-button super mart built to simulate a fairyland. This is where all the people spend many happy hours a week strolling and buying to their heart's content. In this paradise of high-velocity selling, there are no jangling cash registers to disrupt the holiday mood. Instead, the shopping couples—with their five children trailing behind, each pushing their own shopping cart—gaily wave their lifetime electronic cards in front of a recording eye. Each child has his own card, which was issued to him at birth ... Cornucopia City's marvelous mart is open around the clock, Sundays included.

Is Cornucopia City to become not a feverish dream, but instead, an extreme prototype for the City of Tomorrow?[15] Without presuming to answer such a rhetorical and by now anachronistic question, I am going to present some paragraphs of a marketing trend text found during my search for references to the word "Hygeia," in which it appears that the marketing nightmare of Vance Packard might today be interpreted in a dematerialized form as "city as a marketplace." The following is from trendwatching.com "an independent and opinionated trend firm, scanning the globe for the most promising consumer trends, insights and related hands-on business ideas":

> Welcome to HYGIENIA: a marketplace inhabited by mature consumers from South Korea to Brazil, from Australia to Canada, who can instantly and expertly point out the various hygiene factors for each and every good, service and experience on offer. They base their knowledge on many years of self-training in hyper-consumption, and on the now almost biblical flood of new-style, readily available information sources and filters helping them to track down the Best of the Best, the Cheapest of the Cheapest, the First of the First.
>
> If they know, so should you. Best of breed, best practices, it's nothing new, but never before has everything been so in the open, and never before have consumers done their homework so much more diligently than corporations. In fact, armed with all this information, consumers will increasingly be confident enough to try out completely unknown brands, bypassing "trusted" incumbents.
>
> All in all, it leaves no excuses for brands not to know the new HYGIENIA standards.[16]

Such an evolutionary development of "mature" consumers was already forecast by Packard who suggested that "the only sure way to meet all the demands may be to create a brand new breed of super customers."[17] It is in such emphasis toward the consumer rather then the product that we find evidence of the critique leveled by Packard towards both Frederick's manifesto "Selling Mrs. Consumer" and of the early techniques of persuasions piloted by Edward Bernays.

If the nineteenth-century challenge was to perfect the apparatus of industrial production and attend to its development in the most efficient way, by the time Packard wrote *The Waste Makers* it was obvious for most marketing experts that priming the consumer to act in a way that would create demand was the future. As one marketing consultant of that era suggested, "our enormously productive economy":

> demands that we make consumption our way of life, that we convert the buying and use of goods into rituals, that we seek our spiritual satisfactions, our ego satisfactions, in consumption ... We need things consumed, burned up, worn out, replaced, and discarded at an ever increasing rate.[18]

This is a most explicit demand to *consume*, equated as it is with an imperative to dispose at ever-faster rates, and by inference to *waste* at an ever-higher rate. No clearer suggestion could be made to transform the category of waste from a problem to be solved to a necessity to be promoted as socially constructive, and economically necessary. A positively described and socially engineered entropy was to be embraced and the dismissal of preexisting priorities was deemed necessary, through the embrace and the promotion of a new ethic, that of consumerism.

The rise and diffusion of this new consumer ethic, the emphasis on the creation of a new breed of customers and the establishment of a deeper subset of subject-object relations are the most explicit clues to the transformation of the significance of discourses of waste from the nineteenth to the twentieth century as posited at the beginning of this paper. Richardson's utopia of citizenship in the imaginary sanitary city where the provision of public services forms the core of its planning, fades in the distance with the appearance of the new consumerist ethic. The citizen-subject of Cornucopia City displays its citizenship only through its consuming habits, its purchasing ethic, without expecting any provision other than that of being able of extending its own desires into the social realm.

Through such manipulations of cultural semantic operations on the *value of wasting* a series of polarities were confused and repositioned; waste and value were not anymore opposites, but indeed mutual aides in the pursuit of capital interests whilst scarcity and excess were simply indexes of market demand and offer in the terms defined by the relation between production and consumption. These maneuvers had the effect of shifting the reality of waste and its agents of pollution into an abstract dimension, a rarefied secondary space from which only later it would emerge when it could no longer be diluted and dispersed. It is this "suicidal tendency" of modernity—as Zygmunt Bauman describes it—that appears at the centre of Italo Calvino's tale of the city of Leonia.[19]

Five

The last of the scenario I have selected, then, is that of Italo Calvino's imagined community of Leonia, in *Invisible Cities.*[20] The mood is set from the outset; Calvino opens with the words: *The city of Leonia refashions itself every day.* Leonia's citizens are constantly occupying themselves with the daily exchange of the old with the new, which they do in such a methodical manner that it compels Calvino to wonder, "if Leonia's true passion is really as they say the enjoyment of new things, and not instead, the joy of expelling, discarding, cleansing itself of a recurrent impurity."[21] Calvino's dream-like narrative echoes the concerns of the many theories of the twentieth century regarding the relations between economic expansion, consumption and waste as essential for urban development. It is clear that his imagined Leonia is a city where the message of pundits such as Frederick and Bernays, had slowly set in and had been internalized by its inhabitants for whom—as an example—the "process of Creative Destruction" posited in 1950 by Joseph Schumpeter was now a daily ritual.[22] Something akin to the paroxysm of a daily fashion shift, engulfs them in their reckless habits, compels them to joyfully dispose of everything as if following an intrinsic dictum written in the objects themselves, forgetting that yesterday the same objects spoke to them in a different language, a language of seduction and desire.

But where does all the refuse "encased in spotless plastic bags" go?[23] In Leonia, nobody wonders. We learn from the narrator, that "a fortress of indestructible leftovers surrounds Leonia, dominating it on every side, like a chain of mountains."[24] Calvino's landfill of Leonia is a 360-degree mountain range represented as an act of pure accumulation, a preservation "all of itself in its only definitive form: yesterday's sweepings piled up on the sweepings of the day before yesterday and of all its days and years and decades."[25] At times, when the stench from its surroundings reaches the shopping area of the city, the Leonians seem to realize that those solid, indestructible heaps of their spent desires will ultimately survive them; they are the most resilient testimony of their lives, the self-made monument of their decay. As this realization sets in, the seemingly self-satisfied practices of Leonia, its success in turning the new into old, day after day, assumes a menacing tone, a threatening accent. As we read: "The greater the height grows, the more the danger of a landslide looms," the danger being that of literally "submerging the city in its own past, which it had tried in vain to reject."[26] There is no exorcism or feat of engineering powerful enough to dispel such threat, but this does not stop the Leonians from repeating *ad libitum* their daily ritual, the almost magical transformation from beauty to ugliness, from desire to disposal. Prisoners of their own destiny, Leonia's citizens simply wake, use, dispose and slumber, whilst "in the nearby cities they are all ready, waiting with bulldozers to flatten the terrain, to push into the new territory, expand, and drive the new street cleaners farther out."[27]

Six

This is how the tale of Leonia ends, with a prospect of "driving the street cleaners farther out" and this also seems a very appropriate point for us to conclude this literary-historical foray into Waste and Utopia and enter into a specific contemporary situation, that of Cardiff itself and its paradigm of economic growth currently suspended in between two future projects. These are the building of its new Shopping Centre (St. David's 2) and the increasingly difficult search for an alternative location of its last operative landfill, Lamby Way, forecasted at the current rates of disposal to reach saturation in late 2008.

In order to discuss this further I will now introduce the Lamby Way Time Capsule project by presenting an alternative definition of "sanitation" proposed by the artist Mierle Laderman Ukeles in 1990, in her "Sanitation Manifesto":

> Sanitation is the working out of the human design to accept, confront, manage even use decay in urban life.
>
> Sanitation is the city's first cultural system, not its displaced-housekeeper caste system.
>
> To do sanitation is to husband the city as home.
>
> I think it can serve as a model for democratic imagination as follows: sanitation serves everyone, it starts from that premise; it accepts that everyone must be served in a democracy and the city must be maintained in working works everywhere, no matter what socio-economic culture.
>
> Sanitation, in democracy, implies the possibility of a public-social-contract operating laterally, not upstairs-downstairs, but equally between the servers and the served.
>
> Out of these humble circumstances, we can begin to erect a symbol of communality.[28]

Seven

The Lamby Way Time Capsule is a public art project. It chooses to focus its attention on one of the most "public" sites in the city of Cardiff, its own landfill, its own "social sculpture": the Lamby Way landfill waste management site. Lamby Way is at present the only operative landfill in the city. Its full capacity will be reached in a period of time of between two to three years. As I write, no alternative site has been found within the city limits, mostly due to the fact that within the growth pattern now established in Cardiff, rising land values make increasingly difficult to locate operational structures that are not considered revenue generating, no matter how essential they might be.

Already two articles in the recent local press have leaked attempts at finding a replacement site for Lamby Way and have promoted the kind of collective NIMBYism one could expect in these cases, albeit bringing an unexpected twist in the history of environmental justice, by highlighting the implausibility of such a civic site next to the highly coveted homes of BBC executives and footballers.[29] Once Lamby Way closes, it is likely that the visibility of waste disposal operations will move further away from the city itself. Plans for a "world-class pay-and-play golf course with hotel and leisure facilities" are currently being developed, with the intention of having the whole site built, run and managed by a private developer on the ex-landfill.[30]

In the future transition from public to private management of the land lies an interesting aspect for the Time Capsule project, since the invisible character of its presence, buried underneath the development, will act both as a reminder of the site's previous identity and of its public service.

Eight

The project addresses the collective cultural determination of society's own system of values and signification to be tested and reaffirmed. Through the mechanism of a public survey it asks the largest possible number of citizens in Cardiff to respond to a simple question: *What would you say is 'rubbish' in our world today?*

If the landfill is the place where things go when we define them as "rubbish" by our very act of disposal, then the question as posed demands that those who are willing (or able) to engage with it, contribute to the exercise of the power of "naming things" which culture grants us all. The power of "naming things," is also, in an individual form, that bestowed upon the post-Duchampian artist who has the power of "labelling" or "designating" what "art" is, providing that such acts of designation take place within established structures. But semantic authority is also crucial in governmental waste strategies for which definitions of what constitutes "waste" as opposed to "resource" are instrumental in attributing the preferred legal status to each material undergoing processes of transformation.

The project's question was asked to all the households of Cardiff by inserting the question cards into the council's regular door-to-door distribution chain of bin liners, and the specific answers are not made public until the Time Capsule is found and opened. The use of the word *rubbish* in the question is given its broadest significance of "low or no value" enlarged to actions, events, objects, ideas, institutions, and individuals. Reworking the normal function of time capsules by *deliberately setting aside what a hypothetical future finder is anticipated to regard as evidence of the present*, the Lamby Way Time Capsule project combines the expected function of a time capsule, that of preserving items considered of value (only now it is our opinions as citizens of a determined area which are to be preserved), with a location (the landfill which serves that area) considered the low end of land values within a consumptive parameter.[31] The attempt here is for notions of *conservation* and *disposal* to become involved in a dialectic of land use that acknowledges the dilemmas often posed by the conflictual relationships between ethical and pragmatic positions. In the spatial politics of urban developments, the accumulative logic of a waste landfill (much like that of capital, as Bataille reminds us in *The Accursed Share*) sets temporary boundaries to the mires of development expansion precisely because of its purpose and toxicity, whilst it mirrors the sites of accumulative consumption often placed right next to them, in this case the large Tesco store opposite Lamby Way.[32]

Through the set up of the deliberate contrivance of its play between conservation and disposal, the project attempts to shift both the significance of the landfill within the city and its content, and to create a framework for an alternative contribution to its growth by its constituency, this being: *the disposal of "cultural waste" as symbolic refuse*. It is the multifaceted casting of public opinions on what such cultural waste might consist of, that will make up the content of the time capsule. The fact that the question alludes to the "world" at large rather than its immanent cultural and geographic context is an allusion to a reality of waste disposal; whilst this takes place in a specific locality, the origins of its composition are to be found all over the world, such is the amount of imported goods which are available to us as consumers these days. Made all over the world, traded from all over the world, but dumped in a specific location. It is the project's intention not to establish a scheduled retrieval and opening time frame for the Lamby Way Time Capsule; it will be found whenever the landfill ceases to be biologically active and the land becomes once more subjected to reorganization.

Just like Utopia, defined by Benjamin Ward Richardson in *Hygeia* as "another word for time," the Lamby Way Time Capsule will await with confidence its own becoming, with its status as an invisible trigger of discourses around the landfill site; it will act as a rumour, awaiting the future—like a virus in the midst of our waste.

Nine

The viral expectations of the Lamby Way Time Capsule and its deferred agency, point at an unspecified future and attaches to its burial ceremony hopes for the unknown and trust to its trajectory, together with a sense of responsibility that the present projects in its becoming. In doing so, it attempts to materialize the concept that many futures are here with us in the present, constructed in our immanence, hence engendering a responsibility for the "not yet" despite our "not knowing." Therefore in the Lamby Way Time Capsule project, the significance of the virus is constructed in a benign sense, even when surrounded by the potentially much more maligned microbiological activity upon which our hopes of exercising full control are partly linked to our faith in technology and partly discharged within the unpredictability that characterizes any landfill waste mass and its virtually unknown composition.[33] The Time Capsule is proposed here as a *reversed* viral presence, enacting its infectious power within the local discourses of the landfill, inserting a new, unexpected element in the narrative of its contentious geography and its possible future development whilst reinstating the collective ownership of its content—until then separated by the boundaries of its public management entrusted to preside over discarded public properties.

The transformation of the cultural connotation related to viruses that the project is attempting to enact, is also important as it relates to what has been discussed in this paper in terms of historical evidence of nineteenth-century utopia of urban planning.

The nineteenth-century obsessive insistence upon cleanliness as a part of its sanitary project of ordering both nature and society, encountered a moment of epiphany in the isolation and visualization of germs and the confirmation of the "germ theory of disease" attributed to Louis Pasteur. As the microscopic realities of the world were progressively revealed, they acquired an important aesthetic dimension and became one of the primary modes of legitimacy of scientific authority and its tools. It is this authority that underpins Richardson's "sanitary science" at work in his plans for Hygeia and in more general terms such authority underwrites the alliance between scientific discovery and social policies that consolidated much of the impetus of the nineteenth-century utopian reformist agenda.[34] Once visualized and popularized, the germ, the microbe, the virus, became objects of fascination and horror, highly prided natural science evidence harking back to an almost prehistoric past, pre-modern creatures which stood in the way of modernity and its colonizing ambitions upon the future. Their resilience to the dramatic changes that were taking place in our world was the subject of admiration as much as a resolution to extrapolate them out of their invisibility, classify them and whenever possible isolate them at safe distance.

Such drive is explicit in Richardson's *Hygeia*, ensuing a warfare in which the microbe is seen as an enemy of modernity and one of its outcasts, whose only refuge is in the waste and filth spread all across the city. It is no surprise that a great deal of Richardson's urban planning is motivated by distancing the matter in which such dangers can lurk and infect our midst. With such cultural-historical precedent in mind, the word "virus" as associated to the Lamby Way Time Capsule, is significant of an intentionality to reframe such discourses and open up alternative socio-semantic possibilities; rather than pointing at a past order, the notion of "virus" becomes related to a future organization, the function of the virus being that of a carrier of information whose significance might be more appropriately valued in a "not yet" context.

Notes

[1] Zygmunt Bauman, *Wasted Lives: Modernity and Its Outcasts* (Oxford: Polity Press, 2004).
[2] Thomas Beddoes, *Hygëia: Or, Essays Moral and Medical on the Causes Affecting the Personal State of Our Middling and Affluent Classes* (London: R. Phillips, 1802).
[3] Robert Mitchell, in Beddoes, *Hygëia,* p. 2.
[4] *Ibid.*, p. 3.
[5] *Ibid.*
[6] Robert Pemberton and Benjamin Ward Richardson, *The Happy Colony: Hygeia, a City of Health* (New York and London: Garland, 1985), p. 3.
[7] *Ibid.*, p. 44.
[8] See Justus Liebig, *Die Chemie in Ihrer Anwendung auf Agricultur und Physiologie [Organic Chemistry in its Application to Agriculture and Physiology]*, Vol.1 (Braunschweig: Friedrich Vieweg, 1841). The quotes from Liebig's text repro-

duced here are from John Bellamy Foster, "Marx's Ecology and its Historical Significance," in *International Handbook of Environmental Sociology*, second edition. Eds. Michael Redclift and Graham Woodgate (Northampton, MA: Edward Elgar, 2010), p. 107.
[9] Edward L. Bernays, *Propaganda* (Port Washington: Kennikat Press, 1972).
[10] *Ibid.*, p.71.
[11] Christine Frederick, *Selling Mrs. Consumer* (New York: The Business bourse, 1929)(available at: http://memory.loc.gov)
[12] *Ibid.*, p. 246.
[13] *Ibid.*, p. 252.
[14] Vance Packard, *The Waste Makers* (Longmans, Green: London, 1961), p. 6.
[15] *Ibid.*, p. 5.
[16] See: www.trendwatching.com/trends/hygienia.htm
[17] Packard, *The Waste Makers,* p. 11.
[18] These are words of Victor Lebow, a marketing consultant, from an article titled "Price Competition in 1955," from the *Journal of Retailing,* as quoted in Packard, *The Waste Makers,* p. 24.
[19] Zygmunt Bauman, "Wasteful Planet," *Kainos-Rifiuti*, 4-5 (2004) (available at: www.kainos.it), p. 3. The translation from the Italian is mine.
[20] Italo Calvino, *Invisible Cities*, trans. William Weaver (London: Secker & Warburg, 1974), pp. 114-116.
[21] *Ibid.*, p. 114.
[22] Joseph Schumpeter, *Capitalism, Socialism and Democracy* (George Allen & Unwin, London, 1950).
[23] Calvino, *Invisible Cities,* p. 114.
[24] *Ibid.*, p. 115.
[25] *Ibid.*, p. 115.
[26] *Ibid.*, p. 116.
[27] *Ibid.*, p. 116.
[28] Mierle Laderman Ukeles, "Sanitation Manifesto! Why Sanitation can be used as a Model for Public Art," *The Act*, 2: 1 (1990), pp. 84-5.
[29] Phillip Nifield, "Millionaire Row Rubbish Tip Plan," *South Wales Echo*, 03 May 2006.
[30] "Lamby Way Golf & Leisure Development Plan," Cardiff County Council Financial Services Team (tender documents), 21 May 2005.
[31] This and other attempts of definitions can be found in: Brian Durrans, "Posterity and Paradox, Some Uses of Time Capsules," in *Contemporary Futures: Perspectives from Social Anthropology,* ed. Sandra Wallman (London: Routledge, 1992), p.51-67.
[32] Georges Bataille, *The Accursed Share: An Essay on General Economy* (New York, Zone Books, 1988). Most Anglo-American readers know Bataille as a novelist. The Accursed Share provides an excellent introduction to Bataille the philosopher. Here he uses his unique economic theory as the basis for an incisive inquiry into the very nature of civilization. Unlike conventional economic models based on notions of scarcity, Bataille's theory develops the concept of excess: a civilization, he argues, reveals its order most clearly in the treatment of its surplus. The result is

a brilliant blend of ethics, aesthetics, and cultural anthropology that challenges both mainstream economics and ethnology.

[33] An exception to this state of not knowing the content of landfills is the 'Garbage Project' a cross-disciplinary research unit based at the University of Arizona begun in 1972 and still active today. A full description of the project and some of its results can be found in William Rathje and Cullen Murphy, *Rubbish!: The Archaeology of Garbage* (New York, HarperCollins Publishers, 1992)

[34] For an overview of the subject see Roy Porter, *The Greatest Benefit to Mankind: A Medical History of Humanity* (London: Fontana Press, 1999)

CHAPTER EIGHT

RUINS OF THE FUTURE

WILLIAM VINEY

This chapter considers the role that architectural ruins play in the way that we imagine the future, and suggests that they mark both an end and a continuity; the end to use and its muted remainder. Whilst future ruins frequently suggest the termination of some time, people or structure, they also have a lingering or remaining sense of time, a time that is particular to the condition of being "leftover." Future ruins are the ruins of the present. The time of the ruin, which marks the termination of use *and* its cindering persistence, means that projected ruins represent a disrupted continuation of present events. Projecting ruins discloses the duration and shape of time and dramatizes a conflict between material permanence and material transience. This conflict between continuity and cessation makes the ruin an end that remains, an end that is imperfect or unreliable; a sense of termination that has not quite come to its end.

Projected future ruins provide a means to approach the unthinkable. Jacques Derrida observed that an apocalyptic future is always a question of narrative, a textual negotiation with a "phenomenon whose essential nature is that of being fabulously textual, through and through."[1] The ruins of the future give a means to envision a story that both locates a possible landscape and relates that landscape to present surroundings. This is not only done in order to imagine what the future might look like but, as we shall see throughout this chapter, provides an opportunity to re-think the qualities of the now. As Paul Ricoeur writes, "In reading the ending in the beginning and the beginning in the ending, we also learn to read time itself backwards, as the recapitulation of the initial conditions of a course of action in its terminal consequences."[2] In fiction, film and the painterly arts, the anticipation of a ruinous end is frequently a narratological means by which to return to and make sense of the present.

When in *The Planet of the Apes* (1968) Charlton Heston's character confronts the toppled remnant of The Statue of Liberty, he exclaims, "Oh my God! I'm back. I'm home." In the comparative exercise of managing the disrupted continuity of the future ruin, between imaginary times and familiar places, we encounter not only a future in itself but also a future in which the present has been abandoned, cut off, terminated or forsaken. This is the broken continuity, the interrupted endurance that the future ruin offers. The consummation that a ruin might represent does not necessarily become imminent to the present, but ruins are felt to become immanent in present events, thus, we come to see the ruinous potential of things that have not yet been discarded and identify the long held association between ruins and *vanitas*. It is precisely this uncanny distance, achieved through a sense of arriving at and travelling to the estranged familiar, that gives the future ruin the power to relay the present and the future as an object that persists by passing away.

Apocalypse, Then

Confronting the future in ruins is by no means a formulaic exercise. Visions of catastrophe and apocalypse demand interpretative gaps that can energize the exigency of this material waste. Particularly in their painterly and cinematic manifestations, the ruins of the future frequently leave out how or when or for what reason these structures have reached their terminal condition. For instance, in the paintings of Hubert Robert and Joseph Gandy we are given no explicit explanation for why the structures they depict have fallen into ruin; their visual impact plays upon the disjuncture felt between the building existent and the future ruin represented. Robert's *Design for the Grand Gallery in the Louvre* and his *An Imaginary View of the Ruins of the Grand Gallery in Ruins*, manipulate a dual vision of the future: one image presents a new and ideal view of a Republican art institution, the other represents a more ruinous and terminal condition of this institution. As such, the work has often been read as an ambiguous commentary on France's new and emerging attitude to the public control of artworks and, by implication, condemns the futility of the Revolution.[3] For Daniel Brewer, Robert seems to dramatize "the inevitable transience of precisely the institution whose current function is to preserve the artwork from physical [deterioration]."[4] Furthermore, viewing the images together—as they were at the 1796 *salon*—emphasizes the transition between architectural shelter and exposure, museum and ruin. But the nature of this transition, the causal events by which one moves from one condition to another, is a transition rich in absence and enigma. Time has been accelerated in *An Imaginary View*, leaving the Louvre at both the end of time and at a place where time seems to have resumed. The inclusion of the Apollo

Belvedere, seen in the foreground of *An Imaginary View of the Ruins of the Grand Gallery in Ruins*, is just one indication of the intervening years and the exigent potential of imagining ruins. The statue came to the Louvre with Napoleon a year after Robert pre-emptively included it in this painting. In this small detail we see how Robert does not simply give an image of his present as ruin, but a particular, albeit elliptical history by which the future ruin is contextualized. The presence of the Apollo Belvedere is one indication of this causal absence, which the ruins of the future envelop.

The English parallel to Robert's work might be found in another double vision of future ruins. Joseph Gandy's *A Vision of Sir John Soane's Design for the Rotunda of the Bank of England as a Ruin* (1798) and *Soane's Bank of England as a Ruin* (1830), constitute attempts to imagine ruin in construction, ruin which marks the conception and completion of Soane's bank. In his painting of 1798 Gandy was commissioned by Soane to project the building into a future state of ruin, and in the work draws directly from the picturesque style of Robert to give the material foundation of the bank a corresponding ruin. The painting's companion piece, which Brian Lukacher has described as a "Piranesian ruinscape," was commissioned by Soane when the building of the bank was nearing its completion in 1830.[5]

Just as Robert's images of the Louvre were viewed together to emphasize their chronological yet antagonistic pairing—in which ruin is conceived as the beginning and the end of a creative process—so Gandy's paintings of the Bank of England were displayed together at the Royal Academy in 1832.[6] The genesis of Soane's bank, its design, construction and eventual use, was surrounded by images of its dissolution; the imagination of its present was permeated by the state of ruin it would fall into. Again, the idea of the ruin is used to fashion the future and narrate the trajectory of built environments. But, whilst these images of future ruin invite us to consider the idea of waste, the *events* of waste and the particular sequence of events whereby the useful is transformed into the derelict is a transition frequently left to our imaginations. In the paintings by Joseph Gandy, for instance, a subtle exchange occurs between image and viewer, where the temporal absence implied by the image is shaped by the perspective on and texture of the ruin represented. The abundant vegetation contained in *A Vision* and the rustic scene occurring within the sanctuary of the rotunda, means that we are led to assume that this future ruin has come about slowly or without catastrophe, left to collapse gently like the ruins of antiquity. In *Soane's Bank of England as a Ruin*, on the other hand, with its elevated perspective, the bare and uninhabitable interior and the confrontational figure in the lower right hand corner, suggests fire, earthquake or holocaust; in short, a cataclysmic event that has brought the bank to ruin. In this respect, ruins of the future provide a means to visualize consequences before knowing precisely what their cause might be. They are images of the future that stand prior to our reconstructive acts of narration, yet their ruins shape and inform our interpretative responses. The future ruins of Robert and Gandy provide us with objects within incomplete narratives, narratives with absent middles. Whilst suggesting a material continuity between the contemporary and the futural, the future ruin can also appear as cast off from any neat or continuous reconstruction of events.

So far we have considered the narratological effects of the future ruins but these compensatory acts of narration implicitly search for ends, to the temporal terminus that ruins can promise and frustrate. Michel Serres has argued that historical thought searches for this impossible zero point in time, "That point inaccessible—it is a point of accumulation; another point always interpolates itself, iteratively, in front of it."[7] This is because, for Serres, historical thought "seems linear, as if progressional, as if it followed the current of time."[8] Though Serres is describing the activity of working backwards through time, towards the zero point of an origin, we might take this characterisation of historicism and consider how the future ruin confirms the accumulative, progressional nature of future thinking; that is, the way in which ruins help us to think towards an end, the end, our end. In a way not dissimilar to the ever-receding vanishing point of Robert's paintings, Serres gives us the vocabulary to consider the impossible and ever-receding point of the end; just as we might think that, finally, we have determined the point of termination we find another, a time yet more remote that frustrates our neat conclusions.[9] In their capacity to stand as a consummation and continuity of the present, future ruins form part of a visual and conceptual repertoire of an incomplete or provisional end. As both an end and the contingency of that end, a form of material and temporal punctuation; the already-and-not-yet of waste's dissolution. This makes the ruin, particularly the future ruin, a very odd object for historical contemplation. Historiography, argues Serres, takes for granted a transition from the indeterminate to determinate, "indetermination precedes the determinate."[10] That is to say that the historian's explicatory task is to transform the indeterminate into the knowable, the scripted and the sequential. What, then, is the historical quality of the future ruin? The ruin of the future neither fulfills this trajectory from indeterminacy to determinacy, nor flatly contradicts this progression. As Robert's works demonstrate, there is an implicit relation between the Louvre represented as a functioning space for the public display of objects and its consequent condition as a ruin. In one sense the ruin is always that which comes afterwards, it is always the ruin *of* something. Ruins, then, are the outcome of a linear, progressional transition from use to waste; future ruins are an outcome of the present. On the other hand, they are frequently the outcome of events that we do not witness, the cast offs from a time that is yet to occur. In this respect, the fragile terminus of the future ruin is laid bare; the indeterminate does not necessarily precede the determinate when the future ruin can frustrate the continuity of thought that seeks to bridge the determinate present and indeterminate future.

Why should ruin be so frequently used to challenge the stability of human existence? Hannah Arendt once wrote that the "reality and reliability of the human world rest primarily on the fact that we are surrounded by things more permanent than the activity by which they are produced."[11] Arendt is sensitive to the way in which we measure time according to the things that we make and use, doing so within the temporal strictures of human mortality. The time we associate with architecture seems particularly relevant in this case, where the disparity can be great between the time it takes to build and the duration of "use-time" expected from that building. If we are to speak of the permanency of things *vis-à-vis* the relative duration of human labour, then buildings represent some of the more enduring objects that humans create. We might reapproach Arendt's statement in light of the foregoing analysis; we do not simply contrast a building with the activity required to build it but against the ongoing activity that keeps it in the purposive and proleptic time of use, the time in which it is dwelt. In contrast to the great majority of modern electronics, clothing, transportation, ornament or fuel, buildings are not expected to fail before we do. Shoes, for instance, are acquired with the expectation that they will wear out, discarded, replaced; they are objects enveloped within the temporal horizons of mortality. That we expect the usefulness of a building to outlast the demands of an individual user only seems to intensify the disparity felt between the time of buildings and the duration of human life. This rather odd, co-dependent yet divergent relationship between the time of human existence and the time of architecture proves crucial for any assessment of ruins because it explains a building's potential to both metonymically represent human endeavour, to intercede and intermediate between the architectural and the human, and to appear indifferent to the rhythms of human activity. Whilst we maintain the use-time of buildings by networking them within the time of our tasks and activities, doing so does not exhaust their propensity for action. In so far as buildings participate in and anticipate human needs, the time of buildings is human-dependent, yet a building's propensity to participate in this way frequently exceeds any given network or relation. For David Lowenthal and a vast number of writers, artists and travellers:

> The dominant association with [architectural] decay is our own transience. Marks of dissolution figure as *memento mori*, compelling reminders of death's imminence [...] remainders of the evanescence of life and the futility of effort.[12]

And yet ruins represent death only in so far as they contravene the material expectations we have of buildings, the time we assign them and the time we assign ourselves. This failure to act under implicit obligations can take two forms: firstly, that buildings no longer show themselves to comply with the futural, projective tasks that we set for them; and secondly, that some buildings should cease before we do. Should they fail to participate in the "reality and reliability of the human world" we are reminded of the bearings we take from buildings, from the sense of spatial organization and temporal assembly that functioning architecture gives us. The future ruin depends upon our expectation that the usefulness of a building should outlast the demands of an individual user, intensifying the disparity we feel between the time of buildings and the duration of human life, as well as increasing a building's propensity to stand for collective use. The sense of the end that the future ruin generates plays upon the way in which buildings are felt to outlast these multiple relationships with individual users, reflecting an accumulation of uses and users. In doing so, the most pertinent notion that is associated with the future ruin comes to light: the denial of progress and the advent of apocalypse.

The Ozymandias Complex: Monuments to Our Passing

A notion of progress seems hollow without including a sense of historical transition and, with it, a sense of material transmission. The purposive, active, future-orientated temporality of useful things—the "*towards-which*" and "*for-which*" of Heidggerian tool-being—can readily be aligned to ideas of progress: to a time synonymous with the instrumentalization of things.[13] The permanence and continuity of change that characterises notions of progress means that the time of use, which gives provides distinction to objects and architectures of waste, can now take on a wider application to questions of collectives, societies, nations and empires. Temporally congruent with the increasingly efficient production and consumption of resources, with an increasingly rational distribution of resources and rapid rates of technological innovation, a determined and projective relationship to the future is readily appropriated by the imperatives of progress. Whilst progress might denote continuous change—and "creative destruction"—it only does so provided that *things are being put to use*. In short, social progress in its most simple manifestation requires the time of use to maintain a teleological drive towards improving living conditions; the useful employment of things is a sign of social health.[14] And, just as use-time has given shape and clarity to the time of waste, so it is that the progress associated with use-time gives clarity to the decline and cessation of this progress, to the dissolution, decay and waste of things.

Not an explicitly apocalyptic text, Percy Bysshe Shelley's "Ozymandias" exemplifies how questions of progress, time and the monumental become inextricably linked by a particular gesture to and exchange with an apocalyptic future. One might an antiquarian work like "Ozymandias" to speak more of the past than of the future. However, it is precisely the futural dimension of the work that creates its moral ambivalence towards and communicative oscillation between monuments of permanence and monuments to transience. It is, in short, a study of ruin's relationship to the monumental—and the future of that monumental gesture—that rescues the poem from mere antiquarianism or a simple condemnation of the past, opening the work to explore both past and futural aspects of the ruin.

> I met a traveller from an antique land,
> Who said—"Two vast and trunkless legs of stone
> Stand in the desert … near them on the sand,
> Half sunk a shattered visage lies, whose frown,
> And wrinkled lip, and sneer of cold command
> Tell that its sculptor well those passions read
> Which yet survive, stamped on these lifeless things,
> The hand that mocked them, and the heart that fed;
> And on the pedestal these words appear:
> My name is Ozymandias, King of Kings,
> Look on my Works ye Mighty, and despair!
> Nothing beside remains. Round the decay
> Of that colossal Wreck, boundless and bare
> The lone and level sands stretch far away"[15]

What I term the "Ozymandias Complex" attempts to gather together the twofold gesture of consummation and dissolution that the "King of Kings" presents to the "traveller."[16] At one level, the tyrant's claims to permanence and omnipotence are affirmed—the invocation to stand before and look upon his works, which, of course, includes the invocation itself, is dutifully acknowledged not just by the traveller but by the enunciating "I" of the poem. Although far from complete, the statue continues to communicate the features and passions of its subject, as well as the skill of those that sculpted it. A more traditional reading, however, points to how Ozymandias' boasting pedestal is inverted by the dilapidation to which it has fallen. For Kelvin Everest, the ruinous condition of the statue means, "Ozymandias's message to posterity has ended up articulating just exactly the opposite to what was intended", suggesting how the colossal wreck of the statue highlights "the deluded vanity of thwarted human ambition."[17] Herein lie two sides to an "Ozymandias Complex": ruin can be judged as both a monument to endurance and a monument to transience, a testimony to progress or its refutation. Against these dichotomies lies a notion of an ideal monument with an infinite potential to transmit into the future, a monument that can outlast and remain beyond all that surrounds it. In the ode that begins, "Exegi monumentum aere perennius," Horace suggests a definition of the perfect monument as one that does not ruin, one that "cannot be destroyed by gnawing rain, or wild north wind, by the procession / of unnumbered years or by the flight of time" (*Odes* III.30).[18] The dream, here, is of an object so unchanging that it becomes simultaneous with all time. Horace describes a monument that is timeless, in so far as decay no longer gives it time. Whilst these varying positions on the relationship between ruination, progress and the monumental might seem mutually exclusive, all depend on the anticipation of an evaluating subject, a "message to posterity."

Does the Ozymandias statue remain a testimony to the king's "Works" or a refutation of them? Interestingly, much depends upon how we read the word "remains." As both substantive and verb, a tension exists as to whether we understand "remains" to describe the "lone and level sands" or the statue itself. It is a tension that exaggerates the conditional and incomplete endings found in the ruin. Does what remain of the statue represent an outpost of last resistance against the corrosive demands of time, or, do the sands show the eroded future of the stones? The answer to this question should account for how the temporal permutations of waste—as a false ending, as a monument to consummation and transience—organizes the temporal ambivalence we encounter. This is an ambivalence that is mirrored by the sculpture's relationship to the work of the poem. The poem itself is, after all, an act of memorialization and monumentalization: a testimony to a remainder. However fictional the enunciating 'I' of the opening line or his wandering interlocutor might be, the poem finds metafictional resonance with the legend inscribed on the statue. Everest characterises this resonance between poem and the statue described in the poem as one predicated on the capacity for both to endure:

> "Ozymandias," the poem, offers an instance, cognate with that offered by the ruined statue we read of in the poem, of a human activity which is genuinely able to transcend time, to outlast the conditions and occasions of its making.[19]

This textual and evidential quality of the work serves to accentuate the temporal quality of these "remains," calling attention to the way in which remainders develop over time and attract the judgement of posterity. Moreover, the expressive, outward and forward-looking constitution of the Ozymandias statue and, in consequence, the various responses that the poem might elicit, depends upon the narrative structures discussed above: a sense of an end, the unexplained cause of decay or "absent middle," and the ruin's relationship to the individuals and the collectives it is taken to represent.

The power of ruins to architecturally foreclose the present, to void aspirations or expectations of progress and to punctuate the present with an immanent sense of the end, has provided modern and contemporary fiction and film with an enduring dystopian trope. Whether it be the half-submerged figure of the Statue of Liberty found in the closing scenes of *The Planet of the Apes*, the urban wastes of *Blade Runner* (1982) and *Terminator* (1984; 1991) or the dank ruins of "The Zone" in Andrei Tarkovsky's *Stalker* (1979), the futural figure of ruin serves to provide a disrupted continuity between the now and the then. H. G. Wells's Time Traveller journeys to a future somewhere near Richmond, carrying with him the expectations of his age, "the absurd assumption that the men of the Future would certainly be infinitely ahead of ourselves in all their appliances."[20] Temporal advancement, it is assumed, brings with it social advancement. But it is ruin that stands as the negation of such beliefs. In *The Time Machine* (1898), Wells emphasizes the dilapidated condition of the future by characterizing the condition of the world as ruinous, indeed, ruins are the foremost puzzle that he attempts to unravel: "I was watchful for every impression that could possibly help to explain the condition of ruinous splendour in which I found the world—for ruinous it was."[21] Elsewhere, the Time Traveller remarks upon the "big abundant ruins," and, later, he visits the dusty and corroded remains of a large museum, "the ruins of some latter-day South Kensington!"[22] The traveller then wanders through "gallery after gallery, dusty, silent, often ruinous, the exhibits sometimes mere heaps of rust and lignite."[23] By dint of repetition, the sheer scale and number of ruins only seems to reinforce the uniformity of decline; like the inhabitants of this future world, evolution through time does not necessarily assume the "improvement" of material conditions. Here, the certainties of a progressive, preservationist, technologically advanced Victorian England has entered into regression; it is ruin that gives this sense of regression an unequivocal, terminal image.

In later life, and under the influence of Edward Gibbon's *The Decline and Fall of the Roman Empire* and Oswald Spengler's *The Decline of the West*, Wells makes this association between social and architectural disintegration even more explicit: "Mankind which began in a cave and behind a windbreak will end in the disease-soaked ruins of a slum."[24] In the full Judeo-Christian tradition of manipulating waste in order to describe ends, the Wellsian ruin marks the material end of humanity and the decline of human progress. For George Orwell on the other hand, the ruinous future was closer to hand and provided a means to highlight a material continuity rather than an end. The "rotting nineteenth-century houses" of *Nineteen Eighty-Four*, "where the plaster dust swirled in the air and the willow herb straggled over the heaps of rubble," do not represent future ruins in the same way that Wells imagined buildings *becoming* or *having become* ruin.[25] Instead, Orwell gives the inverse—his ruins are the ruins of the present, the ruins of the Blitz perpetuating into the stasis of a dystopian future. And yet the effect of these wastes is the same; progress and the ruin maintain their close relationship. Although *The Time Machine* imagined ruins as the mark of social decline, representing a change for the worse, Orwell's *Nineteen Eighty-Four* imagines ruins as a mark of social decline by *not* representing change and allowing squalid, post-war conditions to continue where one might expect them to have disappeared. The new and the used are, in both texts, seen to be antithetical to notions of progress.

One further example, taken from the many apocalyptic and dystopian works in which ruins figure, should help to give added emphasis to role that ruins play in these fictions. Reported through a dry and unrelenting epistle, Paul Auster's *In the Country of Last Things* (1988) provides an inscrutable image of the near future, assumed to be less than a generation's remove from our own. Narrator-protagonist Anna Blume describes how the city in which she lives is dominated by the death of human and thing: "Things fall apart and vanish, and nothing new is made. People die, and babies refuse to be born."[26] The various ways the city's population die only seems to underline the material parallels Auster draws between the built environment and the human body. So-called "Leapers," who jump from buildings, and "Runners," who career through the streets until they die of exhaustion, suggest falling bodies that approximate and mimic the city's architecture; a fall to earth that litters the streets with the corpses of bodies and buildings. The streets are "reduced to rubble," and, "Dead bodies are therefore everywhere you turn–on the sidewalk, in the doorways, in the street itself."[27] The urban environment is one dominated by the sterility of waste. Katherine Washburn has described this condition in both the terms of periods and cultures, a society at the "terminal stage of collapse and ruin," and, as in Wells' *Time Machine*, a novel portraying "the wreck of Western culture."[28] And yet the novel suggests something rather more sophisticated than this "terminal" view:

> The closer you come to the end, the more there is to say. The end is only imaginary, a destination you invent to keep yourself going, but a point comes when you realize you will never get there. You have to stop, but that is only because you have run out of time. You stop, but that does not mean you have come to the end.[29]

The created and creative elements of the end, as well as the historiographic paradoxes at stake, are well framed in this passage. "Life as we know it," writes Anna elsewhere, "has ended, and yet no one is able to grasp what has taken its place."[30] Anna is conscious that, whilst she writes for and about an end, it is a condition she herself can never experience. Whilst the ruinous landscapes she moves through provide an image or sign of the end, her descriptions of them suggest that this end is being pushed further into the future by her act of description. Ruins do not mark an end here, an absolute apocalypse, but signal how the end is being deferred, supplemented by the material chronography of things falling out of use.

Having fleshed out the way in which future ruins suggest the negation of progress to function as "monuments to transience," we can now turn to the other side of the Ozymandias Complex; the ruinous yet enduring monuments to past success. Born from a more cyclical and determined view of history, these projected ruins stand as a testimony to the grandeur of fallen empire and fulfil the familiar trajectory of famed groups, dynasties and diasporas. One of the central paradoxes of these projected ruins is the way in which, as against Horacian monument, decay serves to secure the awestruck appreciation of future generations. A contemporary of Shelley, Joseph Bounden, sums up this view of ruin succinctly: "In splendid ruin still alive to flame / Thy giant wreck shall long preserve thy fame!"[31] The destruction and dissolution of buildings does not so much negate the values of those who no longer inhabit them, but, instead, ruin takes on an important communicative function; the projection of ruins claims a particular image of the future.

The ruin does not necessarily provide the means to imagine an end but a means to perpetuate and safeguard its future. Imagining future ruins in this way can prove to be counterintuitive with respect to our ideas of the monumental, particularly if we fail to keep in mind the dual nature of the ruinous monument. Writing in 1903, Alois Riegl describes the efficacy of monuments in terms of their relationship to waste:

> The historical value of a monument arises from the particular, individual stage it represents in the development of human activity in a certain field [...] The more faithfully a monument's original state is preserved, the greater its historical value: disfiguration and decay detract from it.[32]

But, as our analysis of Shelley's "Ozymandias" has shown, the historical value of the monument cannot be simply be opposed to but frequently evolves with the process of its dissolution. Moreover, Riegel fails to account for how ruin has provided some to imagine their material endurance, a means to resist visions of erasure by communicating with the unknown.

Invested in the control and manipulation of both the present and the future, Adolf Hitler's architect, Albert Speer, was tasked with influencing the judgments of those thousands of years into posterity. "Hitler liked to say that the purpose of his building," wrote Speer, "was to transmit his time and its spirit to posterity. Ultimately, all that remained to remind men of the great epochs of history was their monumental architecture."[33] Speer, troubled by the sight of debris created whilst clearing the site for the Nuremburg Zeppelin Field—"It was hard to imagine that rusting heaps of rubble could communicate these heroic inspirations which Hitler admired in the monuments of the past"—came up with what he termed his "Theory of Ruin Value."[34] Speer's ambition was to construct buildings that would ruin "effectively" and safeguard a positive image of the Reich for those of the future. In order to do this, he avoided all the typical innovations of modern construction—especially steel girders and reinforced concrete—that might belie the solid, neo-Classical and upstanding self-image that the Reich gave itself. Walls were built so wide that, even if roofs and ceilings collapsed, the walls would remain.[35] To convince his colleagues, Speer had a "romantic drawing prepared" showing the Zeppelin Field in a ruinous condition, "after generations of neglect, overgrown with ivy, its columns fallen, the walls crumbling here and there, but the outlines still clearly recognizable."[36] Whilst many found this image of the future in bad taste, Hitler accepted Speer's theory of ruins as a "logical" stage in his conception of Nazi imperialism.[37] Based on an admiring view of Classical architecture, ruins become the central object by which a regime, empire or society can narrate, perpetuate and safeguard its future. Indeed, by assuming that one's architecture will be recognized in the same attitude as eighteenth and nineteenth-century Europeans acknowledged the ruins of antiquity is to place oneself at the foundations of the future; ruins do not form the end but the basis for a new beginning.

With Speer we have reached a particularly clear example of how ruins operate as an object of politics. Choosing between the two sides of the Ozymandias Complex—deciding what can be done about the ruins of the future—is a decision fraught with the politics of remains, their creation, projection and interpretation. This is true of the dystopian ruination of the future as an indictment of the present, as well as the celebration of ruin as the glorious conclusion to and recognition of civilization. This is a politics of duration and persistence, disappearance and recognition, a politics of traces and their management. Architectural movements in the twentieth century were never far away from defining themselves according to their propensity to waste, to attempt to erase, endure and leave no remainders. This seems particularly relevant when considering the radical aspirations of so-called Modernist architects and agitators. In Antonio Sant'Elia's manifesto, *Futurist Architecture* (1914), to take just one example, it is claimed that the "Futurist architecture will be fundamentally short-lived and transitory."[38] Rather than the durable and solid, Futurism should pursue architecture that could sweep aside all that was old, to build upon the understanding that the next generation will be just as radical, allowing the "constant renewal of the architectonic environment [...] Our houses will last less time than we do. Every generation will have to make its own city anew."[39] As Owen Hatherley observes, this is a theory of architecture with both a fetish for inbuilt obsolescence and a desire to erase every trace of the old—"outrunning the old world before it has the chance to catch up on you."[40] As with Speer, the mandate assumed for intervention is one founded upon the destruction of present conditions on the one hand, and their replacement with an architecture that will last for a requisite period on the other. For Speer, this anticipated period is many generations long and involved the participation of waste, for Sant'Elia it is for just one generation and depended upon waste's erasure. Whilst Speer and Sant'Elia might appear at opposite end of an Ozymandian spectrum, the figure of future ruin stands as their common object, the condition by which they measure their progress, the condition by which and from which they give and take time.

Encountering Last Things

In considering the examples above we have completed a circle that has taken in two sides of the Ozymandias Complex, where ruin has been seen to represent an ambivalence of ending, as both an object of transience and endurance, failure and consummation. But we should not overlook that

Shelley was not the only poet writing about Ozymandias in 1819, Horace Smith also wrote for *The Examiner* on the theme of ruin:

> In Egypt's sandy silence, all alone,
> Stands a gigantic Leg, which far off throws
> The only shadow that the Desart knows:—
> "I am great OZYMANDIAS," saith the stone,
> "The King of Kings; this mighty City shows
> "The wonders of my hand."—The City's gone,—
> Nought but the Leg remaining to disclose
> The site of this forgotten Babylon.
>
> We wonder,—and some Hunter may express
> Wonder like ours, when thro' the wilderness
> Where London stood, holding the Wolf in chace,
> He meets some fragment huge, and stops to guess
> What powerful but unrecorded race
> Once dwelt in that annihilated place.

The productive oscillation between permanence and decay is evident again, with questions of showing, disappearing, disclosing and forgetting underpinning the rhyme scheme of the second quatrain.[41] However, the sonnet's volta introduces a departure from Shelley and provides a fresh theme for our consideration of future ruin, as the confrontation with ruin provokes a moralizing image of London in an imagined state of ruin, seen through the eyes of "some Hunter." Whilst Shelley's "Ozyamandias" left us curious as to the precise status of the remains described, Smith's makes explicit the synonymous relationship between the ruins of the past and the ruins of the future. Smith's poem, then, includes a view of Ozymandias and a more knowing, projective and empathetic consideration of what it means to be an Ozymandias in the making. What accompanies this circular movement from empire building, to ruination and eventual discovery is a sequence of disappearance and assessment; in short, another means to fabricate self-distance through ruin. And, unlike the ruins of the near future, such as those described in the fiction of J. G. Ballard, W. G. Sebald or the essays of Robert Smithson, these are ruins of the distant future precisely because they are discovered by an individual entirely alien to our present, an outsider that can look upon the works of the present and judge their worth with an objectivity unavailable to those of the present.[42] This technique, in which the outsider's view is proleptically incorporated within the self-regard of the present, provides an uncanny vantage point. As a kind of archaeology of the future, the evidential quality of the ruin and the experience of testimony and remembrance that they might inspire becomes a central (and often paranoid) preoccupation. Of course, anxiety arises not just about what kind of testimony might be told by this fictional figure, but whether there will remain any trace of the past to be viewed. We have entered into a final phase in both our analysis of the future ruin, and in our broader attempt to consider the temporal problem of waste. What remains is precisely a question of remainders and their capacity to be interpreted, their capacity to make and take time.

David Skilton has traced the representation of tourists visiting a ruinous London to the letters of Horace Walpole and the poetry of Thomas Lyttelton, but elsewhere acknowledges that this figure, particularly that of the "New Zealander," raised to its extraordinary popularity during the mid-nineteenth century.[43] The trope of the New Zealander entered into popular discourse with Thomas Macaulay, who concluded an article in the *Edinburgh Review* with the prediction that Catholicism:

> may still exist in undiminished vigour when some traveller from New Zealand shall, in the midst of a vast solitude, take his stand on a broken arch of London Bridge to sketch the ruins of St. Paul's.[44]

So prevalent did Macaulay's idea of the inquisitive and judgmental New Zealander become, capable of measuring the duration of things, that by 1865 *Punch* had placed it at the top of their list of rhetorical devices considered "used up, exhausted, threadbare, stale and hackneyed."[45] Appearing in every imaginable kind of publication, the New Zealander was still visiting future ruins over thirty years after Macaulay's initial invocation, and it was left to Gustave Doré to seal the New Zealander's reputation by publishing a famous etching. By evaluating the fantastical products of a ruined present, the New Zealander—invariably male, typically of Maori rather than European descent—dramatizes both a familiar perspective and a perspective entirely remote to us, an impossible view.[46] As Skilton writes,

> although Britain seems to have reached a state of power and prosperity unparalleled in the history of the world, it is in the nature of things to be impermanent. The New Zealander is less a harbinger of doom than a prompt to self-awareness and self-criticism.[47]

It is, however, rarely noted how this effort towards self-awareness and self-criticism involves the New Zealander's own act of artistic production. Unlike Smith's hunter, who views the fragment of a "powerful but unrecorded race," both Macaulay and Doré have their New Zealander carefully recording whatever is available to him, visually producing and reproducing a view. As well as giving an explicit sense of the traveler's labour of assessment, his interpretative response, the emphasis on the visual reproduction of future ruins means that the legacy of British imperialism is artistically perpetuated and critical appraised as a legacy in ruin. In a gesture that might recall the figure towards the centre of Robert's *An Imaginary View of the Grand Gallery in Ruins*, in which artistic endeavor is seen as an intrinsic companion to material decay, the New Zealander enacts a form of Grand Tourism that displaces the Classical world with the modern. And, not unlike Speer's ruin theory of value, the allusion to a Classical precedent means that the New Zealander also perpetuates a circularity of travel, transcription and return. If, as Anne Janowitz has argued, "the national images of both England and of Great Britain are built on Roman ruins," then these national images are also built on the classical ruins that Britons were projecting for and of themselves.[48] One function of the New Zealander, then, aims to reconcile the spatial and temporal foreignness of ruin and the material continuity implied between one experience of ruin and another; mediating one experience of empire's ruin with another. The New Zealander also provides a broader corrective to Walter Benjamin's suggestion that the turbulent momentum of history, "which keeps piling wreckage upon wreckage" at the feet of the Angel of History, is not simply experienced in retrospect.[49] Benjamin describes how his angel is "irresistibly propel[led] *into the future to which his back is turned*, while the pile of debris before him grows skyward."[50] The historiographic device of the New Zealander shows how we also gain perspective on the rubble that history creates by facing towards what has not yet occurred, searching for ways to punctuate and divide time by turning ourselves towards the prospect of future ruin.

"It is not unimaginable that our future archaeologist may have no other means of reconstructing our civilization," wrote Montgomery Schuyler in 1883, deploying the New Zealander to consider the technological grandeur of New York's Brooklyn Bridge, "than that which is furnished him by the tower on which he stands. What will his judgment of us be?"[51] One of the principal functions of the New Zealander is to ask such rhetorical questions, to provide an opportunity to evaluate and *reconstruct* the contemporary. Explicit here is an anxiety about leftovers, about what those leftovers might look like and where they might reside. That is, these future ruins and the viewers they might attract must, above all, be *visible*. Contemporary with Schuyler is Richard Jefferies' *After London; or, Wild England* (1885), a novel that mobilizes a kind of New Zealander figure, a disenchanted wanderer called Felix. Jefferies goes to great lengths to invert the visibility of ruin associated with Macaulay, Doré and Schuyler by describing the almost absolute disappearance of London, as a city nearing intangibility: "The very largest of the buildings fell in, and there was nothing visible but trees and hawthorns on the upper lands, flags, reeds, and rushes on the lower."[52] When Felix enters the further reaches of the Great Lake and encounters the "deserted and utterly extinct city of London [...] under his feet," the grainy substance of the past neither inspires solitary contemplation nor artistic reproduction:

> he came to walls like those of houses, such as would be left if the roofs fell in. He carefully avoided touching them, for they seemed as brittle as glass, and merely a white powder having no consistency at all. As these remnants increased in number, so that he had to wind in and out and around them. In some places the crystallized wall had fallen of itself, and he could see down into the cavern [...] Whether the walls had been of bricks or stone or other material he could not tell; they were now like salt.[53]

These are future ruins verging on the indistinction of dust, towards the dissolution of the idea of ruin into anonymous matter. Georg Simmel observed that:

> the metaphysical-aesthetic charm of the ruin disappears when not enough of it remains to let us feel the upward-leading tendency. The stumps of the pillars of the Forum Romanum are simply ugly and nothing else, while a pillar crumbled—say, halfway down—can generate a maximum charm.[54]

But to judge whether a ruin is "charming" or "ugly," "something" or "nothing else" depends on the visibility and availability of ruin, its resistance to absolute erasure. Jefferies' *After London* suggests the limits of the Ozymandias Complex and the temporal questions it poses, reaching into the sands that surround Shelley's statue to articulate the line between the intelligible materialization of time through waste and the dematerialization of time only made intelligible through the amorphous indistinction brought by dust. As Shelley and Jefferies make us aware, dust or salt, powder or sand frequently marks the disappearance of ruins and all the intimations of human action that can be made by them. Carolyn Steedman has argued that:

> Dust is the opposite thing to Waste, or at least, the opposite principle to Waste. It is about circularity, the impossibility of things disappearing, or going away, or being gone.[55]

But if dust suggests the impossibility of things disappearing, going away or being gone, then it erodes many of the temporal qualities of waste which deign to hold time up, to slow time down, to arrest time's relentless passing. As dust, the ruin no longer provides us with any sense of ending, incomplete or otherwise. Although dust might be the terminal condition of the ruin, it is also marks the termination of ruin's chronographic potential. Steven Connor has described dust as quasichoate, omnivalent and minimally cohesive, dust is too diffuse to chart a transition between use and waste.[56] The temporal qualities of waste, which structure and give license to our reconstructive impulses and provide the basis of ruin's exigency, loses their power to signify. If, as Steedman goes on to suggest, dust represents the "grand circularity" of matter, a time might unfold without any of the temporal punctuation that waste gives us, motivated by a continuous flow of material that neither retains the past nor heralds any particular future.[57] Just as Horace imagined a monument capable of eroding the "now" and the "then" by being impervious to erosion, so, in the monumental potential of dust we are left unable to distinguish this dust from any other, this time from any other. Surrounded by dust, Felix can only note the sheer absence of Ozymandian complexity:

> He looked around, supposing that he might see the gleaming head and shoulders of the half-buried giant, of which he recollected he had been told […] The learned thought, if such a thing really existed, that it might be the upper part of an ancient brazen statue, kept bright by the action of acid in the atmosphere, and shining with reflected light. Felix did not see it.[58]

Concluding with Ruins

Ruins are governed by the way in which action and functionality have time, give time and receive time. But without the material trace of these actions the shadowy absence that becomes absolute. The Ozymandias Complex is snuffed out; waste must communicate an effect, and it needs complicit effects. The material might linger but its momentary associations, which make past action knowable, remain unobtainable.

One might expect a chapter about a ruinous future to conclude with apocalyptic visions of nuclear or some other condition of planetary annihilation. I have avoided these events because, by their nature, we cannot truly approach them. It is the vision of ruin, among other things, that allow us to imagine a future. I have preferred to explore the disappearance of effective waste things instead of the disappearance of individuals capable of narrating their effects. No doubt that the relative comprehensibility of dust depends on the level of analysis being employed; a microscopic analysis of the salty, silted material Felix discovers in *After London* may well have given him a better impression of the city beneath his feet. But this conjecture, at the level of competence and expertise, does not contradict the necessary truth that, if we cannot recognize past action, then reflections upon the recursive nature of that object will slip from view. These are the wider limits of the supplementary logic of ruin that has emerged in this chapter, a logic that includes a commingling of tense, continuities and discontinuities, presence and absence, collation and dispersal, all of which are central to the system of contradiction that makes waste an articulated and articulate thing in time.

Notes

[1] Jacques Derrida, "No Apocalypse, Not Now (full steam ahead, seven missiles, seven missives)," trans. Catherine Porter and Philip Lewis, in *Diacritics* 14 (1984), p. 23.
[2] Paul Ricoeur, *Time and Narrative, Vol. 1.*, trans. Kathleen McLaughlin and David Pellauer (Chicago, IL: University of Chicago Press, 1984), pp. 67-68.
[3] Paula Rea Radisch, *Hubert Robert: Painted Spaces of the Enlightenment* (Cambridge: Cambridge University Press, 1998), pp. 129-133.
[4] Daniel Brewer, *The Enlightenment Past: Reconstructing Eighteenth-Century French Thought* (Cambridge: Cambridge University Press, 2008), p. 192.
[5] Brian Lukacher, *Joseph Gandy: An Architectural Visionary in Georgian England* (London: Thames and Hudson), p. 162.
[6] *Ibid.,* p. 163.
[7] Michel Serres, *Rome: The Book of Foundations,* trans. Felicia McCarren (Stanford, CA: Stanford University Press), p. 29.
[8] *Ibid.,* p. 42.
[9] For further commentary on the visual techniques employed in Robert's painting, see Brewer, *The Enlightenment Past,* p. 189.
[10] Serres, *Rome,* p. 29.
[11] Hannah Arendt, *The Human Condition* (Chicago, IL: University of Chicago Press, 1958), pp. 195-196.
[12] David Lowenthal, *The Past is a Foreign Country* (Cambridge: Cambridge University Press, 1985), pp. 175, 148.
[13] Martin Heidegger, *Being and Time*, trans. John Macquarrie and Edward Robinson (Oxford: Blackwell), pp. 98-101.
[14] See Teodor Shanin, "The Idea of Progress" in *The Post-Development Reader*, ed. Majid Rahmena and Victoria Bawtree (London: Zed, 1997), pp. 65-71; and *Progress and Its Discontents*, ed. Gabriel A. Almond, Marvin Chodorow and Roy Harvey Pearce (Berkeley, CA: University of California Press, 1985).
[15] Percy Bysshe Shelley, "Ozymandias," in *The Poems of Shelley, Vol. 2: 1817-19*, ed. Kelvin Everest and Geoffrey Matthews (Harlow: Pearson, 2000), pp. 310-311.
[16] I have borrowed the phrase "Ozymandias Complex" from Christopher Woodard's *In Ruins* (London: Vintage, 2002), pp. 178– 204. It is unclear what Woodward means by the phrase (it serves as a chapter heading) and, whilst he is

concerned by the complex relationship between empire and the ruin, he does not explore this complexity as a temporal symptom of waste.

[17] Kelvin Everest, "'Ozymandias': The Text in Time," in *Percy Bysshe Shelley: Bicentenary Essays*, ed. Kelvin Everest (Cambridge: D. S. Brewer, 1992), pp. 26, 30.

[18] Horace, *The Complete Odes and Epodes*, trans. David West (Oxford: Oxford University Press, 2000), p. 108

[19] Everest, "'Ozymandias'", pp. 32-33.

[20] H. G. Wells, *The Time Machine and Other Stories* (London: Ernest Benn, 1927), p. 61.

[21] Wells, *The Time Machine and Other Stories*, p. 32.

[22] *Ibid.,* pp. 37, 71.

[23] *Ibid.,* p. 76.

[24] H. G. Wells, *The Fate of Man: An Unemotional Statement of the Things that are Happening to Him Now, And of the immediate Possibilities Confronting Him* (London: Longmans, 1939), p. 247; See also, Patrick Parrinder, *Shadows of the Future: H. G. Wells, Science Fiction, and Prophecy* (Liverpool: Liverpool University Press, 1995), pp. 65–79.

[25] George Orwell, *Nineteen Eighty-Four* (London: Penguin), p. 5.

[26] Paul Auster, *Collected Novels, Vol. 1* (London: Faber and Faber, 2004), p. 283.

[27] *Ibid.,* p. 289.

[28] Katherine Washburn, "A Book At the End of the World: Paul Auster's *In the Country of Last Things*," in *Paul Auster,* ed. Harold Bloom (Broomall, PA: Chelsea House, 2004), pp. 165-169.

[29] Auster, *Collected Novels, Vol. 1,* p. 401.

[30] *Ibid.*, p. 291.

[31] Joseph Bounden, *The Deserted City; Eva, A Tale in Two Cantos; and Other Poems* (London: Longman, Hurst, Rees, Orme, Brown and Green, 1824), p. 2.

[32] Alois Riegl, "The Modern Cult of Monuments: Its Character and Its Origin," trans. Kurt W. Forster and Diane Ghirardo, in *Oppositions* 25 (1982), p. 34.

[33] Albert Speer, *Inside the Third Reich,* trans. Richard and Clara Winston (London: Cardinal, 1975), p. 96.

[34] *Ibid.,* p. 97.

[35] *Ibid.,* p. 702 (note 2).

[36] *Ibid.,* p. 97.

[37] *Ibid.,* p. 98.

[38] Antonio Sant'Elia, "Futurist Architecture," in *Futurism: An Anthology,* ed. Lawrence Rainey, Christine Poggi and Laura Wittman (New Haven, CT and London: Yale University Press), p. 201.

[39] *Ibid.,* p. 201.

[40] Owen Hatherley, *Militant Modernism* (Winchester: Zero Books, 2009), p. 5.

[41] For further comparisons between Shelley and Horace Smith see, Eugene M. Waith, "Ozymandias: Shelley, Horace Smith, and Denon," in *Keats-Shelley Journal* 44 (1995): 22-28; and M. K. Bequette, "Shelley and Smith: Two Sonnets on Ozymandias" in *Keats-Shelley Journal* 26 (1977), pp. 29–31.

[42] See J. G. Ballard, *High-Rise* (London: Flamingo, 2000); J.G. Ballard, "The Ultimate City," in *Complete Short Stories* (London: Flamingo, 2001). W. G. Sebald, *The Rings of Saturn* (London: Vintage, 2002); Robert Smithson, "A Tour of the Monuments of Passaic, New Jersey," in *Complete Writings of Robert Smithson* ed. Jack Flam (Berkeley, CA: University of California Press, 1996).
[43] See David Skilton, "Contemplating the Ruins of London: Macaulay's New Zealander and Others" *Literary London: Interdisciplinary Studies in the Representation of London* 2:1 (March 2004). (www.literarylondon.org/london-journal/march2004/skilton.html). Accessed 15th July 2010; and, David Skilton, "Tourists at the Ruins of London: The Metropolis and the Struggle for Empire," *Cercles* 17 (2007), pp. 93–119. Skilton claims that the figure of the New Zealander "become[s] shorthand for the "coming man," and designate[s] the entire phenomenon of the anticipated ruin of the city and empire," *Cercles* 17 (2007), p. 118.
[44] Thomas Babington Macaulay, "Review of Leopold von Ranke's *The Ecclesiastical and Political History of the Popes During the Sixteenth and Seventeenth Centuries*, trans. Sarah Austin," *Edinburgh Review* 72 (October 1840), p. 228.
[45] "A Proclamation," *Punch* 48 (7 January 1865), p. 9.
[46] See Michael Bright, "Macaulay's New Zealander", *The Arnoldian: a Review of Mid-Victorian Culture* 10 (Winter 1982), pp. 8–27; and Robert Dingley, "The Ruins of the Future: Macaulay's New Zealander and the Spirit of the Age" in *Histories of the Future. Studies in Fact, Fantasy and Science Fiction*, ed. Alan Sandison and Robert Dingley (London: Palgrave, 2000), pp. 15–33.
[47] Skilton, "Tourists at the Ruins of London", pp. 116-117.
[48] Anne Janowitz, *England's Ruins: Poetic Purpose and National Landscape* (Oxford: Blackwell, 1990), p. 20.
[49] Walter Benjamin, "Theses on the Philosophy of History," in *Illuminations*, trans. Harry Zohn (London: Pimlico, 1999), p. 249. My italics.
[50] *Ibid.*
[51] Montgomery Schuyler, "The Bridge as a Monument," *Harper's Weekly* (26 May 1883), p. 326.
[52] Richard Jefferies, *After London; or, Wild England* (London: Cassell, 1885), p. 68.
[53] *Ibid.*, pp. 375-376.
[54] Simmel, "The Ruin", in *Georg Simmel 1858–1918: A Collection of Essays, with Translations and a Bibliography*, trans. Kurt H. Wolff (Columbus: Ohio UP, 1959), p. 265.
[55] Carolyn Steedman, *Dust* (Manchester: Manchester University Press, 2001), p. 164.
[56] Steven Connor, "Pulverulence", in *Cabinet* 35 (2009): 73.
[57] Steedman, *Dust,* p. 166.
[58] Jefferies, *After London,* p. 300.

Chapter Nine

Cézanne and Van Gogh's Quarries: Wasted Landscapes as Modern Art

Maura Coughlin

This essay discusses the waste and reclamation of land, broadly construed: considering "landscape" in both visual representation and material form. Although images of actively exploited quarries are common in French, Dutch and British art after the seventeenth century, a depleted quarry, once abandoned, was rarely painted as a landscape. Like a mineshaft, a quarry is a potentially troublesome hole: it can collapse or fill up with debris and run-off water; it can breed noxious vapors, or disease-bearing insects.[1] A disused quarry is a ruinous wasteland that sits in an uncomfortable reltionship with modern consumption as a reminder of its cost. In the early 1970s, land artist Robert Smithson was drawn to similarly problematic sites (that are neither pristine natural, nor encultured urban, spaces) as "infernal regions—slag heaps, strip mines, and polluted rivers. Because of the great tendency toward idealism, both pure and abstract, society is confused as to what to do with such places. Nobody wants to go on vacation to a garbage dump."[2]

Smithson's thoughts on wasted landscape have shaped my approach to late nineteenth-century paintings of quarries in rural Provence by Vincent Van Gogh and Paul Cézanne. Two interdependent aspects of modern art equally inform this essay: aesthetic modernism and the modernity of bourgeois capitalism.[3] Consequently, I read these paintings as formally and stylistically "modernist" and argue that they trade on exhausted rural landscapes, the ruins of bourgeois capitalism. For my argument, these terms are also always mutually implicated. By relocating this complex form of modernity to the provincial margins, I hope to show that formerly firm oppositions of centralized urban modernity and rural stasis, of culture and nature, corruption and purity are dissolved in certain landscapes—variously devastated and remade in modern terms. Additionally, I pose as counterpart or non-site to these painted rural wastelands the land reclaimed as an aestheticized urban spectacle in the Second Empire: the Buttes Chaumont Park in Paris. In doing so, I follow the advice of literary critic Andreas Huyssen, who states that artistic modernism is never a monolithic discourse, rather it is an elastic category containing both the euphoria of modernization and "some of ... [its] starkest critiques."[4]

In France, in the immediate aftermath of the 1848 revolution, at a time when paintings of peasant labour were commonly taken to be radical references to the newly enfranchised vote, backbreaking labour in the stone quarries that ringed the city of Paris was artistic fodder for mid-nineteenth-century realism.[5] *The Quarriers* (*c.* 1847, Toledo Museum of Art, Ohio) by Jean-François Millet depicts recent peasant immigrants at work on the city's outskirts, pitting their bodies against the resistance of stone. In a similar vein, Gustave Courbet paints rural *Stonebreakers* (1849, formerly Dresden, destroyed 1945) reduced to the mechanical process of crushing rocks for modern roads.[6] This peasant labor is not rural; it is not timeless; it is not explained by reference to the fatalistic rhythms of the seasons or the demands of the earth. Breaking rocks for roads is a modern cliché of grinding toil, of the chain gang, of immigrant, waged slavery.

In nineteenth-century France, the bodies of rural immigrants were worked and worn out like beasts of burden to build the infrastructures of modern life. Poet and folklorist Max Buchon describes Courbet's *Stonebreakers* as "the alpha and omega, the sunrise and sunset of that life of drudgery."[7] Radical Republican author Auguste Luchet, in an article (*Le Vote Universel,* 13 February 1851) dramatically titled, "Those who die that we may live: the Quarriers of the Fontainebleau Forest" (*Ce qui meurent pour nous faire vivre: les carriers de la forêt de Fontainbleau*), makes an explicit link between the city's modernity and the exploitation of stone workers on its outskirts. Because newly paved urban roads are relatively easy to clean, Luchet explains, the urban improvements of the past several years have kept the city free of infection. Yet this benefit has come at a high cost to the unseen quarriers on the city's outskirts who live savage, impoverished and brief existences (in the same rural setting, he caustically adds, that delights Parisian painters). The very stoneworkers who brought health to the city's streets were thought to be at great risk from the unhealthy vapours they encountered in stone quarries and many took precautionary measures such as the wearing of prophylactic sachets of camphor and garlic to protect against subterranean fumes.[8]

Although the demands of urban construction in Paris voraciously consumed outlying raw materials such as stone, sand and clay dug from the ground, the city's growth also depended on the incorporation of various forms of waste into its very fabric. Landscape was moved, removed, mined and filled; wastelands were created, capped, ignored and sometimes represented in the process. Under Napoleon III's right-hand man, Baron Georges-Eugène Haussmann, prefect of the Seine, Second Empire Paris came into being as a modern feat of consumption. It used up newly quarried materials on its outskirts, reclaimed and incorporated formerly wasted space within the city, which had arisen from the demolition of medieval neighborhoods, and assimilated the *terrain vague* of its outskirts.[9] In northeast Paris, the denuded hills of Belleville (not far from the slaughterhouses of La Villette)—known as *Les Monts Chauves*—had served for many centuries as a repository for some of the most abject urban substances. From the thirteen to the seventeenth centuries, this was the site of the notorious Royal Gibbet of Montfaucon. One could ordinarily find there up to sixty corpses swinging in the wind. The scraped and scarred land was pocked with exhausted gypsum and lime quarries that served as hideouts for French Foreign Legion defectors and gypsies. Haussmann's landscape designer dubbed the residents of this place "the worst bohemians of Paris," effectively excluding and compressing culturally marginal lifestyles into marginalized and wasted space.[10]

Long after the resources of the hills had been spent, generations of Paris's trash and sanitary sewage were dumped in its open pits.[11] Many of its holes served as mass graves for those executed at the gallows and for agitators killed during times of civil unrest.[12] Alain Corbin's magisterial social history of smell, *The Foul and the Fragrant*, dwells at great length on Montfaucon as the "epicenter of stench in Paris," whose amalgamated waste led to an extreme fear of the soil for many of the city's inhabitants.[13] Literary critic Anne McClintock comments that such abjection draws cultural power from its ability to dissolve boundaries formerly assumed secure, "inhabiting the cusp of domesticity and market, industry and empire, the abject returns to haunt modernity as its constitutive, inner repudiation: the rejected from which one does not part."[14]

Centuries of accumulated death and decay on this dump and potter's field were efficiently and imaginatively repressed in time for the Paris Exposition of 1867: the event that Walter Benjamin dubbed the most "radiant unfolding" of "the phantasmagoria of capitalist culture."[15] The brand new twenty-three hectare Buttes Chaumont Park, the jewel in the crown of Haussmann's plans for the restructuring of Paris, completely transformed the wasteland to an urban pleasure spot. For many historians of landscape design, this is the "most dramatic early example of the art of landscape to re-create shape and form from apparent waste."[16] Displacing land on a grand scale, its landscaping denied the accretion of corpses, sewage, and miscellaneous trash at its core and went hand in hand with Haussmann's displacement of 350,000 lower-class Parisians to the city's margins. Capping a wasted landscape saturated with abjection with carted-in topsoil and a romantic fantasy of the picturesque was thoroughly in line with the Baron's symbolic reordering of Paris.[17]

Although half the size, the park cost twice that of the Bois de Boulogne and took three years to build; all stops were pulled in the inclusion of a lake, a waterfall within a grotto clad with false stalactites, winding woodland paths, a pseudo-Grecian temple and a suspension bridge from which many suicides have been launched.[18] (It is no wonder that this simulacrum of nature was later a favourite Surrealist haunt). Art historian Nicholas Green views the invention of the Buttes Chaumont park as part of the "official legitimation of the wholesale *reinvention* of nature which had been distinctively articulated in and around Paris of the 1830s and 1840s," but landscape historian Elizabeth Meyer persuasively argues that this was conceived as a showcase of modern technological ingenuity rather than as a nostalgic replica of nature.[19] Using modern steel-reinforced cast concrete to refer to temple gardens of the past, engineer and landscape designer Adolphe Alphand wedded an urban architecture of present and past. Like François Mitterrand's *grands projets* of 1980s, the Second Empire spectacular conversion of waste to pleasurable landscape was a monument to the modern industry and artistry of the French.[20]

Like the Hills of Belleville, the substructure of central Paris was extensively quarried (leaving behind the catacombs as future receptacles of bodily remains); because this practice caused the ground above to collapse occasionally, it was halted in the late eighteenth century. Further anxiety about the unruly underground was generated by failures of the city's very fabric to absorb refuse: to cite a few horrors, the overcrowded mass grave of the Cemetery of the Innocents in central Paris burst its walls in 1780 (into the cellar of a nearby home) and the pre-Haussmann inefficient sewers occasionally failed, such as the one known as the "Stink Hole" whose rupture in 1802 flooded the streets of downtown Paris. Corbin cites an eighteenth-century report that warned of a similar problem in the neighborhood of the future park, that:

> near the refuse dumps of Montfaucon there was already a danger of "streams forming under the earth that were large enough and continuous enough to infect the wells in the neighborhoods and suburbs, and to damage the strata of the earth or the foundation of dwellings"; and great risk that "fetid material from the sluices" would penetrate the soil and infect the sites of future buildings.[21]

Within Buttes Chaumont we find one (repressed) reference to its mingled subcurrents: this is a limestone cavern or grotto that connects the present fantasy to the site's geologic and cultural history. Whereas abandoned urban quarries were eyesores, the romantic form of the natural or artificial grotto was said to have a universal appeal to picturesque taste. In his memoir, Haussmann explains that because the public demanded the "inevitable grotto" in each major park of Paris he had Alphand build them in the Bois de Vincennes and the Parc Monceau, with the most imposing in Buttes Chaumont. The logic of the latter park, according to Meyer, is not wed to the ordering, regularization and geometry of Haussman's plan for Paris, but it rather follows from a process of sedimentation in which urban space was recycled, civilized and then re-imposed upon the existing structure of the city. The case of the grotto is exemplary for it simultaneously reused the void of the former quarry and denied its former function as a receptacle of waste.

Because the site's groundwater was corrupted, the new artificial lake was lined with concrete to block the upward seepage of unthinkably putrescent fluids, and a canal was diverted to feed its waterfall. The grotto, as depicted in Alphand's 1885 study of parks, artificially evokes the majesty of a pure natural spring.[22] The illustration resembles Gustave Courbet's *Source of the Loue* paintings (1864, National Gallery, Washington; 1864, Albright-Knox Art Gallery, Buffalo) that locate a pure and natural water source powerfully emerging from stony land: the artist's own place of origin and wellspring of his "authentic" peasant genius. Courbet's *Source* paintings are often discussed as overtly vaginal, a point made unmistakable in dialogue with contemporary artist Kiki Smith's print *Fountainhead* (1991, Museum of Modern Art, New York) that depicts menstrual flow on a monumental scale. This blunt analogy may well serve to inform the cleaned-up abjection of the Buttes Chaumont's grotto that inadvertently recalls the bodily waste deposited in its former quarry pits.[23] Somewhere in between a modern monument to engineering and a ruin of nature, the sanitized weirdness of the Buttes Chaumont Park evokes Benjamin's conclusion to "Paris, Capital of the Nineteenth Century" that "with the destabilizing of the market economy we begin to recognize the monuments of the bourgeoisie as ruins even before they have crumbled."[24]

Unembellished wasted landscapes feature prominently in the work of Post-Impressionist painters Van Gogh and Cézanne. Before he came to Paris, Vincent Van Gogh ministered to a community of miners in the Belgian region of le Borinage and though his abilities to sermonize were inadequate to the task, he gained a lifelong compassion for underground workers of the earth.[25] While living in Paris in 1886, Van Gogh skipped the

Haussmannized pleasures of the city centre for its as yet formless fringe, the waste zone of the *banlieue* that Victor Hugo had called a "kind of bastard countryside, somewhat ugly but bizarre, made up of two different natures, which surrounds great cities."[26] He painted desolate works there like *The Hill of Montmartre with Stone Quarry* (1886, Van Gogh Museum, Amsterdam): an image of urban expansion in which stone blocks lie as yet unused in the foreground as the city consumes its landscape.

Paul Cézanne lived in Paris on and off for many decades; he too had little interest in celebrating its pleasurable modernity. Only occasionally did he work from the city around him and chose sites that were scruffy, awkward and brutal, such as a back view of a *Wine Merchant's Depot* (1872, Portland Museum of Art). Cézanne's rural landscapes from this period are no more idealized, as seen in the oddly scarred landscape of the *Railway Cutting* (1870, Neue Pinakothek, Munich). Like Van Gogh's Paris quarries, the association is clear: just as stone builds the city, cut earth hastens the train. Later in their careers, when both painted disused quarries in Provence as landscape motifs, their practices inverted or quite literally trashed many long-held academic principles regarding the making of an ideal landscape.

In early nineteenth-century French painting, artists who subscribed to the academic hierarchy of genres sought to elevate landscape above the mere imitation of nature by the inclusion of small-scale mythological narrative and the quotation of classical architectural forms. French Academician Pierre-Henri Valenciennes advocated that aspiring painters make open air sketches of the Italian countryside near Rome to serve as the ideal backdrop for classical narratives. Picturesque travel was thus at the heart of the classical painter's education: many generations of Northern European artists made pilgrimages to the Mediterranean to make contact with the historical ruins of the ancient world.[27] The resulting historical landscapes, churned out by French Academicians for many years following their Grand Tours, were fantasy pastiches of mismatched parts, rather like the sculpted land of Buttes Chaumont.

In the 1830s, a few artists rejected this practice of making idealized constructions, and instead sought out French architectural and topographic landmarks.[28] Little is known about Georges Michel and Paul Huet, who painted on the grimy outskirts of French cities, but legend has it they sometimes set their easels on rubbish dumps like those of Les Monts Chauves (the trash never makes it into the picture).[29] Courbet, in his realist landscapes from the 1850s and 60s (titled with local sites previously unheard of in urban art centres), had trowelled on paint that evoked the visceral materiality of stone, moss, water, foliage and speak of the physical character of the land of his native region—the Franche Comté. His technique was often compared to the mortar application of a stonemason, yet he made this crudity his signature mark of artistic individuality and a declaration of an artistic practice rooted in a provincial place.[30] In the next few decades, the Impressionists painted a range of French landscapes: urban, suburban and rural. Yet with the exception of Cézanne who showed twice with them, they rarely sought out ugly, ruined and non-tourist places.

Neither Van Gogh nor Cézanne bother much with ancient or medieval ruins or other picturesque ways to deal with place, even when surrounded by Roman arenas and arches and Romanesque churches in Provence.[31] Instead, they both turn to their own "inevitable grottoes" in Provence: these were its ubiquitous quarries, some of which were first worked in Roman times. The theme of the quarry provided a particularly unsentimental way to deal with a rural worked landscape and one that carried on a dialogue with realist landscapes of mid-century. Peasant labour as pictorial subject and as a metaphor of artistic practice had forged the public personas of Millet and Courbet, who had been alternately mocked or celebrated as provincial peasants, worker-painters or crude stone masons. Both Van Gogh and Cézanne had forged their own identities in emulation of these realist "peasant painters." As Linda Nochlin notes,

> at the root of realism is this innate 'materialism' of the man of the country, who is at home with mud and manual labor, and is aware, from his earliest days, of the physical contact of man and the material world.[32]

Whereas abandoned urban quarries were eyesores in need of decontamination, urban planning and "greening," similar rural sites were easily forgotten or overlooked. The quarries chosen by Van Gogh and Cézanne are unremarkable sites about 75 km apart: prosaic and forlorn absences in the landscape that had served to build nearby towns.[33] They are depicted neither as refuse dumps nor sites of industry: these desolate ruins of the landscape have been left to the natural processes of entropy. Trees grow improbably from rock; in time, their roots will crack through and further reclaim the site. Unlike the *Quarriers* of Millet or the *Stonebreakers* of Courbet that provoke a dialectical relationship of city and country, of labour and consumption, the later quarry landscapes are not spectacles of human exploitation or laments for lives gone to waste. Instead, these may be read as dialogical meditations on worked earth that re-imagine these sites of cultural intervention as rural spaces that obliquely reference the forms, or senses of place, that their removed matter has given rise to elsewhere.

Van Gogh painted two quarry images in the summer and fall of 1889; this was a prolific time in his short career, when his signature style of calligraphic brushstrokes (that later became a foundational credo of expressionism) was fully developed. His worked up, tactile impasto was inspired by his interests in the textures of rural material culture; he employed his animated paint surfaces to describe phenomenological qualities of rock hit by the scorching sun, pools of shadow in the overhangs, and scrubby swirling underbrush. Van Gogh countered his rough provincial aesthetic to overly refined urban taste in a letter from the previous summer:

> What a mistake Parisians make in not having a palate for crude things … for common earthenware. But there, one must not lose heart because Utopia is not coming true. It is only that what I learned in Paris is leaving me, and I am returning to the ideas I had in the country before I knew the impressionists.[34]

Writing on the first version of *Entrance to a Quarry* (1889, Van Gogh Museum, Amsterdam), Van Gogh notes the emotional effects of its colour:

> to my taste the sombre greens go well with the ochre tones; there is something sad in it which is healthy, and that is why it does not bore me.[35]

As an appeal to melancholy, this quarry painting is much less heavy-handed than his earlier landscapes of the grungy Parisian *banlieue,* that, as T. J. Clark mentions, are a bit too trite, *triste* and over-determined by realist precedent to succeed as modern paintings.[36] Previously, he had also sought a "healthy" approach to rural life in the strangely soiled and equally excessive *Potato Eaters* (1885, Amsterdam, Van Gogh Museum), noting that:

> it would be wrong … to give a peasant picture conventional smoothness. If a peasant picture smells of bacon, smoke, potato steam- all right, that's not unhealthy … If the field has an odour of ripe corn or potatoes or guano or manure—that's healthy, especially for city people … To be perfumed is not what a peasant picture needs.[37]

When the Dutchman was in Provence, as art historian Griselda Pollock tells us, he never wanted to think of himself as a tourist, although he brought with him a vast array of cultural baggage, from the landscape formulas of Dutch landscape painting to the emulated aesthetics of Japanese woodblock prints.[38] The latter, in particular, seem to inform his second quarry painting, *Entrance to a Quarry near Saint-Rémy* (1889, Bellagio Gallery of Fine Art, Las Vegas) that he describes to painter Emile

Bernard as "pale lilac rocks in reddish fields, as in certain Japanese drawings."[39] Just like the dirt or smells of the *Potato Eaters,* the abject sadness of a worked and deserted quarry (translated through his various artistic interests) was precisely the thing that needed to be put under the nose of an urban viewer. As a parallel icon of modernist painting, the swirling brushwork of *Starry Night* (1889, Museum of Modern Art, New York) comes from this period of Van Gogh's work when he was seeking out naturally occurring "monuments" that typified the cultural landscape of Provence. In Roman times, cypress trees were planted on graves as long-living obelisks. Provençal olive groves had also seen centuries of history; together as natural and enduring monuments they symbolically and literally linked past and present.[40]

Cézanne worked at Bibemus Quarry on the outskirts of Aix-en-Provence several times from 1895-1902. Instead of observing a scar in the earth from afar, as in the awkward early *Railway Cutting*, Cézanne's quarry paintings in Provence ventured below ground level. Down in Bibemus, he reveals an intimately known Provençal landscape: a negative architectural form "found" in the chasm below ground that could easily be overlooked or passed by. Without any horizon but the pit that surrounds us, we look up the crazy surfaces of the subterranean quarry wall, stepped and cantilevered by its absent stone. It is more or less parallel with the picture plane, not foreshortened or reclining before us; not quite behaving as flat land should. Instead of the idealized, rationalized historical landscape that Valenciennes prescribed, it courts associations with built structures like moats, tanks, graves, tombs, or cellar holes. The rock wall stands as evidence to past activity, an unintentional, un-built earthwork: the indigestible remains of devoured land. The most intriguing of these, *Bibemus Quarry* (1895; Museum Folkwang, Essen), was painted in a style that has been credited as the origin of cubism; it was bought by a German art collector who had visited Cézanne in Aix.[41] From the viewer, it elicits a response that exceeds the purely visual or intellectual. The longer one looks, the greater grows an illogical, phenomenological sense of the looming, engulfing void, a bodily, visceral pull of the land that reads as a presence, an environment.

If, as Mary Douglas tells us, a cultural perception of dirt is signified by matter out of place, can matter displaced on a scale like this signify differently?[42] Can it create its own sense of place? The modernity of Cézanne's Essen painting is not merely a set of aesthetic implications of brushstrokes that announced their flatness or their disregard of subject matter that excited the would-be Cubists. This painting's strange engagement with the consumption of a rural landscape manages to speak to the demands of urban centres past and present, and to demonstrate the sedimentation of their entropic wastes. As a spatial dialogue of centre and periphery, it combines an emphatically modern painting style and an anti-nostalgic take on landscape.

Because Cézanne painted there, Bibemus quarry and its unintentionally sculpted rock forms were located and documented. As part of an indexical treasure hunt pursued in Provence in the early twentieth century, art historian John Rewald sought out many of Cézanne's motifs, photographing the quarry just before it was reworked. Van Gogh's remnants of rock have also been found and catalogued.[43] But these traces tell us very little; I prefer to read these sites metaphorically and I have found the language for this in earthworks of the 1960s and 70s. Van Gogh's notion of the cypress as a natural monument cannily anticipates Land Artist Robert Smithson's revolutionary proposition that the cultural value of a "monument" is determined by the spectator alone. In his infamous essay of 1967 on industrial wasteland in New Jersey, "A Tour of the Monuments of Passaic," Smithson asks, "has Passaic replaced Rome as the eternal City?" Could a quarry also read as a monument to the continual use of the landscape? As a *boring into* the earth rather than as a spiral form *rising from* it? Cézanne's subject, Bibemus quarry, had been named by its use as a drinking spot for hunters.[44] The ongoing exploitation of this quarry's reddish-gold sandstone to build the city of Aix had inadvertently created this found "monument" to male bonding outside of cultural confines. That is not to imply a tidy symmetry between these paintings and earthworks, although some of Smithson's metaphors and juxtapositions could well apply to these chaotic, painted, landscape junkyards.

In the "Monuments of Passaic" essay, Smithson seemingly free-associates all that crosses his path during one day of anti-picturesque tourism.[45] Details of his account underscore this as a parodic Grand Tour, but unlike any other. A clipping from a copy of the *New York Times*, carried with him on the bus that day, reads, "Art Themes and the Usual Variations" above a reproduction of an idealized *Allegorical Landscape*, then on view at an art dealer. This landscape (like Valenciennes' imaginary construct), painted by failed nineteenth-century American painter-turned-inventor Samuel F. B. Morse, strategically enters into dialogue with the banal "entropic ruins" that Smithson visits. Critic Lawrence Alloway summed up Smithson's New Jersey sites as "one of the places where the geological network of faults and the human network of waste penetrate each other to form a solitary landscape."[46] Smithson's "mesh of collapsing systems"[47] broke down a dual way of thinking about the world out there vs. the world of the art object into a reconceived experience of a site and its layers of accumulated history. These ideas informed his practice of making site selection studies that included photographs, maps and samples from quarries, mines and gravel pits; the entropic mess that is the desolate byproduct of modern consumption. In thinking about these quarry paintings as landscapes of modernism's wastes, it has been productive to engage Smithson's concept of the entropic ruin which he first used to contrast New Jersey's construction sites and holes to Manhattan's skyscrapers. Elizabeth Meyer, writing on the limitations of the binary mode of thought endemic to discussions of landscape, such as nature/culture, female/male, figure/ground, proposes that, like Smithson, we think in engaged and compound terms such as figured ground, articulated space, and so on. "It is the back and forth thing" was the succinct way that Smithson put it.

Reading these painted quarries as entropic monuments, through Smithson's terms, can position them as the encultured, natural non-sites to human practices of waste: historical and modern, metropolitan and provincial. The paintings neither heal the scarred landscape nor do they advocate for social intervention. Like the park of Buttes Chaumont that revealed modern architectural technology through its simulation of the natural, the quarry paintings of Van Gogh and Cézanne are complex "ruins in reverse": [48] aestheticized modern monuments constructed by the painters, monuments not to accomplishment, but to consumption.

Notes

[1] Alain Corbin, *The Foul and the Fragrant: Odor and the French Social Imagination* (Cambridge, MA: Harvard University Press, 1986), p. 23.
[2] Robert Smithson "Cultural Confinement" (1972), in *Robert Smithson: The Collected Writings* (Berkeley, CA: University of California Press, 1996), p. 155.
[3] These terms come from Matei Calinescu, *Five Faces of Modernity: Modernism, Avant-Garde, Decadence, Kitsch, Postmodernism* (Durham, NC: Duke University Press, 1987), pp. 41-47.
[4]Andreas Huyssen, *After the Great Divide: Modernism, Mass Culture, Postmodernism.* (Bloomington, IN: Indiana University Press, 1986), p. 186.
[5] Linda Nochlin, *Realism and Tradition in Art, 1848–1900: Sources and Documents* (Englewood Cliffs, NJ: Prentice-Hall, 1966), p. 33; T. J. Clark, *The Absolute Bourgeois: Artists and Politics in France, 1848-1851* (Princeton, NJ: Princeton University Press, 1982), p. 78.
[6] T. J. Clark, *Image of the People: Gustave Courbet and the 1848 Revolution* (Princeton NJ: Princeton University Press, 1982), pp. 79-80; Nochlin, *Realism and Tradition in Art*, pp. 50-51.

[7] For the full text of Buchon's Advertisement for Courbet's Exhibition in Dijon (1850), from which this quote comes, see Clark, *Image of the People,* pp. 162-64.
[8] Corbin, *The Foul and the Fragrant,* p. 64.
[9] See T. J. Clark, "The Environs of Paris," *The Painting of Modern Life: Paris in the Age of Manet and his Followers* (New York, 1984), pp. 147-204.
[10] "On sait que le parc des Buttes-Chaumont occupe l'emplacement du gibet légendaire de Montfaucon, de sa voirie, et des carriers à plâtre voisines, jadis le repaire favori des pires bohèmes Parisiens." See Alfred Auguste Ernouf and Adolphe Alphand, *L'art des Jardins. Parcs, Promenades* (Paris: J. Rothschild, 1886), pp. 314-15. On the cultural construction of Bohemia, see *Modernity and Modernism: French Painting in the Nineteenth Century*, ed. Francis Frascina (New Haven, CT: Yale University Press, 1993), pp. 81-111. See also Marilyn R. Brown, "Barbizon and Myths of Bohemianism," in *Barbizon: Malerei der Natur – Natur der Malerei.* (Munich: Klinkhardt & Biermann, 1999), pp. 457-58.
[11] This practice continued until 1849. See David H. Pinkney, *Napoleon III and the Rebuilding of Paris* (Princeton, NJ: Princeton University Press, 1958), p. 101.
[12] Elizabeth K. Meyer, "The Public Park as Avant-Garde (Landscape) Architecture: a Comparative Interpretation of two Parisian Parks, Parc de la Villette (1983-1990) and Parc des Buttes-Chaumont (1864-1867)," *Landscape Journal,* Vol. 10 (1991), p. 19. Also see Corbin, *The Foul and the Fragrant,* pp. 29-31.
[13] Corbin, *The Foul and the Fragrant,* p. 31.
[14] Anne McClintock, *Imperial Leather: Race, Gender, and Sexuality in the Colonial Conquest* (New York: Routledge, 1995), p. 72. Also see David Sibley, *Geographies of Exclusion* (New York: Routledge, 1995), p. 8.
[15] Walter Benjamin, "Paris: Capital of the Nineteenth Century," *Perspecta,* Vol. 12 (1969), p. 188.
[16] Susan and Geoffrey Jellicoe, *The Landscape of Man: Shaping the Environment from Prehistory to the Present Day* (London: Thames & Hudson, 1987), p. 257.
[17] Clark, "The Environs of Paris," p. 50.
[18] Pinkney, *Napoleon III and the Rebuilding of Paris,* p. 101.
[19] Nicholas Green, *The Spectacle of Nature: Landscape and Bourgeois Culture in Nineteenth-Century France* (Manchester and New York: Manchester University Press and St Martin's Press, 1990), p. 69.
[20] Meyer, "The Public Park as Avant-Garde (Landscape) Architecture," p. 20. Also see Ann Komara, "Concrete and the Engineered Picturesque: The Parc des Buttes Chaumont (Paris, 1867)," *Journal of Architectural Education* (2004), pp. 5-12.
[21] Corbin, *The Foul and the Fragrant,* p. 24.
[22] Alphand in Ernouf and Alphand, *L'art des Jardins*, p. 173.
[23] "Buttes Chaumont is a mechanical, as well as biological, green lung for Napoleon's urban organism. Its forms, materials and iconography testify to the ability of science to transform the city, to lead it out of the underworld. In Alphand's words, his role was to 'convert nature, modify it, urbanize it by taking its naturalness to extremes, to make it exploitable in the urban setting, to replace it by technique.'" Meyer, "The Public Park as Avant-Garde (Landscape) Architecture," pp. 21-22.
[24] Benjamin, "Paris: Capital of the Nineteenth Century," p. 172.

[25] See Fred Orton and Griselda Pollock, "Rooted in the Earth: A Van Gogh Primer," *Avant-Gardes and Partisans Reviewed* (Manchester and New York: Manchester University Press and St. Martin's Press, 1996), p. 9.
[26] Clark, "The Environs of Paris," p. 26.
[27] Renzo Dubbini, *Geography of the Gaze: Urban and Rural Vision in Early Modern Europe,* trans. Lydia G. Cochrane (Chicago, IL: University of Chicago Press, 2002), pp. 99-105. Also see Michael Marlais, John Varriano and Wendy M. Watson, *Valenciennes, Daubigny, and the Origins of French Landscape Painting* (South Hadley, MA: Mount Holyoke College Art Museum, 2004).
[28] Greg M. Thomas, "The Topographical Aesthetic in French Tourism and Landscape," *Nineteenth-Century Art Worldwide* (Spring 2002) (available at: www.19thc-artworldwide.org).
[29] Steven Adams, *The Barbizon School and the Origins of Impressionism* (London: Phaidon, 1994), pp. 45-48.
[30] Thomas, *ibid.*, writes of Courbet's technique as a metaphorical performance of rooted identity.
[31] Vojtech Jirat-Wasiutynski, "Van Gogh in the South: Antimodernism and Exoticism in the Arlesian Paintings," in *Antimodernism and Artistic Experience: Policing the Boundaries of Modernity,* ed. Lynda Jessup (Toronto: University of Toronto Press, 2001), p. 182.
[32] Linda Nochlin, *Gustave Courbet: A Study of Style and Society* (New York: Garland, 1976), pp. 37-8.
[33] Nina Kallmyer mentions that the yellow-red stone from Bibemus had built the townhouses of Aix-en-Provence, thereby contributing to its local identity and sense of place. See Nina Kallmyer, *Cézanne and Provence: The Painter in his Culture* (Chicago, IL: University of Chicago Press, 2003), p. 171.
[34] Letter to Theo, 11 August 1888. Cited in Ronald Pickvance, *Van Gogh in Saint-Rémy and Auvers* (New York: Metropolitan Museum of Art, 1986), p. 119.
[35] Letter from Vincent van Gogh to Theo van Gogh Saint-Rémy, 19 September 1889. *Vincent Van Gogh—The Letters* (available at: www.vggallery.com).
[36] Clark, "The Environs of Paris," p. 30.
[37] Judy Sund, *Van Gogh* (London and New York: Phaidon, 2002), p. 95.
[38] Griselda Pollock, "On not Seeing Provence: Van Gogh and the Landscape of Consolation 1888-9," in *Framing France: The Representation of Landscape in France 1870-1914,* ed. Richard Thomson (Manchester and New York, 1998), pp. 97-101.
[39] Letter to Emile Bernard. Saint Rémy, 8 October 1889; cited in Pickvance, *Van Gogh in Saint-Rémy and Auvers,* p. 149.
[40] Vojtech Jirat-Wasiutynski, "Vincent van Gogh's Paintings of Olive Trees and Cypresses from St.-Remy," *The Art Bulletin,* Vol. 75 (1993), p. 657.
[41] *Bibemus Quarry* was in the Folkwang museum collection in Essen in the early twentieth century but was deaccessioned by the Nazis as was all "degenerate" modernism, and later was repurchased by the museum.
[42] Mary Douglas, *Purity and Danger: an Analysis of Concepts of Pollution and Taboo* (1966; London: Routledge, 1984), p. 36.

[43] See *Cézanne: the Late Work,* ed. William Rubin (New York: Museum of Modern Art, 1977), p. 94; John Rewald, *The Paintings of Paul Cézanne: a Catalogue Raisonné* (New York: Harry N. Abrams, 1996), p. 502; and Pickvance, *Van Gogh in Saint-Rémy and Auvers,* p. 150.
[44] Kallmyer, *Cézanne and Provence,* p. 170.
[45] Robert Smithson, "A Tour of the Monuments of Passaic, New Jersey" (1967), in *Robert Smithson: The Collected Writings* (Berkeley, CA: University of California Press, 1996), pp. 68-74.
[46] See Lawrence Alloway, "Robert Smithson's development," *Artforum*, Vol. 11: 3 (November 1972), pp. 52-61. Reprinted in *Art in the Land: a Critical Anthology of Environmental Art,* ed. Alan Sonfist (New York: Dutton, 1983), p. 128.
[47] Alloway, p. 128.
[48] Smithson describes elements of the industrial fringe of Manhattan that "seemed to contain ruins in reverse, that is—all the new construction that would eventually be built. This is the opposite of the 'romantic ruin' because the buildings don't fall into ruin after they are built but rather rise into ruin before they are built." Smithson, "A Tour of the Monuments of Passaic, New Jersey," p. 72.

Chapter Ten

Arman's Spectacles of Waste

Jaimey Hamilton

We often talk of waste as the uncanny double of the commodity system, the spectre that haunts our concept of modernity. Waste is usually thought of as concealed, but its presence as always somehow palpable; it seems to skulk in waste dumps or incinerate into invisible particles that permeate our atmosphere. Because of this, we tend to think of it as the abject or dialectical other to our systems of efficiency. Perhaps this comes from our anxiety that waste does not simply go away. It tenaciously persists as obdurate materiality, resisting transformation into wealth or pure energy—a reminder of the darker side to the conveniences of contemporary consumer culture.

Perhaps this impulse to represent waste as the other of modernity exposes a different sort of cultural repression—a denial of our fascination with garbage and excess. We commonly assume a dialectical relationship between efficiency and waste or between production and excess. But these linear trajectories of new stuff to used-up stuff may not completely characterize the complicated circulation of material in our consumer system and our relationship to its excesses. Waste, as it turns out, can easily be framed, aestheticized, and commoditized—made visible and integral to modern commerce, however much we hate to admit it. There are, in fact, many cases in which the spectre of waste is constructed as spectacle. And the notion of excess is actually much more integral to "efficient" production than we perhaps realize. Waste is not always simply pushed into the trash heap, but sometimes is actually reinserted into modes of production and consumption in order to keep the commodity system moving forward. I would like to sketch out a few instances of this in the postwar art of the French sculptor Arman [Fernandez] in order ultimately to answer this question: what function could the spectacularization of waste possibly serve in the larger economy of modern consumption? How can we talk about waste, not only as abject, but also as sometimes useful?

Arman's series of trash sculptures from the 1960s are perhaps some of the most literal manifestations of waste as spectacle in the recent history of contemporary art. Following a tradition of the found object, Arman was one of a number of artists, including Americans Robert Rauschenberg, Richard Stankiewicz, Lee Bontecou, and Europeans César, and Jean Tinguely, who were using the waste of an increasingly prolific postwar mass production system in their work.[1] Working in France as it was dramatically affected by business, manufacturing, and consumer models coming from the United States, Arman was fascinated early on by the cycles of production, consumption, and waste of the postwar markets. In his first statement, published in the avant-garde art journal *Zero* in 1960, Arman wrote,

> Useless objects, as a medium of expression, particularly attract my attention and treatment … [My] procedure [of accumulating consumer objects] correlates with actual methods: automation, assembly lines, and the repeated discarding [of trash], creating strata and geologic layers full up with the force of the real.[2]

Arman developed his interest in consumer goods and waste early in his career through his friendship with fellow Nouveau Réaliste, Yves Klein. They had met as teenagers in a judo class when they were both living in Nice. For a time, they followed different paths toward the Parisian art world. But as they both developed their artistic careers in the context of a hyper-emotive abstract *art informel* painting scene, they intuited the increasingly market-oriented gallery scene, as well as the importance of a new kind of "signature style" if they were to succeed within its structure. They formulated artistic strategies that would both distinguish them from the proliferation of abstract painters and play upon the new consumer-oriented bent of the market. Yves Klein suggested that they "divide up reality" and make art about their particular "realms."[3] Klein took the "natural" world, going on to describe it with his famous Blue Monochromes made with his patented International Klein Blue pigment. Arman took the "man-made," creating sculptural expressions of the new industrial civilization in which he lived. Both artists were eventually convinced by Parisian critic Pierre Restany, who also believed in the importance of "packaging" as the way to break through the dominance of the *informel* scene (and who was taken with Klein's charismatic persona), to become part of his Nouveau Réaliste movement, established in 1960.[4] In an essay the next year, Restany attempted to articulate a common aesthetic for the motley group of artists he had convinced to sign his manifesto:

> Through [their] specific images the entire sociological reality, the common food of human activity, the large republic of our social exchanges, of our commerce in society, is summoned to appear. [5]

Taking up Restany's call to be a contemporary ethnographic "witness" to the new wave of postwar French industrialization, Arman devoted himself to recording the material effects of consumerism. This took the form of a few series of found object sculptures from the early sixties, including the *Poubelles*, essentially clear glass "trashcans" that visualize the waste of consumer society, and the *Accumulations*, clear glass boxes filled ad infinitum with mostly newer consumer items of the same kind. One of Arman's first *Poubelles*, *Petits Déchets bourgeois*, from 1959, acts as a specimen of French postwar bourgeois consumption. It includes lightbulb packages, cheese wrappers, tea cartons, film, wine corks, string, and some old embroidered clothes. This sedimenting material could be likened to an archeological trash midden, a core sample of modern daily life. Indeed, Arman borrowed from archeological, anthropological, and sociological approaches in order to capture a long view of living culture that seemed on the verge of consuming itself. "It has always been obvious that society feeds its sense of security with a pack-rat instinct demonstrated in its window displays, its assembly lines, its garbage piles," he said:

> As a witness of my society, I have always been very much involved in the pseudo-biological cycle of production, consumption, and destruction. And for a long time, I have been anguished by the fact that one of the more conspicuous material results is the flooding of our world with junk and rejected odd objects.[6]

In order to capture the rising tide of consumer objects, Arman continued constructing the *Poubelles* for over ten years. If *Petits Déchets bourgeois* exhibits postwar France's consumer habits, another core sample of French trash, *La Grande Bouffe* (The Big Feast), 1973, constructed fourteen years later, registers an increasing amount of packaging and plastic. In addition to the bits of string and cheese wrappers, there are 7-Up bottles, disposable milk cartons, and Heineken beer packing. Playing up his role as "archeologist of the present," Arman explained his mounting evidence: "we were passing from local consumption of the small grocery store and the butcher to the supermarket, to packaging, to plastic."[7]

The changing content of each sculpture indicated the rapid growth of the European economy revitalized by Marshall Plan aid and other US-sponsored economic programmes. There was a reorganization of French business practices, a boom in suburbanization, household modernization, and a new consumer culture based on American-style marketing. Along with this came a rapid transformation of the socio-economic structure; agriculture declined rapidly, while the service industry and middle class surged.[8] This transition represented what came to be known in postwar France by the title of Jean-Jacques Servan-Scheiber's 1967 best-seller, *The American Challenge*.[9] That "challenge" was caught up in what cultural critics, theorists and artists identified as a new culture of the spectacle.[10]

The term "spectacle" was often used simply to denote the culture of advertising, movies, television, and shop windows that had eventually transformed America and that, by 1960, seemed to be colonizing the French landscape through "le marketing." But Guy Debord, the primary theorist of the spectacle, argued for a more specific historical meaning within the context of his Marxist critique of capitalism, *Society of the Spectacle* (1967). In his terms, the spectacle described a culture in which the reign of the commodity was finally accepted as a natural turn of events: social relationships were increasingly mediated and reified by images, and any object became subsumed by a larger system of commodity value:

> The spectacle corresponds to the historical moment at which the commodity completes its colonization of social life. It is not just that the relationship to commodities is now plain to see … the world we see is the world of the commodity.[11]

Debord's activist agenda, developed through the Situationist International (a reformulation of the Lettrist International from which he broke, and a continuation of the unfinished Surrealist International, itself modeled on the Communist International), sought to combat the spectacle's implicit capitalist cultural colonization by fighting it with his own multi-national Marxist art movement. Though he did not condemn the "society of the spectacle" as a specifically American importation, it is fairly obvious that by the sixties, he took this as a given. The American challenge had presented itself; the gauntlet had been thrown down.

Debord was critical of Arman's work, as well as the rest of the Nouveau Réalistes because he thought they simply reproduced the conditions of spectacle rather than critiquing them (we shall get to a fuller articulation of his complaint a bit later).[12] This is interesting since many others have seen Arman's *Poubelles* as exaggerated dialectical reversals of our efficient system of capitalism, exposed through abject remainder.[13] Certainly, the sculptures are aggressive, agonistic, excessive objects, especially when seen in the context of the history of art. To pit Arman's pile of refuse against centuries of marble sculptures, which embodied concepts of beauty, order, and transcendence, is to upset, in a very visceral and dialogic way, a whole system of aesthetics. Even in comparison to the modernist assemblages of Picasso and the ready-mades of Duchamp, they seem to lack a certain amount of arrangement and choice, as they simply present a random and untransformed state of found filth placed on a pedestal. Even the Surrealists, who talked about tapping into a language of the abject, of filth, and the unconscious, still held onto the notion of poetically arranging a few outmoded objects. By comparison, Arman's *Poubelles* might better visualize surrealist philosopher Georges Bataille's description of the accursed share—a notion of excess that articulates all that refuses our modern logic systems. (Indeed Bataille had a famous falling out with the official surrealist circle because he thought they did not push far enough to offend bourgeois tastes and notions of rationality.) In "The Notion of Expenditure," written in 1933, Bataille specifically proposed that excessive waste could be used to combat the more refined and stylized expenditure of luxury, including art. He claimed that we needed "a potlatch in which real refuse and revealed moral filth entered into a rivalry of horrible grandeur with everything in the world that was rich, pure, and brilliant."[14]

This image of the impoverished rising up against the wealthy through an agonistic potlatch of trash could easily describe one aspect of Arman's *Poubelles* as they counter the very discourse of art and its rapid spectacularization in the 1960s. In fact, to put trash in the place of a finely crafted art object, or even of a Duchampian conceptual challenge, was undoubtedly to upset not only aesthetics, but also a system of fine art consumption. Arman's *Poubelles*, like a lot of junk art from that period, came at the very moment when art galleries were more and more influenced by new marketing and business models. He, as well as other junk assemblagists, exploited the growing commercial identity of the gallery space, with its new emphasis on hobnobbing events, rapid turn-over of shows, and the cultivation of *nouveau rich* patrons. By simply moving trashy objects onto the pedestal, junk artists exposed the uneasy coexistence of a new throwaway culture with the implied immortality of luxury objects. Their strategy implied that not even art was immune to the driving force of the culture of obsolescence. More than one critic picked up on this, but *New York Times* critic Stuart Preston characterized it best. Junk art, he complained, flashes "across the horizon and then, like last year's car … [is] discarded."[15]

This exposure of the instability of cultural value did not preclude junk art, including Arman's sculptures, from becoming the hot new commodity. Indeed, things get more complicated, the dialectical opposition between efficiency and waste break down, as we register the ways in which Arman's appropriated trash did not totally disrupt the smoothness of commodity spectacle because of the very fact that it was a commodity itself. For even though Arman was throwing trash at art patrons, that trash *was* aestheticized, systematized, and eventually accepted by avant-garde patrons as an art object as early as 1960. There was a calculated semblance of chaos in each *Poubelle* carefully arranged by Arman. Despite the contrasts to Picasso and Duchamp, and after the initial shock of seeing pure garbage in the space of the gallery, one can begin to see the formal and symbolic resonances of the material to which Arman carefully attended. This is evident in the *Accumulations*, where variations in colour and texture play against the monotonous repetition of objects, but it can also be found in the *Poubelles*, as clothing is artfully arranged around other fragments of trash, and labels are made conveniently to congregate at the edges of the containers. There are, in fact, a number of documentary photographs of Arman carefully sorting out and arranging the contents of each *Poubelle*. The cheese wrapper in *Petits Déchets bourgeois,* for instance, is nicely arranged among more neutral brown paper and string in the bottom corner so that it sits against the glass and is clearly legible. The composition of the piece as a whole is not symmetrically balanced, but rather arranged to give off the sense that the identifiable objects and packaging at the bottom of the container are being crushed under the weight of the more amorphous material at the top. So even though there was some chaotic piling, there were also some considered compositional tactics involved in the *Poubelles'* construction that made them more compelling visual objects, and thus more readable and desirable to patrons. Arman's composed detritus at first had only a few enthusiastic supporters beyond the Nouveau Réalistes themselves. Pierre Restany, of course, championed them and gallerist Alfred Schmela exhibited them in Germany in early 1960, but the Parisian dealers were initially reluctant to accept them into the gallery. Arman eventually convinced Iris Clert to allow an exhibition of the most chaotic and aggressive, but also the most composed and spectacular, use of trash to date.

In 1960 Arman created *Le Plein*, in which he literally filled the whole space of the Clert Gallery with trash. Arman had proposed the idea of *Le Plein* right after Yves Klein's 1958 show of *Le Vide*, an exhibition of a completely empty Clert gallery. With hyped-up invitations and purposeful pomp-and-circumstance surrounding the opening, Klein exposed the whole of art to be nothing but capitalist marketing and art world consensus building. The buzz about Klein's invisible art was infectious and aided in the sales of his Blue Monochromes.[16] Arman tried to convince Clert that his project was as spectacle-worthy, but for two years she resisted the idea of filling her gallery with foul-smelling detritus. Arman finally convinced her to present a "full-up" gallery that included two hundred and fifty pounds of pure garbage, two hundred pounds of old records, two and a half tons of scrap metal, one hundred and eighty birdcages, and three cubic yards of used light bulbs.[17] On 25 October 1960, when the show opened, the gesture of accumulation, as Restany noted, was "elevated to an architectural dimension."[18]

Arman claims that the original idea was to hire city garbage truck drivers simply to unload their garbage into the gallery, but the city would not agree. This was replaced (as with most of his work) with a more premeditated alternative. Arman enacted the role of the garbage man by scavenging the trash bins of Paris himself.[19] He also wryly added all of the leftover invitations to past Clert openings to the piles of detritus. He even piled in the paintings that were on display from the show before *Le Plein*, noting them in his catalogue of collected waste as "fifty small paintings from the Micro salon exhibition at the Iris Clert by artists ranging from Picasso to Hartung to Manzoni" as if they were equal to the "twenty olive oil bottles" and "seventy pounds of old curtains." But while it seems that Arman was clearly de-valuing art by allowing trash to overtake its domain, he was also making concerted efforts to aestheticize and quantify the excesses of trash. As with his *Accumulations*, similar objects were organized together and spread out so that they would take up more volume and present themselves in an artful chaos. Records were placed with other records, light bulbs with other light bulbs, and so on. He also made a list of all the objects in the gallery and documented his diligent stacking of fruit baskets, and jumbled arrangements of bicycle wheels. With the aid of the list, of course, the paintings could be read as equivalent to more objectionable material, but more importantly perhaps, everything he collected, all the waste, was subjected to a scientific, observational, and rational system of knowledge. The point of listing and arranging may have been to revel in just how much junk he stuffed into the gallery, but it also had the effect of making the formlessness of the mess a known quantity, a calculated *representation* of waste.

This is ultimately how Arman's agonistic potlatch became a marketing spectacle—horribly beautiful, abjection contained. The gallery, not filled with saleable items (or a more stately "void"), perhaps seemed to make a mockery of the art-commodity system. But what one could not sell, perhaps one could make into a marketing event. This is exactly what the five hundred invitations for *Le Plein* indicated. Iris Clert had taken charge of getting these made. Each was a used sardine can that contained a random selection of trash (metro tickets, matches, etc.) and a card stating: "Iris Clert asks you to come to contemplate in 'Le Plein' the power of the real, condensed into a critical mass."[20] This too is its own example of packaged trash—used detritus organized in order to entice patrons to an even more spectacular display of "the real." The stuff, despite possibly being abject, became a focal point, like Klein's *Le Vide*, of art world social events. Everyone wanted to come and see the unusual window display that Arman had concocted.

In this respect, *Le Plein* immensely complicates the underlying impulse of junk art to expose the marketing procedures of the art commodity system. In fact, Arman's *Le Plein* should be read in relation not only to Klein's *Le Vide* but also to the work of Rauschenberg, Jasper Johns, and Stankiewicz, who had all shown in Paris in 1959.[21] Critic Françoise Choay picked up on this when she identified *Le Plein* as a possible response to Martha Jackson's 1960 *New Media-New Forms I: Junk Culture* show in which Rauschenberg and Stankiewicz's works were key contributions.[22] Arman most likely learned about the show from Rauschenberg and Stankiewicz personally, when they were in Paris the year before. In comparing *New Media* with *Le Plein*, Choay wrote that the Americans were too serious, too bent on their principled provocation in comparison to Arman's much more savvy display of trash.[23] Choay's comparison is perceptive, especially as the New York junk aesthetic was taken to environmental levels in Allan Kaprow's trash-aided installations, *Subway,* 1960 and *Yard*, 1961 (shown a year later in the same gallery).

In Kaprow's environments, visitors were encouraged to enter into the chaos, stumbling over trashy objects and getting dirty in order to walk away transformed by the multi-sensory experience. Arman's *Le Plein*, on the other hand, was "full of irony."[24] Not side-tracked by the alternative repurposing of trash in order to disrupt our habitual patterns and mindset, Arman instead tempted viewers with what could have been a repulsive experience (if one could get close enough to smelling the molding cheeses hidden in the garbage) but then redirected them to see the "culture of debris" mainly as spectacle. Viewers simply could not enter into the full-up gallery. Instead, Arman forced the viewer to look in on the layers of garbage either from a platform just inside the maintenance entrance or (for most) through the street-front window of the gallery. In this way, the artist played upon the power of the spectacle rather than attempting to subvert it.

Arman purposely created a consumable *image* of the entropic state of capitalist trash. Refuse, like every other cultural object that he accumulated, was contained by the glass vitrine and structured by its signifying logic. This is exactly how despicable and ostensibly abject objects get reincorporated back into the capitalist system—by becoming representations.

Debord's opening words of *Society of the Spectacle* capture the experience of *Le Plein* well:

> The whole life of those societies in which modern conditions of production prevail presents itself as an immense accumulation of *spectacles*. All that was once directly lived has become mere representation.[25]

Instead of trying to provoke an "authentic" and possibly unsettling experience with trash, Arman sensed that there was no real way to counter the power of spectacle. For no matter how close one got to the trash in junk art, it was already "framed" as an art commodity, a representation of the experience of our throwaway culture. Even Kaprow's work, despite its early radicalism, was ultimately subsumed within this paradox. Grappling with the way that the window vitrine of any gallery could directly incorporate capitalist excess back into the system, Arman chose to outdo his American counterparts by exposing the irony at the heart of the "junk aesthetic."

Though art critic Françoise Choay claimed that this was an exhibition from which "no collector could profit" because no one could buy *Le Plein*, patrons *could* purchase a smaller version of it as a *Poubelle* or *Accumulation,* on display in Clert's office.[26] In relationship to the overwhelming architectural scale of *Le Plein*, two by three foot cubes of trash may have started to seem more acceptable and even quaint mementos of the amazing experience. This marketing strategy is an important aspect of *Le Plein* that often goes unacknowledged, but one that Clert and Arman both realized straight away. In fact, Clert chronicled the success of *Le Plein* as it began to generate interest in Arman's smaller, now, comparatively aesthetic and arty sculptures. She eventually lured collectors to purchase a good many of them.[27] Thus the "critical mass" of trash indicated in the invitation became an image and concept, a digestible and marketable part of the functioning system of art capital—in a word, a spectacle.

Though Arman did not consciously play up all of these contradictory aspects of his work's production, display, and eventual sale, it is consistent with his sociological position of witnessing, through literally enacting, the way that the consumer system had become so pervasive that it could successfully incorporate garbage into its logic. Guy Debord eventually took Arman to task for this, explaining that Arman's sociological approach of merely recording, instead of critiquing, our need to consume, was too fuzzy and ironic. Debord saw such explorations of spectacle as turning the "radical" dada past into a "reactionary advertisement."[28] But this contradiction did not in the least bother Arman, who, in his somewhat cynical stance, saw the re-absorption of his trashy artworks as inevitable, and necessary to his own livelihood.[29] Indeed, he was savvy enough to recognize the fact that his trash sculptures would never escape capitalist marketing. Even in his first artist's statement, Arman was attentive to the contradictions of his position:

> I hope to express the anxieties that come from the reduction of spaces and surfaces and from the invasiveness of our industrial secretions. I would like to stop speed, the explosion, the bursting, pieces brought back from the past, precise accidents or hazards as always the same and already repeating. I see myself as a product of our civilization and formed by our rhetoric. I do not think I bring anything new but just my little fragment of reality through the prism of all arts of all times.[30]

Instead of fighting spectacle culture as Debord did, Arman courted its rhetoric and tested its mettle by throwing as much material as he could into its machinery.

While *Le Plein* is the most visible and extreme visualization of the power of spectacle culture to turn everything—including notions of waste, destruction, and disintegration—into meaningful components of the capitalist system, it is also evidenced in Arman's *Colères*. For this series, Arman destroyed status-bearing imitation antiques found in the flea market—usually period furniture and musical instruments such as cellos and violins. *Faux Louis xvi*, 1961, a destroyed imitation table mounted on a wood ground, speaks of the destruction of history by spectacle culture. This piece can be usefully read through Jean Baudrillard's commentary on postwar collectables from his 1967 book *System of Objects*:

> The way in which antiques refer to the past gives them an exclusively mythological character. The antique object no longer has any practical application, its role being merely to signify.[31]

In other words, the commodity status of the antique in spectacle culture trumped its use-value, and even its historic and heirloom value. Authenticity, ancestry, and a vague connection to French nobility became signifiers that could be purchased. There was a booming postwar market in faux antiques, and particularly in reproductions such as the imitation Louis xvi table. In fact, Arman's father sold such pieces in his furniture stalls "Les interieurs rustiques des Provinces de France" and "Au Foyer" in Nice, which specialized in *brocante*—less valuable and cheaply made second-hand pieces.[32] Consumers scoured stalls, like Arman's father's for imitation provincial period styles, and if they were lucky, haute époque or French regency styles that corresponded to the models pictured in magazines such as *La Maison Française, Mobilier et Décoration,* and even *Elle*. The growing professional middle class wanted the status that such objects signified, but did not have the budget to buy the original or the pedigree to have inherited it.

As with *Le Plein*, there is more than a bit of irony in the over-performance of spectacle logic *Faux Louis xvi*. Its title emphasized its fakeness and its over-the-top destruction obviously made the logic of the system visible. The simple way that Arman cut to the chase exhibits a hint of iconoclasm as he enacts the way that spectacle can symbolically destroy a rich European past. He also very obviously destroyed a mass-produced object. In its splintered state, the table reveals the cheapness of the wood under the finish and a stamp that clearly identifies its manufacturer. The object's destruction (no less than its title) exposes its position within a pervasive system of other mass-produced objects. Yet despite this excessiveness, or perhaps even because of it, the strategy was fully integrated into the condition of the spectacle. He simply took away its functionality so that it could exist purely as a visual signifier: décor that could now be mounted on a wall in a central location instead of sitting in some corner. Whether smashed or set on fire, the end result of the *Colères*, like the *Accumulations*, was an aesthetic jumble of pieces stilled and frozen in a way that corresponded to its increasingly fixed position within the consumption of a lifestyle. Its symbolic status as part of symbolic décor subsumed any articulation of historical memory.

In conjunction with *Le Plein* and the *Poubelles,* the *Colères* expose the way in which seemingly outmoded objects were as susceptible to spectacularization as new ones. As Debord argued and as Arman performed, spectacle "buries history," remolding present time by "preserving the old culture in a congealed form, going so far as to recuperate and re-diffuse even its negative manifestations."[33] Faced with the "American Challenge," Arman elaborately demonstrated the ways in which spectacle could be seen to penetrate every aspect of French material culture. His interventions in marketing strategy expose the increasingly tangled relationship between France and the United States to the point where the immediacy and pervasiveness of a more generalized "spectacle culture" had overtaken much of Europe. The "pseudo-cyclical time" of consumption, as Debord called it—of season finale sales, department store and even flea market hours of operation—became the only kind of time that mattered. Mass production and rote accumulation, even of faux Louis XVI furniture, had triumphed by metamorphosing a complicated past into a present continuum of things. The differences between trash and slightly used objects of Arman's *Poubelles* and *Colères* became incorporated into the evenness, immediacy, and increasingly, the internationality, of commodity culture and its marketing practices.

Arman's works were not the only ones to participate in this spectacularization of waste. Arman's Nouveau Réaliste cohorts were especially attentive to the circulatory logic of capitalism. Klein took advantage of it, as we have already seen, with *Le Vide*. In addition, César's crushed cars, Jacques de la Villeglé's torn posters, and Tinguely's self-destroying machines, all performed the increasing rapidity and aggressiveness of capitalism's production-consumption-destruction cycles in humorous and unsettling ways. Tinguely's self-destroying machines, for example, could be seen as models for the self-enclosed but vertiginous system of consumption. The mechanisms of his machines always seemed to be in chaotic, excessive movement, but their excessiveness was precisely what was consistent about them. The machines would continually "produce" destruction, and then their destroyed parts would be used to build a new machine: "one continuous process—out of the chaos of the dump and back again."[34] As Tinguely put it, "the constant of movement, of disintegration, of change, and of construction is static."[35]

Tinguely's *Rotozaza iii* in particular captures the specific challenge of making ruin and demolition a successful marketing gimmick. Conceived as early as 1950, Tinguely first proposed the sculpture for the store window of Neiman-Marcus in Texas. In a letter to the retailer, Tinguely described the relief in terms of American marketing extravaganzas: "It would be a fantastic relief, which would joyfully destroy the same object everyday, making a mystifying (and I hope fascinating) spectacle."[36] Though Mr. Marcus turned down Tinguely's proposal, *Rotozaza iii* was finally built in 1969 for the Swiss department store window of Victor Loeb. Equipped with conveyor belt and hammer, it destroyed objects (mostly china plates and teapots) on sale in the store with "mechanical reliability."[37] Tinguely solved the problem of capitalism's overflowing excess the same way Arman did—by making it into a spectacle that functioned to drive the imperative to consume; the Swiss store owner, understanding the value of such outlandish marketing tactics, reportedly posted an invitation in the store window next to *Rotozaza iii*, "come in and buy before Tinguely smashes everything."[38]

The ultimate point here is that while Tinguely's machines and Arman's *Le Plein* might be "contestation[s] of the rule of efficient production," they are also (and this is the "fullness" of their irony) hyperbolic performances of spectacle logic that get subsumed by the capitalist machine.[39] The shock of trash, overproduction, or destruction is displayed through the store window, in which there is a consistently safe "view." The impermeability of the glass screen reintegrates useless, abject, or overproduced things, extending their lives in order to express the new profligate wealth of capitalist enterprises—ones that could spare engine parts for Tinguely's machines, that could afford to destroy plates, or throw away one hundred and eighty bird cages, and, most notably, capitalist enterprises that could celebrate the leisure-time consumption of a new junky "art culture" that these Nouveau Réaliste spectacles represented.

Arman and Tinguely's projects present the full cycle of potlatch—not just the vision of excess that Bataille had examined in his early text, but also the paradoxical social value of such agonistic displays. Bataille actually acknowledged this second aspect of potlatch in his later, more thorough, and more refined text, called *The Accursed Share*, in 1949.[40] Potlatch, he reminds us, is really a momentary ritual in which destruction or disintegration actually expressed prestige and a transition in social order.[41] Whereas traditional ceremonies of potlatch, as practised by the Native American cultures such as the Tlingit in which tribal leaders ceremoniously destroy their wealth, were unrelated to capitalist accumulation and expenditure, they nevertheless find their echo in Arman and Tinguely's performative sculptures. The Tlingit chief's ceremonial expenditure is now replaced with the figure of the wastrel as the figure that represents the immense wealth of the contemporary economy. Jean Baudrillard, even more forthrightly, articulates this connection:

> The principal basis of this [capitalist] system would appear to be the official, obligatory and supervised demise of the objects that it comprises: a gigantic collective "happening" whereby the death of the group itself is celebrated through the euphoric destruction or ritualistic devouring of objects and gestures.[42]

Arman and Tinguely's work directly addresses, through enactment, this "euphoric destruction or ritualistic devouring" as they paradoxically *reaffirm* the need to produce more and celebrate the obligation to consume more.

For this reason, Arman's accumulations of outmoded objects present themselves as useful allegories for the paradoxical relationship between commodities and waste. Through their various manifestations, from consumer-ready object to trash bin to the museum's repository, they can be read to embody the extreme contingency of value within the capitalist system as well as our contradictory attraction and repulsion to our prodigious system of objects. In their intertextual references and performative displays, Arman's works expose the power of spectacle, consumerism, and the capitalist system in general. In fact, they point to the idea that perhaps modernity, if it is embodied in mass production and consumption, is not the opposite of excess, but is excessive itself—not in the way that it excludes or suppresses our trashy excretions, but in the way that its prodigious machinery (especially the machinations of spectacle) keeps expanding and colonizing to *include* such seemingly anomalous items. Capitalism's packaging and systemizing of waste, as Arman exposes, spur us to admit that even though we may be appalled by the spectre of waste, we cannot turn our heads away: we have become more than a little bit enthralled by it.

Notes

[1] See my forthcoming book project, "Strategies of Excess" for an expanded account of this phenomenon. I'd also like to thank the many engaged readers and audiences who have contributed to this essay, in particular Caroline A. Jones, Yve-Alain Bois, Mari Dumet, John X Christ, Emily Gephart, Maura Coughlin and all of the participants of the "Modernity and Waste" conference held at the University of St. Andrews, 2006.

[2] Arman, "Réalisme des Accumulations," *Zéro* v. 3 (1961), reprinted in *Arman: Passage à l'acte*, edited by Tita Reut (Nice: Musée d'art moderne et d'art contemporain, 2001), p. 18. This is Arman's first artistic statement, originally written in 1960.

[3] Arman, interview conducted by Sevim Fesci on 22 April 1968, n.p. Archives of American Art/Smithsonian Institution. On-line transcript: http://tiny.cc/yvs0j

[4] Arman in conversation with Otto Hahn, *Mémoires Accumulés* (Paris: Pierre Belfond, 1992), p. 35. See also Pierre Restany, "The *Nouveaux Réalistes* Declaration of Intention," written in Milan on 16 April 1960, trans., Martha Nichol, *Le Nouveau Réalisme* (Paris: Union Generale, 1978), pp. 281-85.

[5] Restany, *A 40°au-dessus de Dada* (May 1961) cited in *1960, Les Nouveaux Réalistes* (Musée d'Art Moderne de la Ville de Paris, 1986), pp. 266-267.

[6] Arman quoted in Henry Martin, *Arman: or Four and Twenty Blackbirds Baked in a Pie: or Why Settle for Less When You Can Settle for More* (New York: Abrams, 1973), p. 9.

[7] Daniel Abadie and Arman, "L'Archéologie du future," in *Arman* (Paris: Galerie nationale du Jeu du Paume, 1998), p. 55.

[8] These are broad brush strokes, but ones that have been generally acknowledged by a number of historians of twentieth-century France. By no means as quick and pervasive as the American postwar boom, and happening about ten years after, there is nevertheless adequate historical evidence to claim a visible and tangible change in the character of postwar life in France. For more on the transition of postwar French culture, see Kristin Ross, *Fast Cars, Clean Bodies: Decolonization and the Reordering of French Culture* (Cambridge, MA: MIT, 1999); D. L. Hanley, A. P. Kerr, and N. H. Waites, *Contemporary France: Politics and Society*

since 1945 (London: Routledge, 1984); Nicholas Hewitt, ed. *The Cambridge Companion to Modern French Culture* (Cambridge: Cambridge University Press, 2003); Richard Kuisel, *Capitalism and the State in Modern France: Renovations and Economic Management in the Twentieth Century* (Cambridge: Cambridge University Press, 1981).

[9] Jean-Jacques Servan-Schreiber, *The American Challenge*, trans. Ronald Steel (New York: Atheneum, 1968). Originally published in 1967, as *Le Défi américain*, it became one of France's all-time best sellers.

[10] Some called it spectacle, some called it mythification, some the consumer society, but all were describing the new significance of consumer objects and mass media in ideological systems. See Debord, *Society of the Spectacle*, trans. Donald Nicholson-Smith (New York: Zone, 1995); Roland Barthes, *Mythologies*, trans. Annette Lavers (New York: Noonday, 1972); Jean Baudrillard, *The System of Objects,* trans. James Benedict (New York: Verso, 1996); and also Jean Baudrillard, *The Consumer Society: Myths and Structures* (London: Sage, 1998); Henri Lefebvre, *Everyday Life in the Modern World*, trans. Sacha Rabinovitch (New York: Harper and Row, 1971).

[11] Debord, *Society of the Spectacle,* p. 29.

[12] See for instance Guy Debord, "The Situationists and the New Forms of Action in Politics on Art," reprinted in *The Situationist International, 1957-1972: On a Passage of a Few People through a rather Brief Moment in Time,* ed. Elisabeth Sussman (Boston: ICA, 1989), p. 151.

[13] Recent historians who take this view include Jan van der Marck, *Arman* (New York: Abbeville Press, 1984); Tita Reut, *Arman: La Traversée des objects* (Venice: Chateau de Villeneuve, 2000); and Jill Carrick, "Le Nouveau Réalisme: Fetishism and Consumer Spectacle in Post-war France" (PhD Thesis, Bryn Mawr College, 1998).

[14] Georges Bataille, "The Notion of Expenditure," in *Visions of Excess*, ed. Allan Stoekl (Minneapolis, MN, Minnesota University Press, 1985), p. 127.

[15] Stuart Preston, "Divisions of Today," *New York Times* (19 December, 1954), sec. 2, p. 12.

[16] An enthusiastic review of the show appeared in *Combat* and the general public had a positive response according to a surprised Restany. Hannah Weitemeier, *Yves Klein* (New York: Taschen, 2001), pp. 33-34.

[17] Arman dutifully recorded everything that went into the gallery for the installation. It is reproduced in Denyse Durand-Ruel, *Arman: Catalogue Raisonné II 1960 1962* (Paris: La Différence, 1991), p. 46.

[18] Pierre Restany, *Arman* (New York: Abrams-Horay, 1973), reproduced in Durand-Ruel, *Arman: Catalogue Raisonné ii,* p. 48.

[19] Clert, *Iris-Time*: *L'Artventure* (Paris: Denoël, 1978), p. 214. Just as an interesting side note, Iris Clert recounts that she aided in the effort by paying a truck driver one hundred francs to pick up old furniture off the streets because she was afraid that Arman would pull off the proposed excessiveness of his gesture.

[20] Clert recounts the tribulations she encountered over this project: The first hurdle was finding five hundred empty sardine cans. Once she found a supplier, she then

decided that they needed to be filled with garbage, so she dipped into the trash bins outside the metro stations fishing for expired tickets. Finally, she was chastised by the post office and police for sending sardine cans through the mail. Clert, *Iris-Time*, pp. 209-213.

[21] Johns and Rauschenberg participated in the Biennale des jeunes in Paris in 1959 along with Tinguely and Yves Klein and were also included in a show at Galerie Cordier at the end of the year on International Surrealism. Johns showed at the Galerie Rivé Droite in 1959 and he and Stankiewicz showed at the Venice Biennale of 1958 in the international pavilion.

[22] *New Media—New Forms: Junk Culture,* Martha Jackson Gallery, New York, 1960.

[23] Françoise Choay, "Lettre de Paris II: 'culture de débris,' à la Galerie Iris Clert," *Art International,* 4, 9 (Dec. 1960), p. 36.

[24] Choay, "Lettre de Paris II," p. 46.

[25] Debord, *Society of the Spectacle*, p. 1.

[26] Choay, "Lettre de Paris II," p. 36.

[27] Clert, *Iris-Time,* p. 214.

[28] Debord, "The Situationists and the New Forms of Action in Politics on Art," p. 151.

[29] Benjamin Buchloh makes a similar argument, though his concern is to position Arman and Klein's "neo-avant-gardism" in terms of the changing history of the avant-garde. Benjamin Buchloh, "Plenty or Nothing: Yves Klein's *La Vide* and Arman's *Le Plein,*" in *Premises: Invested Spaces in Visual Arts, Architecture and Design from France, 1958-1998*, ed. Bernard Blistène (New York: Guggenheim, 1998), pp. 86-99.

[30] Arman, "Statement," 1964, reprinted in *Arman: Passage à l'acte,* ed. Tita Reut (Nice: Musée d'art moderne et d'art contemporain, 2001), p. 19.

[31] Baudrillard, *The System of Objects*, p. 74.

[32] Marcelin Plenet and Tita Reut, *Arman: La liberté en peinture* (Paris: Au Même Titre, 1999), p. 9.

[33] Debord, *Society of the Spectacle*, p. 136.

[34] Billy Klüver quoted in Calvin Tomkins, *The Bride and the Bachelors* (New York: Penguin, 1987), pp. 170-171.

[35] Jean Tinguely, "Untitled Statement," *Zéro* 3 (1961); reprinted in *Theories and Documents in Contemporary Art,* ed. Kristine Stiles and Peter Selz (Berkeley, CA: University of California Press, 1996), pp. 404-405.

[36] Tinguely, letter to Mr. Stanley Marcus (of Neiman-Marcus), reproduced in Pontus Hulten, *Jean Tinguely: "Meta"* (Boston, MA: New York Graphic Society, 1975), p. 332. It is fascinating to note that the Swiss artist might have been catering to an American sensibility with his argument, but that the spectacle only played itself out in a European department store.

[37] Jean Tinguely quoted in Hulten, *Jean Tinguely,* p. 332.

[38] Hulten, *Jean Tinguely,* p. 332.

[39] Branden W. Joseph, *Random Order: Robert Rauschenberg and the Neo-avant-garde* (Cambridge, MA: MIT Press, 2003), p. 238.

[40] Georges Bataille, *The Accursed Share, Vol. 1*, trans. Robert Hurley (New York: Zone Books, 1991).
[41] Bataille, *The Accursed Share*, pp. 64-77.
[42] Baudrillard, *System of Objects*, p. 132.

CHAPTER ELEVEN

RETREAD AMERICA: POST-WAR RE-ADJUSTMENT, BOREDOM, AND LIFE IN THE "LONELY CROWD"

EDWARD J. K. GITRE

A now well-known photograph entitled "Family Utopia" appeared in *Life* magazine in the autumn of 1946 as part of a special tenth-anniversary issue. Showcasing "what Americans want," it has the look of the commodity-cluttered 1950s, not of this far-less-certain first postwar year. The draft was still in place, the economy was not exactly booming, and the perennial fears of inflation hung thick in the air. Perhaps *Life* was intimating an oblique message through this image—that commodities and technologies of living were the new spoils of modern (postcolonial) wars. "What will the net result [of the Second World War] be?" queried members of the American Historical Association's Historical Service Board in 1945. "Such conveniences as television sets, food freezers, self-contained heating, lighting, and even air conditioning units" would give even the most rural of homeowners "all the conveniences for good living that were formerly associated with city life."[1] *Life* reassured readers, less they think this a mere fantasy:

> The posed scene above is an honest representation of the dream of most US families. Fanciful as it looks, it is based on the hard statistics of consumer demand and manufacturers' unfilled orders.[2]

Life was optimistic, if nothing else. In 1946 not every American was clamouring for helicopters (though ostensibly some were).[3] This photograph is of a fantasy, a dream—perhaps an "honest representation of a dream" but a dream nonetheless.[4] And it was still as much a dream for the vast majority of Americans in the GDP-rich 1950s as it was in 1946. "Reconversion after World War II raised the hopes of Americans of many political persuasions and social statuses that a more prosperous and equitable American society would finally be possible in the mid-twentieth century due to the enormous, and war-proven, capacities of mass production and mass consumption," argues historian Lizabeth Cohen in *A Consumers' Republic*. "The new growth economy of the Consumers' Republic promised more affluence to a great number of Americans than ever before."[5] Reconversion "raised the hopes of Americans" with the promise of "affluence to a greater number"; however, were those dreams realized? This essay suggests that scholars need to rethink postwar American culture, particularly the 1950s, not only by questioning images like "Family Utopia," but also by rethinking the complicated effects of the Second World War—especially the gaps between dreams and realities, and what those gaps reveal—for fantasies are never fully realized in everyday experiences. "Hard statistics of consumer demand and manufacturers' unfilled orders" tell only a partial story (even of *Life*'s sanguine postwar outlook); they are unable to get at the emotional lives of postwar Americans in general and returning veterans in particular.

Like other public institutions, post-Pearl Harbor, the Museum of Modern Art (MoMA) in New York City embraced its wartime "civic duty" of educating the population and bolstering community morale. It also took rather seriously its commitment to the government, serving at least six government departments and agencies in an advisory and technical capacity, including not only the Library of Congress and Office of Civilian Defense but also the War Department and the Office of War Information (the precursor to today's CIA). Staff, space, and time were committed to projects proposed by the government, including exhibitions. On 21 May 1942, MoMA opened to enthusiastic encomium "Road to Victory—a Procession of Photographs of the Nation at War," a photographic exhibition executed by Lt. Cmdr. Edward Steichen, who was at the time directing the us Navy's expansive photography division and had been "especially assigned" by the Navy for the job. (After the war Steichen went on to head MoMA's Photography Division.) Steichen was a very accomplished photographer, yet none of his work appeared in the exhibition. Most, nearly ninety percent, of the 150 photographs came instead from other governmental agencies, such as the Farm Security Administration (FSA) and the Tennessee Valley Authority; and the text was written by Steichen's brother-in-law, famed poet and author Carl Sandburg. The exhibition melded together images of Native Americans, sweeping vistas, mountains and prairies, farmers, factory workers, and soldiers in one great tribute to all things American. "Our purpose in preparing this exhibition was to enable every

American to see himself as a vital and indispensable element of victory," wrote Monroe Wheeler, MoMA's Director of Exhibitions:

> Throughout the giant procession of photographs of our farms and industries and our armed forces in actual combat there recur, again and again, the faces of the fathers and mothers and sons and daughters upon whom everything else depends . . . I think no one can see the exhibition without feeling that he is a part of the power that is America.[6]

The exhibition, which filled the entire second floor of the museum, received enthusiastic and unqualified reviews. Effusive *New York Times* critic Edward Jewell noted that:

> We had been promised "a procession of photographs of the nation at war." A colorless understatement that promise soon appeared … if photo murals of such power and splendor as these have been produced in the past, I have not seen them.[7]

The summer after the exhibition opened, attendance at the museum was up fifty percent. MoMA knew it had a hit: within months plans were well underway to send "Road to Victory" on the road. Seven museums, from Cleveland to Portland, were interested, with "scores of requests from smaller cities" clamouring for a more compact version, Wheeler boasted. In his estimation, the circulating exhibition would attract 750,000 patriotic patrons.[8]

During its one-month stay in the City Art Museum in St. Louis, Missouri, it alone attracted over nineteen thousand visitors. "This reviewer candidly confesses that two years of hard reading of current books on war has failed to give him such a comprehensive view of what America is, what America is fighting for and why America must win, if the hope of the world is not to be extinguished," wrote the critic for St. Louis, Missouri's *Globe-Democrat*.[9] Hollywood mogul David Selznick was so impressed with the exhibit and anxious to use Sandburg's text in a film he was shooting, *Since You Went Away*, that he had his staff rush a wire to Sandburg's representative to get permission. The movie would be a "serious and fine study of the home front," Selznick's staff assured Sanburg, "and your words will greatly help us to keynote the scene."[10] Permission was granted.

Was the exhibition mere propaganda? "Yes, it is propaganda," *Chicago Sun*'s critic admitted:

> I think this exhibition is a must for all who want inspiration and need encouragement as we march along the hard road to victory, might as well use

> that word. It is propaganda if all those mental pictures you have stored up of your childhood home, of the walk to the grocery store, the look of your mother as she pulled the biscuits out of the oven or of your father as he raked the lawn—if such images are propaganda.[11]

One man's propaganda was another's nostalgia. What drew visitors and critics to "Road to Victory" was not simply the war, but the image of Americans the exhibition realized through photography and text. "The drama enacted transcends the picturing of 'a nation at war,' in that it searches and searchingly reveals the very fiber of the nation itself that is now fighting to preserve all it holds most dear," argued Jewell:

> Yes, this panorama as it unwinds goes far beyond the factual account of a tremendous war machine in action or in process of being forged. In its vast sum it constitutes a true, a stirringly articulate portrait of America. It becomes, indeed, as the sequences is the essence of a people, vitally and indissolubly linked with a people's environment, the very soil itself that nourishes and that in a sense shapes a people's life.[12]

In Jewell's eyes, and those of many who passed through MoMA's sacrosanct chambers, this was *the* American story, a "portrait" of the nation.

The men and women of "fiber" in "Road to Victory" have a peculiar photographic history. As noted, many of the images were culled from the large archive of the FSA (whose predecessor was the "Resettlement Administration"). President Roosevelt established the agency in 1937 to assist dislocated Depression-era farmers, especially those that had been hardest hit by the Dustbowl. Its "Historical Section" documented those efforts by recording the need, which was indeed great. Under the direction of Roy Stryker, the pictorial effort far exceeded expectations, producing in a period of ten years over 160,000 black-and-white film negatives and over 100,000 colour transparencies. Cultural historian Alan Trachtenberg has argued that these represent "one of the prime cultural artefacts of the New Deal," embodying "the era's ideology of human history as 'universal' and 'progressive'."[13]

This is the visual archive Steichen tapped (and reinterpreted) to construct "Road to Victory." To illustrate America's fibre, he chose not well-fed Brahmin bankers but instead a simple Mormon farmer and his wife from Utah, and other couples of similar ilk—Depression-era survivors. In so doing, Steichen, Sandburg, and MoMA's staff were not merely painting a photographic portrait of a transcendent, ethereal America (though in the exhibitions there are intimations of this aim). Rather, they were conjuring the far more immediate memory of downtrodden everyday Americans, which visitors would have known well enough. Here, however, all self-doubt is removed: America prevails, victims are now heroes.

Unequivocally this transfiguration was self-conscious. In the exhibition catalogue Wheeler explained:

> The force of the exhibition lies above all in the sequence of the pictures. Each room is a chapter, each photograph a sentence. It moves with magnificent simplicity and a kind of visual eloquence from the landscape of the primeval continent through the folkways of simple Americans, the extraordinary mechanism of peace and war, to the cavalcade of men flying and sailing and motoring and marching to the defense of that continent.[14]

"Nostalgia" transformed stale, soup-kitchen bread into hot-out-of-the-oven biscuits. "The earth is alive. The land laughs. The people laugh. And the fat of the land is here," Sandburg reassured his uneasy compatriots.[15] As a Midwesterner who had lived through the 1930s Dustbowl, of course Sandburg knew that the land was not always so fat. Still, his words were simple yet effective.

America's future success rested on the fat on the land—and the shoulders of its young men and women in uniform. In one room, amidst a collage of war photographs, sat a large white placard, which read:

> Country boys, big city lads, home town fellers, they're in the army now—behind a jeep instead of a plow—engineers bridging a river, chutusts in the sky, skirmishers on the land, troopers on sea transports—into sea fog and land smoke—into the test of fire ... Trouble shooters, millions stepping along now, millions more on the way—killers in khaki riding smoke wagons—ready for long war or short—trouble shooters in the first round-the-world-war.[16]

Depression-era FSA farmers had sired these virile men—and they were men (the women are pictorially absent here)—men who no longer guided ploughs but drove jeeps. And through this act, this exchange—both technical and generational—the Depression (and thus America) was ostensibly redeemed. "Road to Victory" concluded with a "final mighty climax," a six-by-fifteen-foot mural reinforcing this narrative arc. In front of a sea of marching soldiers, Steichen positioned a series of photographs of folk farmers with their wives, with Sandburg's closing tribute tying it all together:

> America, thy seed of fate have borne a fruit of many breeds, many pages of hard work, sorrow and suffering—tough strugglers of oaken men—women of rich torsos—they live on—the fathers and mothers of soldiers, sailors,

> fliers, farmers, builders workers—their sons and daughters take over—tomorrow belongs to the children.[17]

He meant children like the US Naval Academy's newly minted ensign, George T. Weems, whom *Life* magazine featured in a "Close Up" six-page spread soon after the attack against Pearl Harbor. Allaying anxieties about America's capacities for warfare, *Life* featured Weems as the "prototype" of the "ideal" naval officer, the kind of man who would help make the world safe for democracy. On the one hand, one gets the impression from reading *Life*'s superlative biography that this twenty-year-old graduate had descended straight from Zeus:

> Young Weems has the look of a naval officer, the sharp jutting chin, the wide, deep-set eyes, the short hair, the well-braced physique. He is 6 ft. tall, unmistakably American, a good-looking blond extrovert.[18]

Weems had more than a chiselled chin, though: he was an intercollegiate 165-pound wrestling champion, master of the midshipmen yacht crew, member of the varsity lacrosse team, vice president of his class, and president of the Trident Society and French Club.

Despite all these accomplishments—which are as much aesthetic as substantive—apparently Weems was merely "normal," the Navy's average:

> He is not far above the average intelligence of his class [he graduated only in the top quarter (114th)]. Seen in person he does not seem to give off sparks or pulsate with inner cosmic vibrations.

No, Weems has only a "normal" and "healthy" amount of "animal energy." Even his chiselled good looks appear not to be uniquely Weems's: "This is a general appearance with which four years in the Naval Academy seems to endow nearly every one of the heterophysiognomied young men who enter it, along with the marine engineering, navigation, seamanship, ordnance and gunnery it teaches them." Weems was the product of a system, an academy that was "not designed primarily to produce genius"—indeed, the "Navy even prescribes the average"—but rather "to turn our men with similar loyalties, habits of thought, and powers of leadership and character." Upon this discipline the future of democracy would rise or fall. Weems may not have been "No. 1 graduate in his class," yet in the Navy's opinion he was the man most likely become an Admiral of the Fleet.[19]

Henry Luce, the founder of *Life* magazine, would have heartily agreed with Edward Steichen's (and many other Americans') conviction that the war could and would redeem America (and the Depression) and that redemption would be accomplished through America's youth. In the days that followed the Declaration of War in December of 1941, Luce published a special editorial for the magazine, "The Day of Wrath," which tried to put the attack on Pearl Harbor into perspective. For him the attack was a "sign of all the weakness and wrongness of American life in recent years"; the victims of Pearl Harbor were not merely the victims of Japanese "treachery," he rued; rather, they were "the victims also of a weak and faltering America that had lost its way and failed the world in leadership."[20] Luce eulogized the country's former greatness:

> We have come to the end, now, of as pusillanimous an epoch as there ever was in the history of a great people: the twenty years of American history between 1921 and 1941. It is not even possible to call these years tragic, for tragedy implies at least the dignity of fate. And there was no dignity in these years, and nothing of fate that we did not bring upon ourselves. The epoch that is closing was much less tragic than it was shameful.[21]

Luce's embarrassment was buoyed by the one legacy of these "shameful" years: its youth. "Despite the meanness of the spiritual environment which we have recently provided them, the strength of this land is in their hearts."

Not only was the strength of the nation in their hearts: so too was its future. Although all Americans would be in the fight—"for this is indeed total war"—Luce admonished, "never forget that the hardest fighting is done at the front by our soldiers and sailors":

> We have fighting men in the field and on the high seas. Thousands of soldiers and sailors in the Pacific are at this very hour facing death. We tell ourselves how young they are, how gallant, how brave. We salute with our hearts in our throats, tears in our eyes—pray for them too … In the great fog of battle, we cannot see our soldiers, where they fight or how. We know they fight with every ounce of guts a good American has. And that's plenty. But as we cannot see them in the fog of battle, they cannot see us. When the fog lifts, in the days to come, let us see to it that they look upon us, here at home, and can say as honestly to us as we say to them, "Good going, boys." That will be the day of victory and of peace in our hearts.[22]

A week after this call to redemption through arms, *Life* ran its "Close Up" of Ensign Weems, on 29 December 1941 (just a few short weeks after Pearl Harbor). Weems's prototypicality would not only typify what the Navy could produce but also what *Life* displayed every week in the page of its magazines in support of war. "Oaken men" and "rich-torsoed women" in uniform appeared often in Luce's periodical. Over half of *Life*'s first ten years in print were devoted to covering the Second World War, and every issue after the entry of the United States into the War contained at least a minimal amount of, and in many cases expansive amounts of, space dedicated to the war. Henry Luce's and *Life*'s editorial staff's commitment to the common soldier, like Weems, and others, such as "Marine Ace Smith", who donned the 7 December 1942 cover, was, in short, unflagging. Pauline Kael, one-time film critic for the *New Yorker* recalled,

> During the war years, the whole spirit of the country seemed embodied in *Life* magazine. Its covers featured GI Joes, girls, and generals. The GI Joes were always clean-cut, wonderful kids. And so were the girls they dated . . . Everybody was patriotic and shiny-faced. Wiped clean of any personality.[23]

Everything about the common soldier's experience seemed to have the look of a *Life* magazine cover.

Photographs displaying virility were buttressed both textually and graphically, and not simply through close-up biographies and war reportage. Nearly every advertiser in *Life* promoted the war as well. Virile soldiers and wartime support workers performing heroic deeds on behalf of the country were at the heart of Madison Avenue's "big sell"—deeds that included, not least, eating simple, wholesome bread. Consider this Fleischmann's Yeast advert, "This is no time to be FRAIL! The dainty days are done for the duration—likewise the dainty diets":

> So as America goes to war, it also goes for more and more bread. Recent surveys show bread sales increasing steadily ever since war was declared. The reason is that bread is one of our best sources of food-energy. No other food can be eaten with such regularity and still not tire your taste.

Bread was "standard equipment" in the lunch boxes of American workers. Soldiers and war workers—men and women, old and young—were called upon (at least their visage was) to sell everything from vitamins, cigarettes, and flatware, to bars of chocolate.

Consider one final advert for a railroad company, entitled, "The Kid in Upper 4," doubtless the most saccharine advert of the lot:

> It is 3:42 am on a troop train. Men wrapped in blankets are breathing heavily. Two in every lower berth. One in every upper. This is no ordinary trip. It may be their last in the USA till the end of the war. Tomorrow they will be on the high seas. One is wide awake … listening … staring into the blackness. *It is the kid in Upper 4*. Tonight, he knows, he is leaving behind a lot of little things—and big ones. The taste of hamburgers and pop … the

> feel of driving a roadster over a six-lane highway … a dog named Shucks, or Spot, or Barnacle Bill. The pretty girl who writes so often … that gray-haired man, so proud and awkward at the station … the mother who knit the socks he'll wear soon. Tonight he's thinking them over. There's a lump in his throat. And maybe—a tear fills his eye. *It doesn't matter, Kid.* Nobody will see … it's too dark. A couple of thousand miles away, where he's going, they don't know him very well. But people all over the world are waiting, praying for him to come. *And he will come*, this kid in Upper 4. With new hope, peace and freedom for a tired, bleeding world …

The messianic role attributed to even the youngest of soldiers was clear. Essaying *Life* magazine and its adverts, historian Mark Grandstaff concluded, "The war provided the means for the American male as soldier to redeem manhood from the crises of economic modernity; to undergo a trial or an initiation into a new life; to become a transformed being — a man reborn with the ability to redeem or regenerate society." [24]

Dreams and fantasies deconstruct under the bright light of lived experience. As a representation of postwar affluence, *Life*'s "Family Utopia" is a convenient fantasy for scholarly reflection and critique. Yet, *Life* did acknowledge that the spoils of war had come with an incalculable price tag. Recall that "Family Utopia" was published in a special tenth-anniversary issue. The issue showcased the magazine's accomplishments over the decade as well as surveyed the "state of the nation" since the magazine's founding and the war's end. The editors of *Life* promised readers that the magazine would take a "hard, sharp look at the new and still strange US" in which postwar Americans were living, a country that was "older" and "wiser" but in some ways also "sadder."

Acknowledging the toll of war in the state of the nation survey, one segment, "Cost of War: The US Is Still Paying Them," admitted:

> Much of the new strength of America which can be seen in the preceding pages was a by-product of war, which spurred the whole nation to heights of accomplishment it had never dreamed of. Yet the cost, as always in war, was appalling. Life would never again be the same for the families of the servicemen who died in their youth on the battlefields. Nor would it ever be quite the same for boys like Jerome Von Mohr …

Private First Class Jerome Von Mohr was a 22-year-old patient at the Army's Walter Reed Hospital; he lost both his legs in the Philippines when his jeep hit a landmine. During his visit at the hospital, photographer Leonard Mccombe took several pictures of Von Mohr. Unlike this photograph that made it into the magazine, another shows the youth actually smiling. Apparently, though, the more dour private proved more dramatic.

The "readjustment" of servicemen and women like Von Mohr was critical to the nation's postwar transition. Readjustment was a national concern, occupying social scientists, psychologists, university presidents, ministers, doctors, advertisers, local business leaders, and, of course, the GIs and their families. Although the pressure to find jobs for the GI proved tremendous, leading to the "Servicemen's Readjustment Act" of 1944 (better known as the GI Bill of Rights), the concern ran much deeper than the soldier's pocketbook. The fantasy of redemption through war necessitated the messianic GI's facile postwar readjustment. Even before the war had ended, psychologist Jerome Bruner was warning Americans:

> The returning soldier is to the American people more than a fact-to-be. He is at once a symbol and an alarum. As a symbol he is the sentimental cynosure around which our interests in domestic post-war problems center. We talk anxiously of jobs. But 'jobs for the returning soldier' is the expression most fraught with emotion. So too social security for the returning soldier, a decent America for the returning soldier, and the rest. But these are the flag-wavings, the hero-worshipings. The hero gloss will wear off when eleven million ubiquitous Johnnies come marching home, tired, human, and deadly sick of war. In all wars, the soldier has been a hero. But hero worship has never counted for much in assuring the run-of-the-mill heroes a place in the post-war sun … this returning soldier is not just Joe Doakes, but our son, our father, our cousin, or that young fellow next door.[25]

Many wondered (and fretted over) what would become of this, the "silent" generation, and those used-up "retread" GIs. Take a look at the "52-20 club" (52-20 being a reference to the GI's unemployment insurance: $20 a week for 52 weeks); this picture also appears in *Life's* "state of the nation" survey. What would become of this blonde fag-smoking lad in his US Navy hoodie who is either bored out of his mind or caught up in yet another fantasy? The contrast between Ensign Weems, the Kid in Upper Four, and this Navy veteran is striking.

Perhaps the most famous of the Second World War GIs to come of age in the 1950s was Tom Rath from *The Man in the Gray Flannel Suit*, the semi-autobiographical creation of Sloan Wilson, who was himself a World War II veteran. Wilson, like Norman Mailer, the author of the most acclaimed US World War II novel, *The Naked and the Dead* (1948), wanted to pen an American *War and Peace*. Weighing in at a svelte 276 pages, Wilson's production was, however, far more modest than Mailer's (and certainly Tolstoy's). Yet, its mid-century cultural import is equally significant. *Naked and the Dead* captured the unique (if heightened) experiences of war: *The Man in the Gray Flannel Suit*, the banal experiences thereafter. A quick best-seller, *Man in the Gray Flannel Suit* was turned into a Hollywood blockbuster, starring Gregory Peck (a role that, in turn, spawned an advertising campaign and fashion trend in menswear, the "Gregory Gray" suit).[26] Within a year, *The Man in the Gray Flannel Suit* had sold over half a million copies (mostly through book clubs), was translated into over twenty languages; and is still in print fifty years later. Not only did it influence how mid-century Americans thought about themselves, it also markedly moulded how the rest of the world imagined what it must have been like to be a suburban American.

The novel is about Tom and Betsy Rath, a young couple who by appearances should have everything going for them—three children, their own home in Connecticut, a steady if modest income, in short an average middle-class suburban existence. But (of course) not all is well. Housewife Betsy, unlike husband Tom, is a striver who wants more out of suburbia. Tom—a "young man," wrote one book reviewer, "with a slipped disk in the backbone of his ambition"—is caught in the rat race of corporate life with no way out.[27] After the war, he took a job working for a major foundation that does the public good but with paltry pay. Worried like fellow commuters about keeping up with the Joneses, he is lured into taking a public relations job at the United Broadcasting Company (UBC), a New York City broadcasting empire. Once an aspiring artist and writer, Tom finds himself trapped in the world of double speak. He becomes the president's speechwriter, a president who treats Tom like a long lost son yet, aside from lucre, can proffer nothing but overnight business trips, late nights and weekends in the office, a rebellious child, and a broken marriage.

Summing up Tom's life and the world of mid-century malaise in which he lives, one book reviewer concluded, "Tom seems to be a suitable hero for a society of vanishing extremes where minimal comforts are fairly secure and maximal ambitions extend to the next step ahead."[28] In the end, when confronted with a decision between more money and career advancement—and thus more work—and a more-secure 9-to-5 routine, Tom chooses the latter. And why does he do this, when asked by the company president?

> I'm just not the kind of guy who can work evenings and week ends and all the rest of it forever ... I'm not the kind of person who can get all wrapped up in a job—I can't get myself convinced that my work is the most important thing in the world. I've been through one war. Maybe another one's coming. If one is, I want to be able to look back and figure I spent the time between wars with my family, the way it should have been. Regardless of

war, I want to get the most out of the years I've got left. Maybe that sounds silly. It's just that if I have to bury myself in a job every minute of my life, I don't see any point to it.[29]

One course of conformity is chosen over another, the comfort and security of the family over the comfort and security of the organization. Nevertheless, notice how the decision is framed: against the war. The novelist Jonathan Franzen has it exactly right when he writes,

At the moral and emotional core of *The Man in the Gray Flannel Suit* are Tom's four-plus years of military service. Whether he was killing enemy soldiers or falling in love with an orphaned Italian teenager, Tom Rath as a soldier felt intensely alive in the present.[30]

Everything in Rath's life is measured against wartime experiences. And the world around him, the world of UBC double speak, wants nothing but to forget that such a war ever happened—it is "unfashionable." Betty Basye Hutchinson, a World War ii plastic surgery nurse, decades after the war recalled:

It was like we put blinders on the past. When we won, we believed it. It was the end. That's the way we lived in suburbia, raising our children, not telling them about war. I don't think it was just me. It was everybody. You wouldn't fill your children full of these horror stories, would you?[31]

The incongruity of wartime and postwar experiences is at the heart of the grey flannel suit experience. In one of the book's key passages, Rath ponders its consequences:

They ought to begin wars with a course in basic training and end them with a course in basic forgetting. The trick is to learn to believe that it's a disconnected world, a lunatic world where what is true now was not true then; where Thou Shalt Not Kill and the fact that one has killed a great many men mean nothing, absolutely nothing, for now is the time to raise legitimate children, and make money, and dress properly, and be kind to one's wife, and admire one's boss, and learn not to worry, and think of oneself as what? That makes no difference, he thought—I'm just a man in a gray flannel suit.[32]

To argue that the man in the grey flannel suit's curse is suburban inanity, and that alone, is to miss the message of this novel and its popular appeal. As well, it is to gloss over the complexity of postwar American culture.

The success of *The Man in the Gray Flannel Suit* cannot be attributed solely to its literary merits (quite to the contrary) or to the persuasiveness of Gregory Peck's laconic cinematic performance. Tom Rath is a fictional rendering of a pervasive and contested sociological "type," the "other-directed" "organization man." Numerous 1950s best-sellers warned Americans that materialism and affluence were sapping the country's moral marrow, the most important of which were Erich Fromm's *Sane Society* (1955), William Whyte Jr.'s *The Organization Man* (1956), and David Riesman's runaway hit, *Lonely Crowd: A Study of the Changing American Character* (1950). Setting a precedent for future scholars working in a critical vein, like Lizabeth Cohen, the *Lonely Crowd* literature focused most of its attention on suburban postwar affluence, on *Life* magazine's "Family Utopia." "Affluence" was and is an easy target. That the Kid in Upper 4 might have matured into a 1950s Man in a Gray Flannel Suit is nevertheless an uncomfortable, complicated outcome not simply of affluence but of war as well—its reality in the 1940s and its future possibility in the Cold War. The absence of the maladjusted GI from histories from, *and* about, this period has skewed portrayals of postwar American culture.

Though absent from the *Lonely Crowd* literature, ex-GIs—as products of war — had a remarkable influence upon postwar popular culture. Indeed, some of the period's most memorable cultural figures represented, or were, these "maladjusted" Americans—52-20 club losers. To name but a few: Marlon Brando in his first feature film, *The Men* (1950); Jack Kerouac; Hal Hingham in Alan Harrington's *The Revelations of Dr. Modesto* (1955); Frank Wheeler in Richard Yates's *Revolutionary Road* (1961); Binx Bolling in Walker Percy's *The Moviegoer* (1961), which in 1961 beat out *Revolutionary Road* for the National Book Award; Tayo in *Ceremony* (1977), by Leslie Marmon Silko, one of the, if not the, most important modern Native American novelists; Luke in *Cool Hand Luke* (1967); and R. P. McMurphy (aka 'RPM'), the quintessential unadjusted GI, who gave Jack Nicholson his most famous role in *One Flew Over the Cuckoo's Nest*. The list could go on. These were not merely fictional creations: often enough they were in fact autobiographical.

At base, the long-term effect (and affect) of the war is that in complicated ways war itself became a baseline against which meaningful personal and social experiences would be measured. But for many men and women nothing else could measure up. In other words, the gray flannel suit existence seemed doomed from the starting blocks. Jack Kerouac, an ex-merchant mariner, spoke for many of his generation in *On the Road* and through its narrator, Sal Paradise:

> I wished I were a Denver Mexican, or even a poor overworked Jap, anything but what I was so drearily, a "white man" disillusioned. All my life I'd had white ambitions ... I was only myself, Sal Paradise, sad, strolling in this violet dark, this unbearably sweet night, wishing I could exchange worlds with the happy, true-hearted, ecstatic Negroes of America ... Oh, the sadness of the lights that night! ... It was the Denver Night; all I did was die.[33]

Boredom, as social disengagement or disaffection, was the key emotional register for expressing repulsion against postwar readjustment—boredom and its cognates: apathy, alienation, melancholy, tedium, etc. So widespread was boredom that some intellectuals were convinced America would be destroyed by either it or The Bomb. University of Chicago President Robert Hutchins prophesied in 1946, "if we are not all killed in the next few years we will all be bored to death," adding "the horrors of peace in the atom age are almost as bad as the horrors of war." Nearly two decades later, the future looked all the more bleak. "An epoch has come to an end, an epoch in which our best knowledge has been dimmed with boredom or darkened by destruction," Ruth Anshen eulogized in 1963:

> We have felt for too long that this must be the very nature of life; this is the way life is ... leading us to a deep apostasy of the heart and a crucifixion of our natural aspiration for experience and growth.[34]

If scholars want to understand American Cold War culture, they must come to terms with the relationship between boredom and bombs, the experiences of suburban "peace" within history's most violent century.

Notes

[1] American Historical Association, "GI Roundtable: What Will Your Town Be Like," War Department, AG 300.7 (20 January 1945), p. 17.

[2] "Family Utopia," *Life*, 25 November 1946, pp. 58-59.

[3] The fascination here with helicopters may well have resulted from a spectacular 15 May 1941 news report that Igor Sikorsky had hovered over a Connecticut town for more than an hour in a helicopter, which, according to the aha War Department pamphlet, cited above, turned the public "helicopter slap-happy": "The newspapers and magazines broke out in a rash of futuristic articles forecasting suburbanites with helicopters in every backyard. Cartoonists had a field day depicting young girls eloping out of darkened upstairs windows in hovering helicopters, housewives drying their clothes under others, and bellhops helicopting through hotel lobbies 'paging Mr. Jones'" ("GI Roundtable," p. 16).

[4] For further discussion of *fantasy,* within a psychoanalytic vein, see, for instance, Jacqueline Rose, *States of Fantasy* (Oxford: Oxford University Press, 1996); Slavoj Žižek, *The Plague of Fantasies* (London: Verso, 1997); and Joan W. Scott, "Fantasy Echo: History and the Construction of Identity", *Critical Inquiry* 27 (Winter 2001), pp. 284-304. Although my work here in World War ii and postwar American culture is not psychoanalytic (it is decidedly historical) it does depend upon discourses of moods, states of minds, and emotions (the stuff of psychoanalysis)—apathy, boredom, *jouissance*, etc., as well as their structuration within dreams and fantasies, public and private—as I try to cut across the prevailing winds that guide the period's historiography. Toward this end, I am indebted to Walter Benjamin, in particular his *Arcades Project*, trans. Howard Eiland and Kevin McLaughlin (Cambridge, MA: The Belknap Press of Harvard University Press, 1999), esp. "Convolute D: Boredom, Eternal Return," pp. 101-19.

[5] Lizabeth Cohen, *A Consumers' Republic: The Politics of Mass Consumption in Postwar America* (New York: Knopf, 2003), p. 114.

[6] Monroe Wheeler to Miss Newmeyer, 13 May 1942, Circulating Exhibition II: 98(1), Road to Victory, MoMA Archives.

[7] Edward A. Jewell, "Portrait of the Spirit of a Nation," *New York Times*, 24 May 1942, p. x5.

[8] Monroe Wheeler to Mr. Abbott, 4 August 1942, Circulating Exhibition ii: 98(1), Road to Victory, MoMA Archives.

[9] Quote from *Globe-Democrat* (St. Louis, mo), 16 May 1943, in Circulating Exhibition ii: 98(1), Road to Victory, fol. 1, MoMA Archives.

[10] Margaret McDonnell to Carl Sandburg, c/o Carl Brandt, n.d. [22 December 1943], Circulating Exhibition ii: 98(1), Road to Victory, MoMA Archives.

[11] Quote from *Chicago Sun*, 4 April 1943, Circulating Exhibition II: 98(1), Road to Victory, fol. 1, MoMA Archives.

[12] Jewell, "Portrait of the Spirit of a Nation," p. x5.

[13] Alan Trachtenberg, *Lincoln's Smile and Other Enigmas* (New York: Hill and Wang, 2007), p. 267.

[14] "Road to Victory," *The Bulletin of the Museum of Modern Art,* 9 (June 1942), pp. 19-20.

[15] *Ibid.*, p. 4.

[16] *Ibid.,* p. 12.

[17] *Ibid.,* p. 16.

[18] Oliver Jensen, "Ensign Weems," *Life*, 29 December 1941, p. 56.

[19] *Ibid.,* pp. 55, 56, 56-57, 59, 55.

[20] Henry R. Luce, "The Day of Wrath," *Life*, 22 December 1941, p. 11.

[21] *Ibid.,* p. 11.

[22] *Ibid.,* p. 12.

[23] Studs Terkel, *"The Good War": An Oral History of World War Two* (New York: Ballantine Books, 1984), p. 120.

[24] Mark R. Grandstaff, "Visions of New Men: The Heroic Soldier Narrative in American Advertisements During World War ii," *Advertising and Society Review* 5, 2 (2004),

http://muse.jhu.edu.proxy.libraries.rutgers.edu/journals/advertising_and_society_review/v005/5.2grandstaff.html (accessed 8 June 2006).
[25] Jerome S. Bruner, *Mandate From the People* (New York: Duell, Sloan and Pearce, 1944), p. 210.
[26] Dan Cordtz, "Man in Gray Flannel Gets a New Tie with Merchandising Weave," *Wall Street Journal*, 8 March 1956, p. 1.
[27] "Slipped Disk", *Time*, 18 July 1955,
www.time.com/time/printout/0,8816,866528,00.
html (accessed 19 April 2007).
[28] Isa Kapp, "Wilson, Sloan. *The Man in the Gray Flannel Suit* (Book Review)," *Commentary* 21 (1956), p. 91.
[29] Sloan Wilson, *The Man in the Grey Flannel Suit*, intro. by Jonathan Franzen (1955; New York: Four Walls Eight Windows, 2002), pp. 251-52.
[30] Franzen, intro to Wilson, *The Man in the Gray Flannel Suit.*
[31] Terkel, *The Good War*, p. 131.
[32] *Ibid.*, p. 98.
[33] Jack Kerouac, *On the Road* (1955; New York: Penguin, 2003), p. 180.
[34] Erich Fromm, *Beyond the Chains of Illusion: My Encounter with Marx and Freud*, intro. by Ruth Nanda Anshen (New York: Giant Cardinal, 1963), p. ix.

Chapter Twelve

Gatsby and Garbage

Rex Ferguson

About Halfway between West Egg and New York the motor road hastily joins the railroad and runs beside it for a quarter of a mile, so as to shrink away from a desolate area of land. This is a valley of ashes—a fantastic farm where ashes grow like wheat into ridges and hills and grotesque gardens; where ashes take the form of houses and chimneys and rising smoke and, finally, with a transcendent effort, of ash-grey men who move dimly and already crumbling through the powdery air. Occasionally a line of grey cars crawls along an invisible track, gives out a ghastly creak, comes to rest, and immediately the ash-grey men swarm up with leaden spades and stir up an impenetrable cloud, which screens their obscure operations from your sight.

But above the grey land and the spasms of bleak dust which drift endlessly over it, you perceive, after a moment, the eyes of Doctor T. J. Eckleburg. The eyes of Doctor T. J. Eckleburg are blue and gigantic—their retinas are one yard high. They look out of no face, but, instead, from a pair of enormous yellow spectacles which pass over a nonexistent nose. Evidently some wild wag of an oculist set them there to fatten his practice in the borough of Queens, and then sank down himself into eternal blindness, or forgot them and moved away. But his eyes, dimmed a little by many paintless days, under sun and rain, brood on over the solemn dumping ground.[1]

What is this valley of ashes doing in a novel that famously depicts the decadent lifestyle of the wealthy classes of 1920s America? The answer, of course, lies in the question. It is precisely because the world we all know of in *The Great Gatsby*—the world of huge wealth and extravagant parties—is a world that creates waste. The valley of ashes is, like the aftermath of Gatsby's parties in which eight servants have to toil "all day with mops and scrubbing–brushes and hammers and garden-shears, repairing the ravages of the night before," a metaphor for the wastefulness of the Long Island elite.[2] This is exemplified, no less, in the figures of Tom and Daisy Buchanan.[3] But it is more than this. Garbage is more than a metaphor. It is matter in its most absolute form: materiality in its most indubitable moment. In filth and grime, revulsion and disgust, reality appears just as it is. Jay Gatsby, the ultimate idealist, is thus haunted by the materiality he encounters every day between West Egg and New York: its mute substantiality in stark opposition to the world that his imagination calls into existence. It is the inescapability of matter—the inescapability of garbage—that is truly at the heart of both Gatsby's downfall and *The Great Gatsby* as a novel. That is, to understand why garbage is so prominent in the novel is not a question of understanding what it represents but rather a question of understanding how it is represented. How does its materiality come to infiltrate the ideal of both Gatsby and Fitzgerald?[4]

Civilized Garbage

In this context a simple question immediately presents itself: from where does the rubbish we are reading about come? Let us return to the first sentence of our passage. The valley of ashes, a non-place in itself, exists only in a spot where the "motor road hastily joins the railroad and runs beside it for a quarter of a mile"—it is geographically located in relation to technologies of transit. In much the same way, on the very first page of the novel, Gatsby is compared to a seismograph in his "heightened sensitivity to the promises of life," and throughout the narrative his conduct is mediated by a range of technologies such as the car, the aeroplane and the telephone.[5] These instruments make Gatsby's world archetypally modern and do so specifically because they alter the geography he occupies. In his motor car he travels at great speed, reducing the space of distances to the time taken between points, while in the aeroplane he opens the sky to travel. With the telephone he can even occupy two places at once. Modern America is smaller than it used to be; its landscape has shrunk, become malleable. In his book, *Narratives and Spaces*, David Nye has written that "people use technologies to reshape and re-imagine their material context, and their experience of any space is a complex, mediated encounter"; and that "technologies are always involved in the creation of a landscape."[6] Technology creates a landscape by marking a place as a human possession. In an extension of the Lockean idea that the right to property is fulfilled by the mixing of one's labour with the land, technology (human *techne*) masters the natural world for human use.[7] One of Nye's examples is the way that the American West was conquered by the technology of the railroad.

Just such an emplacement is seen in our passage as the train and the motor car make a place of the "desolate area of land" that is the valley of ashes. Of course, in being located only in relation to modes of transport, the point is emphasized that this is a place in which you do not want to stop. It is simply a place to be got through on the way to either West Egg or—the ultimate technological landscape—New York. Consequently, while it is never made explicit, it seems fair to assume that the ash-piles must be the waste produced by the technology of modern travel and the energy consumption of New York.

Technology's interplay with, and appropriation of, the natural world is also implicit in the images into which the ashes are transformed. They are "fantastic farms" and "grotesque gardens": one an appropriation of nature for the production of consumable goods, the other an appropriation of nature for man's aesthetic pleasure. In a steady gradation, the descriptions come to be of the more recognized forms of man-made civilisation such as "houses" and "chimneys": images and forms in which matter becomes fully humanized and thus part of an un-natural world. From the strictly natural process of "growing like wheat into ridges and hills," the ashes are sculpted into images of the human world, and thus a very modern separation is created. Humanity, reason, civilisation and order are all permanently cut off from untouched nature, barbarism and mess.

While humanity is ordered and nature messy, garbage and waste occupy a different realm. The truly natural has yet to be touched by humanity and so is a mess. Waste, by contrast, is matter that has undergone the specific relation of having been humanly *used*. For matter to be garbage, humanity has to be, in some sense, finished with it; waste is the leftovers of something that has been important to us. While coal is messy, ash is garbage. Waste occurs via human use, and at exactly the same moment as land is claimed by man's labour and *techne*. Technology and garbage thus stand side by side. The gleaming exteriors and cold, clean lines of the New York skyline stand as a monument to the wholly unnatural beauty of modernity, but they cast a huge shadow; for in the process of creating and sustaining a city like New York, human waste appears on a truly grand scale.[8] The forms the ashes metamorphose into are thus the pictures, in effect, of their own creation as humanity appropriates nature, and makes waste, in its own image.

In the novel as a whole, a somewhat similar, if decidedly warped, interpretation of modern experience is made by Tom Buchanan (a fictional creation that Fitzgerald considered one of his best). Tom speaks to Nick early in the novel about his fear for the continued dominance of what he identifies as his own race. Influenced by the book *The Rise of the Col-*

oured Empires by "this man Goddard," Tom declares that they are all Nordics (he, Nick, Daisy) and, as the dominant race who have "produced all the things that go to make civilization,"[9] they must ensure that they are not submerged by the greater numbers of other races, including Mediterraneans, who exist in the United States.[10] In effect, Tom's thoughts are an obsession of purity under threat: not the threat of revolutionary violence or political challenge but simply of contamination. For him, the African and Southern European races are the leftovers of civilisation. They are the ash-piles of civilisation. As Michael Thompson has written:

> The survival of a world view can be ensured only by eliminating, rejecting, or ignoring those intrusive and dangerous elements that preclude the continued coexistence of differing world views. These elements that elicit such responses constitute the cultural category "rubbish."[11]

Tom's fears are, at their most basic, not of other races as individual people or even as indistinguishable groups but simply as rubbish. His inherent anxiety over waste in this context is made apparent in his confrontation with Gatsby in the Plaza Hotel. Tom's initial anger over the blatant relationship of Gatsby and Daisy soon gives way to a general rant along familiar lines:

> "Nowadays people begin by sneering at family life and family institutions, and next they'll throw everything overboard and have intermarriage between black and white."
>
> Flushed with his impassioned gibberish, he saw himself standing alone on the last barrier of civilisation.
>
> "We're all white here," murmured Jordan.
>
> "I know I'm not very popular. I don't give big parties. I suppose you've got to make your house a pigsty in order to have any friends—in the modern world."[12]

Tom's main objection to Gatsby is that he does not belong to the world of West Egg. Intuitively, he discerns Gatsby's poor immigrant roots, and so his permanent inability to fit in. He has not come from the right stock and his money (the dead matter of waste in itself) has come from the wrong places.[13] Of course, it is not that Tom thinks that such a person is unsuitable for an affair; his own relationship with Myrtle Wilson is testament to that. But while Gatsby owns the largest house in West Egg, Myrtle is handily placed within the very confines of the valley of ashes and can easily be discarded at any time amongst the city's other waste. Indeed, it is indicative of Tom's theorising that the valley of ashes is the place where the immoral, the sordid and the violent (in short, all the garbage of civilised behaviour) are permitted.

Modern Garbage

Unsurprisingly, New York looms over this novel, a classic modernist text. Ever since Baudelaire's writings on the *flâneur* in the mid-nineteenth century, modern city life had become not simply a setting for literature but a subject matter in itself: a theme developed in a systematic way particularly in the writings of Georg Simmel and Walter Benjamin.[14] In a famous essay, Simmel wrote that "The psychological basis of the metropolitan type of individuality consists in the *intensification of nervous stimulation* which results from the swift and uninterrupted change of outer and inner stimuli."[15]

Fitzgerald displayed, in his adult life, just such a metropolitan type of individuality. He wrote Gatsby while in Rome and also spent many years in New York and Paris. He repeatedly exposed himself to the unpredictable and ever-changing stimuli that characterized metropolitan experience, and this phantasmagoria found its way into *The Great Gatsby*. Nick Carraway works in the city as a bonds salesman. Coming from small town life in the West, he finds big city life shocking to begin with but soon notes that he:

> began to like New York, the racy, adventurous feel of it at night, and the satisfaction that the constant flicker of men and women and machines gives to the restless eye.[16]

The mention of machinery once again alerts us to the technological landscape in which the modern individual is immersed—but it is the "constant flicker" and the "restless eye" that are really indicative of the city experience.

The flicker is a reference to a novella—*Heart of Darkness* by Joseph Conrad—that inspired several of the formal devices employed by Fitzgerald. Conrad's narrator, Marlow, surveys British imperial history and the history of civilisations, imagining empire as a "running blaze on a plain." Now, he says, "we live in the flicker."[17] In a further layer of intertextuality, Fitzgerald's quote also echoes a passage from *The Soul of London* by Conrad's great friend and collaborator, Ford Madox Ford. Ford writes that "an awakened sense of observation is in London bewildering and nerve-shattering, because there are so many things to see and because these things flicker by so quickly."[18] While Fitzgerald's knowledge of this passage is perhaps uncertain, what is beyond doubt is that his "flicker" is both Conrad's and Ford's. It is the flicker of the shifting visual stimuli produced by the city, observed from a restless eye, and it is also the flicker of dying embers—only they are not the ashes of empire but rather those of modernity itself. *The Great Gatsby* is about living alongside, and even in, a state of decay.[19]

But why is modernity seen as being in a state of decay? Why are we in a world of flickering leftovers and piles of ash? Such questions have answers that can be justified on the basis of ideal reason alone. Images of rubbish and decay are representations of decidedly modern theories of transience, degeneration and decline.[20] While these explanations are clearly of some importance, they, in a sense, make garbage of the waste matter in *The Great Gatsby* for a second time as they attempt to turn rubbish into a kind of nothing. When the ash-piles become a metaphor, they stop being piles of ash and instead are rehabilitated into the world of ideal reason as they explain something that has already been conceived by the human mind. Yet the explanation that garbage represents something else fails to account either for the prominence of that something else or for the fact that it is best represented by an image of decaying waste. I suggest, rather, that we try to understand why material garbage crept into the ideal modernist aesthetic—in other words, in what specific way did garbage flicker in the restless eye of the modern city dweller?

To address this question, we can return to our opening passage, where eyes feature prominently—namely, "the eyes of Doctor T. J. Eckleburg" whose "retinas are one yard high." These are the eyes of the metropolitan individual but they are also the arresting image of a very striking advert. The rise to prominence of advertising in both newspapers and magazines, as well as in eye-catching billboards and posters, is coterminous with the huge explosion in city populations in Europe and North America at the turn of the twentieth century.[21] The city was already, as Benjamin put it, a "dream world of capitalism," teeming with the phantasmagoric images of products for sale.[22] The advert simply extends such phantasmagoria, as the image of the commodity becomes bigger and more accentuated: plastered to a wall while the object itself remains invisible, hidden, somewhere else. In the production of such campaigns, advertising agencies—"modernity's town-criers" as Roland Marchand calls them—responded to an economic crisis of overproduction.[23] The solution, according to Richard Ohmann, was the creation of new styles of products and therefore a new class of consumer.[24] The main feature of this new style was disposability. Any sense of durability being a positive feature for a product had to be ousted, as did any notions of recycling.[25] A short shelf life was the key; the consumer had to be encouraged to buy and consume more. With a higher turnaround of production and sales, it became more important for corporations that customers were loyal to their brand. As Susan Strasser has written in her book *Waste and Want*, more and more producers "promoted not simply the contents of the container but a new kind of product that included the packaging, emblazoned with a brand that could be advertised."[26] The brand itself, rather than the material product, became the item of consumption, and the way that brands advertised themselves created an identification between the individual and the brand that they chose to consume: that choice becoming a statement of their singular individuality, paradoxically made in the form of consuming an item of mass production and marketing. A consumer class was born who bought all kinds of goods and continued to buy and consume them; but, more than buying a product, in the process, they bought themselves a self.[27]

It is interesting to note here that the eyes of Doctor T. J. Eckleburg are just that: eyes. There is no face—even the nose is "nonexistent." The product on show—the spectacles—supersede, in existential terms, the very consumer who would wear them. Or rather, because seeing an advert is an intermediate stage in which the subject is yet actually to consume, they, in a sense, do not properly exist yet. The subject prior to consumption is withdrawn in potentiality and so our advert for spectacles is set amidst the nothingness of dead matter.

The imagery of the advert amongst the waste matter is also a picture of beginnings and endings. In the beginning the product is advertised, in the end it is thrown away. And just as the train and the motor car race through the valley of ashes, so too the product races through its transitory life cycle. More goods and more consumption equal more garbage. Strasser writes:

> during the forty or so years around the turn of the twentieth century, mass production and mass distribution created unprecedented quantities of trash that disturbed private citizens and plagued city administrations.[28]

Rubbish was moving from a matter of invisible private disposal to one of visible public refuse. What was one day the brand new commodity with its prevalent brand name was the next day a rotting piece of garbage. The problem of garbage collection was, first and foremost, a practical one; but its visibility on the streets of the city created more than simply practical considerations. The identification process that consumers had gone through could not be put on hold and thrown out with the rubbish. Consumers continued to identify with their products of choice; only now those products were very quickly becoming visible items, not just of rubbish, but of public refuse. The self that had been bought now lay in an overflowing trash can. The transient lifespan of the product is thus connected to the self's own ephemerality, and rotting garbage prefigures the mortality and eventual death of a degenerating body. The ash-piles on the way to West Egg are not just the garbage that humanity creates but the garbage that humanity becomes—ashes to ashes, dust to dust.

Ideal Garbage

Although the extract from *The Great Gatsby* we have been examining is, indeed, a famous one, Fitzgerald is most often associated with his depiction of the decadent jazz age; and the popular image of the novel is undoubtedly of Gatsby's extravagant parties. The party-goers who descend on his mansion every night are, in effect, the ultimate modern consumers. The whole point of the party is consumption. Not just the food and drink on offer but the music, the conversation, the atmosphere, the outrageous actions are all items to be consumed. And the most important thing is simply to have gone, to have been there, and to have taken it all in.

The ideal commodity—an extravagance offered uniquely as the essence of the party—is Gatsby. Not Jay Gatsby, neighbour of Nick Carraway and lover of Daisy Buchanan, but just Gatsby the name: the unique mystery. It is because of this name—because it is Gatsby—that everyone goes to the party. This Gatsby is phantasmagoric because the mystique which surrounds him in no way relates to the truth, or value, of his background and the source of his wealth, or to any inherent charisma that he possesses. Gatsby is simply a brand name in the capitalist city. Or rather, he is the new packaging on an already known brand. Gatsby, the name, is a product only of Jay Gatsby's imagination—and his wealth, mansion and parties are all simply large-scale advertisements which announce to Daisy that he is a new and improved Jay Gatsby when, beneath the wrapper, he is, and still wants to be, the old one. Even in his business dealings, he is the respectable packaging that fronts the criminal activities of his partner, Wolfsheim—who is, in effect, the ad man to Gatsby's saleable product.

Just like any other product in the consumer culture of modern America, Gatsby's main feature must be disposability. The next summer the party-goers will need another lawn to descend on and the memory of the previous summer will be trashed. Gatsby is discarded once he ceases to provide the possibilities for consumption; only his father and Nick attend his funeral. In his death Gatsby, the name, becomes a waste product in just the same way as Jay Gatsby, the body, becomes the garbage of dead matter. His murder is not so much a misplaced killing as a fateful return to his lowly roots. It is as though garbage rose up, in the figure of George Wilson, and enveloped him in a fatal crush.

Wilson, despite being a central agent in the tragedy, is a figure as indistinct and lifeless as the "ash-grey men" in our extract.[29] He is consistently described as grey in his colouring and is permanently exhausted by his work as a mechanic (technology and rubbish are again set side by side). In a telling moment when he is first presented, he is described as emerging from his garage "wiping his hands on a piece of waste," the implication being that any attempt at cleansing is in vain.[30] There is no way out of garbage for Wilson; just as, in spite of his temporary escape from it, there is none for Gatsby.

The "transcendent" effort with which the ash forms into men in our passage is the echo of the effort with which Wilson manages to rise above the limitations inherent in the matter he has come from, and to kill Gatsby. But it is also the echo of the transcendental in the novel more generally: of the transcendental idealist that is Gatsby. Gatsby, the commodity, is sold to both himself and to others as an ideal. Nick writes that "Jay Gatsby of West Egg, Long Island, sprang from his Platonic conception of himself."[31] Another important philosophical reference occurs when Gatsby and Daisy meet at Nick's house. Rather than staying, and so being able to impart what occurs at their reunion, Nick departs to the garden. While sheltering from the rain, he reports that "there was nothing to look at from under the tree except Gatsby's enormous house, so I stared at it, like Kant at his church steeple."[32] At just the moment Gatsby seems about to realize his idealistic vision, and Nick stares at the monument of his great wealth, Kant, and therefore transcendental idealism, is invoked. The Gatsby experience is predicated upon the vision (the Platonic conception) he has of both himself and of Daisy. This should be understood in the broadest sense possible. Experience for Gatsby is bound to the conditions of advanced modern technology and consumer culture in just as profound a sense as Kantian experience must come bound in categories of space and time. The material world for Gatsby contains not a real world of objects as brute facts but only the potential world that can be rearranged into his own ideal: namely, the past.[33] When Nick points out to him that to repeat the past is impossible, his response is typically idealistic:

> "Can't repeat the past?" he cried incredulously. "Why of course you can!"
>
> He looked around him wildly, as if the past were lurking here in the shadow of his house, just out of reach of his hand."[34]

Repeating the past is more than the goal that Gatsby is working to attain. It is, rather, that the potential for repeating the past—and finding the past in the presence of his material surroundings—is the way that those material soundings and that present must be inhabited.

This turns the everyday world into garbage. Nick is right, of course; the past cannot be repeated. It is as unattainable as the Kantian *noumena* of "things-in-themselves" which lie beyond the *phenomena* of experience. Nietzsche, in his later writings, became increasingly more hostile to such a dichotomy of *appearance* and *thing-in-itself*. In a passage openly critical of Kant, he wrote:

> to talk about "another" world than this is quite pointless, provided that an instinct for slandering, disparaging and accusing life is not strong within us: in the latter case we *revenge* ourselves on life by means of the phantasmagoria of "another," a "better" life.[35]

The Gatsby ideal is, like Gatsby himself, phantasmagoric; it is the "better" life that disparages empirical reality. His past life with Daisy makes waste of everything else: the rest of his life being just the attempted rearrangement of dead matter into the shape of the past. What ultimately leads to the demise of Gatsby's dream is Daisy's inability to share fully in his vision and to deny her life, and her other past, with Tom. Daisy is a materialist in this sense, but her own materiality, as a body, is also crucial. At the reunion, Nick's absence is required because, in a sense, the scene cannot be written. Not even the ideal vision of Fitzgerald's artistry could live up to Gatsby's vision, and the material fact of their reunion is discreetly avoided.[36]

Gatsby has to live through his vision and thus, when the ideal is finally and completely shattered by Daisy's failure to reject Tom, he dies. Despite being in some senses a mistaken—almost accidental—killing, Gatsby's murder follows the logic of his existence. Once the ideal is broken Gatsby has to die, because the Gatsby world no longer exists for him to inhabit. The world catches up with Gatsby. Not the world of the past (which would be a more standard literary trope) but, rather, the world of the present. It is presence in the everyday world—the world of *phenomena*—that bursts in on Gatsby. As the novel draws to its close, Nick speculates on how Gatsby must have felt upon Daisy's rejection, that it must have shattered his world. Everything must have appeared different to him like "a new world, material without being real."[37] The reality of the Gatsby experience would have been lost, gone forever, to be replaced by the material world of concrete existence. Entering this new world would have been, for Gatsby, to be plunged into garbage; in particular it would have meant recognising the trash that Daisy herself had turned out to be, part simply of the "rotten crowd" on which Nick has passed judgment.[38]

In his final definition of Tom and Daisy, Nick characterizes them as follows:

> They were careless people, Tom and Daisy—they smashed up things and creatures and then retreated back into their money or their vast carelessness or whatever it was that kept them together, and let other people clean up the mess they had made . . .[39]

At the end of the novel, it is left to Nick to clean up, as he wanders around the deserted Gatsby mansion where there was "an inexplicable amount of dust everywhere."[40] Indeed, Nick's attempts to tie up all the loose ends of his time in the East are attempts to cleanse himself of his involvement in it: to make it almost as though he had never inhabited West Egg, as if he had never existed. He writes of his breaking up with Jordan: "I wanted to leave things in order and not just trust that obliging and indifferent sea to sweep my refuse away."[41] Nick wants to leave the world of West Egg as he found it, to deny the material fact that he was there. He wants to leave it as Gatsby conceived it, in a pure act of willed idealism.

But Gatsby cannot repeat the past and Nick cannot tidy everything up; he cannot deny his own presence during the sickening events that took place. Even more than Gatsby, Nick was exposed to the valley of ashes, shielded only by the flimsy veneer of a train window, as he travelled from West Egg to New York every day. Nick, who thought Gatsby had "turned out all right at the end," was truly repelled only by "what preyed on Gatsby, what foul dust floated in the wake of his dreams."[42] But what floats in the wake of Nick's narrative, what intrudes on the pristine prose of Fitzgerald?—what else but the "spasms of bleak dust which drift endlessly over it."

Notes

[1] F. Scott Fitzgerald, *The Great Gatsby* (Ware, Hertfordshire: Wordsworth Classics, 2001), p. 16.
[2] *Ibid.*, p. 26.
[3] For two examples of the many critical responses which analyse the scene in metaphorical terms see Tom Burnam, "The Eyes of Dr. Eckleburg: A Re-Examination of *The Great Gatsby*," in *F. Scott Fitzgerald: A Collection of Critical Essays*, ed. Arthur Mizener (Englewood Cliffs, NJ: Prentice-Hall, 1963) and A. Robert Lee, "'A Quality of Distortion': Imagining *The Great Gatsby,*" in *Scott Fitzgerald: The Promises of Life*, ed. A. Robert Lee (London: Vision Press, 1989). Burnam writes that "the Buchanans, Jordan Baker, the obscene barflies who descend in formless

swarms on Gatsby's house, (are) all symbolized by the gritty disorganized ash heaps . . ." p. 110; while Lee notes that the wasteland is "Nick's perfect correlative metaphor for all the 'bleak dust' which will hover ruinously over Gatsby and his dream of possessing Daisy," p. 51.

[4] The ideal of Fitzgerald being his artistic work, as opposed to what he termed the "trash" he produced for *The Saturday Evening Post*. See Robert Roulston, "Rummaging Through F. Scott Fitzgerald's 'Trash': Early Stories in 'The Saturday Evening Post,'" in *Journal of Popular Culture*, 21, 4 (1988: Spring), pp. 151-163.

[5] Fitzgerald, *The Great Gatsby*, p. 3.

[6] David Nye, *Narratives and Spaces: Technology and the Construction of American Culture* (Exeter: University of Exeter Press, 1997), p. 4.

[7] John Locke, *Two Treatises of Government* (Cambridge: Cambridge University Press, 1999), Treatise ii, Section 27, pp. 286-87.

[8] The classic source on the conception of beauty as being wholly man-made as opposed to natural is Baudelaire who wrote that "everything beautiful and noble is the result of reason and calculation." See Charles Baudelaire, "The Painter of Modern Life," trans. by P. E. Charvet, in *Baudelaire: Selected Writings on Art and Artists* (Cambridge: Cambridge University Press, 1972), p. 425.

[9] *Ibid.*, p. 11.

[10] Fitzgerald, *The Great Gatsby*, quotations, p 10. "*The Rise of the Coloured Empires* by 'this man Goddard,'" is either a veiled reference or simply a mistitle and misattribution of *The Rising Tide of Color* by Lothrop Stoddard.

[11] Michael Thompson, *Rubbish Theory: The Creation and Destruction of Value* (Oxford: Oxford University Press, 1979), p. 90.

[12] Fitzgerald, *The Great Gatsby*, p. 83.

[13] See John Scanlan, *On Garbage* (London: Reaktion, 2005), "money can become anything and this makes it nothing—the equivalent of waste matter ready to be refashioned into a new form," p. 15.

[14] See Baudelaire, "The Painter of Modern Life."

[15] Georg Simmel, *The Metropolis and Mental Life*, trans. by Hans Gerth, in *Simmel on Culture,* ed. David Frisby and Mike Featherstone (London: Sage, 1997).

[16] Fitzgerald, *The Great Gatsby*, p. 37.

[17] Joseph Conrad, *Heart of Darkness and Other Tales* (Oxford: Oxford University Press, 1990), p. 139.

[18] Ford Madox Ford, *The Soul of London* (London: Everyman, 1995), p. 96.

[19] There are several interesting parallels in other literature. Eliot's *The Waste Land* (1922) the most prominent among them, with not only the waste land itself but the blind seer Tiresias also being echoed in the oculist who sinks down into "eternal blindness." The disease and decay of Thomas Mann's *The Magic Mountain* (1924) is another source, and for a Victorian echo see the dust-heaps of Dickens' *Our Mutual Friend* (1865).

[20] Influential philosophical-historical theories include Max Nordau's *Degeneration* (1883) and Oswald Spengler's *The Decline of the West* (1918-1922). Incidentally, in an article from 1955 R. W. Stallman quoted from a letter written by Fitzgerald in 1940 in which he remarked "I read [Spengler] the same summer I was writing

The Great Gatsby." However, since then Robert Sklar has noted that Stallman had "not checked Fitzgerald's memory against the fact that the first volume of *The Decline of the West* was not published until 1926 and that Fitzgerald was not a reader of German." Despite this blow to Stallman's argument the contiguity of the works in Fitzgerald's own mind remains worthy of note. See Robert Sklar, *F. Scott Fitzgerald: The Last Laocoön* (New York: Oxford University Press, 1967).

[21] Newspapers and Magazines are prominent in the novel as a whole: *The Saturday Evening Post, Tribune, The Journal* and *Town Tattle* are all mentioned at various points. See also Richard Sennett, *Flesh and Stone: The Body and the City in Western Civilization* (London: Faber and Faber, 1996).

[22] Quoted in David Frisby, *Fragments of Modernity: Theories of Modernity in the Work of Simmel, Kracauer and Benjamin* (Cambridge, MA: MIT Press, 1986), p. 3.

[23] Roland Marchand, *Advertising the American Dream: Making Way for Modernity, 1920-1940* (Berkeley, CA: University of California Press, 1985).

[24] Richard Ohmann, *Selling Culture: Magazines, Markets, and Class at the Turn of the Century* (London: Verso, 1996).

[25] On the prominence of recycling prior to this time see Susan Strasser, *Waste and Want: A Social History of Trash* (New York: Metropolitan Books, 1999)

[26] *Ibid.*, p. 14.

[27] Scanlan, *On Garbage:* "in consumption we buy the self rather than simply buying a product," p. 133.

[28] Strasser, *Waste and Want*, p. 17.

[29] A. E. Dyson has described Wilson as "a representative of the ash-gray men." See A. E. Dyson, "The Great Gatsby: Thirty-Six Years After" in Mizener ed. *F. Scott Fitzgerald*, p. 122. Robert Sklar also has written that "George Wilson is like one of those ash-grey figures, spiritless and anaemic, so covered with dust that he seems to fade into the cement color of his garage walls." Sklar, *F. Scott Fitzgerald*, p. 178.

[30] Fitzgerald, *The Great Gatsby*, p. 17.

[31] *Ibid.*, p. 63.

[32] *Ibid.*, p. 56.

[33] At this point it could be argued that for an idealist Gatsby pursues a very materialistic course. Of course this claim has only limited truth as it is quite clear that material wealth are but a means to and end for Gatsby. In addition Ronald Berman has noted, referencing George Santayana's 1920 essay "Materialism and Idealism in American Life" that "Gatsby is materialistic because Americans do not have many other alternatives. Material life offers one of the few recognized ways in which the American can express his idealism," Ronald Berman, "*The Great Gatsby* and the Twenties," in *The Cambridge Companion to F. Scott Fitzgerald*, ed. Ruth Prigozy (Cambridge: Cambridge University Press, 2002), p. 86.

[34] Fitzgerald, *The Great Gatsby*, p. 70.

[35] Friedrich Nietzsche, *Twilight of the Idols*, trans. by R. J. Hollingdale (Harmondsworth: Penguin, 1990), p. 49. In the previous paragraph to the one quoted Nietzsche writes "… the 'real world' has been constructed out of the contradiction

to the actual world: an apparent world indeed, in so far as it is no more than a *moral-optical* illusion." The image of Dr. Eckleburg eyes and the optical illusion, or actual world, he surveys, contrasted with the "real world" constructed by Gatsby is striking.

[36] The mention of Kant becomes doubly significant here. As Robert Sklar notes, "It was Kant's habit to stare at a church steeple through his window while he mulled over his philosophical concerns. A neighbour's trees grew so tall, however, they hid the steeple from Kant's view. But at Kant's request their tops were cut, and he was able to resume his contemplation. Nature (or the material—my note) resists man's efforts to find transcendent reality behind its earthly face." Sklar, *F. Scott Fitzgerald*, p. 185.

[37] Fitzgerald, *The Great Gatsby*, p. 103.

[38] *Ibid.*, p. 98.

[39] *Ibid.*, p. 114.

[40] *Ibid.*, p. 94.

[41] *Ibid.*, p. 113.

[42] *Ibid.*, p. 4.

Chapter Thirteen

Exteriority: Boredom, Disgust, and the Margins of Humanity

Harvie Ferguson

… this increasingly vacant, forlorn bourgeois age
—Ernst Bloch, *Traces*[1]

A "something" that I do not recognize as a thing. A weight of meaninglessness, about which there is nothing insignificant, and which crushes
—Julia Kristeva, *Powers of Horror*[2]

What is the fascination of waste? In one sense the question is hardly problematic; the *topos* of waste and its many cognates—garbage, rubbish, trash, spoil, and so on—is intellectually interesting in an essential way just because, in spite of everyday familiarity, it resists conceptualization. The intellectual puzzle of waste is how to *think* garbage; how to translate the incoherence of its indubitable reality into a clear and distinct idea. And it is just this paradox—the reckless demand to think the unthinkable—that charges all genuine theoretical reflection with an alluring tension. In this sense the fascination *with* waste is nothing other than the lure of thinking. The more pertinent question is not why the idea of waste is engaging but why this real, putrid, decomposing, stinking, squashy, and utterly banal rubbish should be fascinating? Whence rises our *existential* involvement with waste?

Seizing the existential issue, however, is not to abandon coherent reflection. As a pure phenomenon, considered exclusively in respect of its wastefulness and, for the moment, disregarding the accessory, empirical determinations that characterize any of its particular instances, waste appears distinctively through two modalities. In relation to time, waste is experienced as boredom; and in relation to space it is experienced as disgust. Waste time is formless, empty, and eternal; waste space is formless, filled, and unbound. As waste time, boredom is experience bereft of substance; as waste space, disgust is experience filled with nothing but substance. Boredom and disgust are intimately related as alternative modalities of the experience of waste, which finds its common property in the paradox of experience itself; it is being without form. The fascination of waste lies precisely in the intuition that, in acknowledging its enigmatic presence, we can in some way gain access to being without form; a possibility we find attractive and repulsive in equal measure. Waste is an indeterminate region in which transcendence and immanence meet and merge. Filled with *dread*, it is the anxious moment in which existence at once asserts itself and surrenders to oblivion. Waste thus marks the boundary of humanity; the boundary with the unbound. And it is characteristic that, for contemporary society, humanity is nothing other than this boundary.

The zone of demarcation, no longer enclosing and protecting humanity, becomes identical with and exhausts the human. Waste, far from being the cluttered marginalia, is now the essential text of humanity; and its charm, no longer devalued as kitsch, assumes the awful dignity of truth. Waste is the authenticity of the present; the unbound to which we bind ourselves with desperate resolve. These bare assertions require considerable elaboration and support to become fully intelligible and, more significantly, they need to be located in an appropriate historical and cultural context if they are to be grasped critically as living forms and relations.

An immediate connection between boredom and disgust is available, nonetheless, in the experience of satiation and excess. One aspect of boredom might be described as a surfeit of time; its sheer overwhelming abundance piling up in useless and, in fact, unusable excess. And disgust is roused by over indulgence; the excess not just beyond need, but over and above the most exaggerated appetite. Boredom and disgust, that is to say, are linked by the incommensurability of appetite, interest, or desire to the overwhelming presence of a reality that remains inexhaustible, exterior, and alien.

In a preliminary way, then, the existential import of waste emerges, not as the exhausted residue of a process of production or consumption, but, rather, as that transformation of reality corresponding to loss of appetite, interest, and desire. Waste is the radically inconsumable: the opposite of scarcity, it is not what we want but cannot have; it is what we have and do not want. And in a world made coherent and meaningful through the psychic tension of desire for the unattainable, the scandal of overwhelming abundance is irresistibly repellent.

To gain a fuller insight into the significance of boredom and disgust, and their interrelation through waste, it is important to locate these experiences in the context of developing modernity; that is, in the culture of modern society. Of course, boredom and disgust are not unique to modernity. The possibility of these experiences is common to many, if not all, societies; but the actualization of such a possibility depends on specific features of the cultural field in which it emerges. That is to say, the kinds of experiences that we recognize as boredom or disgust appear elsewhere—but elsewhere are thematized in terms of quite different schemas of interpretation and meaningfulness. Lethargy, lack of interest in life, and an engulfing sense of indifference to, and of, the world have appeared as noteworthy experiences in many societies, but as the experience of boredom they are bound together in a peculiar unity that is distinctive of modern society. Similarly, repulsion, defilement, and fear of contagion are possible experiences in any society but, for modern society, constitute the peculiar aesthetic-moral unity of disgust.

The fascination of waste lies precisely in the lure of the counter-conceptual. Where philosophy fails adequately to grasp, by abstraction, the boundaries of humanity in the context of modern life, waste, by contrast, manages perfectly to actualize these.

Spirit of Boredom

Boredom is a modern phenomenon, and not just a modern term for a universal condition. Boredom and modernity, furthermore, are linked in an essential manner; that is to say, rather than being viewed as a contingent *consequence* of the modern transformation of society, boredom is to be grasped as one of its most distinctive *moods*. In modern society, boredom is *waste time*: time that "arrives" in an unusable form or, rather, fails to arrive at all and remains in formless suspension outside the subject. In boredom time resists the ordinary qualifications of temporality given in the irresistible experience of duration and succession. Boredom, thus, is not to be confused with idleness, tedium, meaninglessness, or emptiness. It is not that "nothing happens" but that, in boredom, time stands aloof from all experiences that ordinarily appear effortlessly in flowing interconnection. In boredom time piles up as so much waste; it is time that cannot be "passed." Boredom is not wasted time, it is *waste time*; time realized as waste.

Genealogy of Despair

In spite of the essential modernity of boredom, an impressive body of contemporary scholarship has traced out a typological lineage of closely related conditions located in ancient, medieval, and early-modern western society. Though misleading if read as developmental history, this literature, in fact, provides many key insights into the wholly modern character of the phenomena of boredom and its interpretation.

Thus, where *acedia* has frequently been treated as the prototype of boredom, it is more properly grasped as an experience peculiar to the religious world in which it emerged. *Acedia* is spiritual lethargy, in contrast to compunction, which is the prick or sting of religious conscience. Compunction effectively motivates and guides action; whereas *acedia* is a condition of spiritual torpor immune to the most fervent wish and determined effort to discharge religious obligations.[3] *Acedia* became prominent in the experience of the Desert Fathers during the early centuries of Christianity. Evagrius Ponticus in the fourth century describes it thus:

> The demon of acedia, also called "noonday demon" is the most oppressive of all demon … First he makes the sun appear sluggish and immobile, as if the day had fifty hours … Moreover, the demon sends his hatred against the place, against life itself, and against the work of his hands, and makes him think he has lost the love among his brethren and that there is none to comfort him.[4]

The "demon of noonday" entered medieval Latin theology through John Cassian's *Conferences*, and subsequently gained general recognition as the sin of sloth.[5]

Acedia is complex and subtle. It is not simple unwillingness, or weakness in the face of a burdensome task. It is not the temptation of the non-spiritual, but a distinctive spiritual temptation that deflects the faithful from the true object of worship and contemplation. The temptation of *acedia*, that is to say, is just the kind of overcoming of earthly preoccupations and surrender to spirit also required of genuine religious faith. The *religious* problem of *acedia*, thus, is to distinguish the legitimate withdrawal from secular life and the mystical experience of the divine it inspires, from exhausted and vulnerable passivity and the demonic and terrifying visions that spring uncontrollably from a fevered imagination. This difference might seem to have a modern resonance. The unstable oscillation between *acetic* faithfulness to authoritative commands, and *acedic* self-projection of monstrous images is central, for example, to Flaubert's unequalled fictional account of spiritual trial, *The Temptation of St Anthony*. But the historic setting of this text, and its eccentricity in relation to the equally incisive and compelling account of modern boredom found in *Madame Bovary*, in fact serves to clarify the contrast between premodern religious torpor and modern secular lethargy.

What connects premodern *acedia* with modern boredom, and both with the question of waste time, lies in the *privilege* these experiences claim as expressions, albeit in pathological form, of the "highest" and most refined human feelings. It relates corporeal exhaustion with frenzied imagination: disembodied images escape and terrorize the exhausted body.[6] *Acedia* is ultimately a sin; a loss of faith and rejection of God, which is experienced at its extreme in a literal falling apart of the body-soul unity. But sin is already a religious condition and, thus, it takes a religiously inspired individual to suffer *acedia*. Later, and in quite a different context, boredom also came to play with this intriguing paradox: the spiritual loss of spirit.

The connection between *acedia* and boredom appears only in an abstract and distant perspective. Grasped contextually and historically, however, a new cultural formation that did have a direct bearing on the formation of boredom, namely melancholy, emerged as a distinctive feature of the late medieval and Renaissance world. For over a century *acedia* and melancholy co-existed as distinctive expressions of spiritual lassitude appropriate to different social groups.[7] Developing outside the church and its religious foundations, melancholy characterized a significant aspect in the experience of a new, secularizing, and humanizing experience of life that did not depend for its inner form and meaning on earlier religious worldviews.

Robert Burton traced the history of melancholy back to classical antiquity, identifying its authorship with Democritus, but it emerged fully in his own work, and in an exemplary fashion, as a specific character type within early modern humanism. Melancholy, in fact, is the psychic cost of humanism. It appears as an unplanned consequence of the determined rejection of Christian transcendentalism and the recovery of a Pagan worldviews that was such a conspicuous aspect of Renaissance thought. Already with Marsilio Ficino the revival of classical learning was accompanied by dull and listless states: a gloominess that frustratingly clung to the otherwise exciting and energizing grasp of reality as a human centred world. Almost every major Renaissance writer and artist struggled with, and ultimately embraced, melancholy as the inevitable accompaniment of their own creativity, truthfulness, and sensitivity—the latter all highly valued qualities of a new age that required and rested on the powers of a uniquely human spirit.[8] The melancholic, regarded initially as a specific humoral character-type, was peculiarly adapted to the new age of humanism, and the malady that Burton, with uncharacteristically brief lucidity in the midst of the torrential outpourings of his famous *Anatomy of Melancholy*, wonderfully characterized as "sorrow without cause," epitomised its ineradicable dark side.

Burton's influential definition was coined in the context of a new physical-psychological view of human functioning. The differentiation of human character types was established on the basis of the relative balance and imbalance of bodily *humours*. The melancholic was preponderantly "cold" and "dry," as distinct from, for example, the "hot" and "dry" or choleric. Imbalance and excess might be seen as the local "cause" of a specific pathological physical or psychological condition. And a purely physical theory of melancholy was popular for a time; the constitution of the melancholic resulting in the production of dark vapours within the body (black bile), that rose and filled the head with a deadening, and obscuring fog. But these notions never supplanted Burton's insightful psychological account; and, significantly, for early-modern writers and artists as varied as Thomas Bright, Michel de Montaigne, and Albrecht Dürer, black bile was also a black mood, and the vapours that rose to the brain, obscuring vision and depressing spirits, were never treated as the simple mechanical cause of a complex psychic state.

Melancholy, like *acedia*, was never common or vulgar; it was found only among the most refined and civilized, and overwhelmingly among men—hence the association of melancholy with aristocracy, with court society, and the creative genius of the new age.[9] Aristocratic disdain for ordinary life, however privileged, and artistic rejection of the commonplace, came together in a new cultural world that valued, above all, human superiority over nature and human autonomy from God.[10] In this sense melancholy became emblematic of the modern world. As such it could not be confined to a particular character type, or social group. Melancholy was gradually democratized and, in a later period, became available to all in a new form as boredom.[11]

The puzzle of melancholy is that, from its inception, the modern, the new, and the novel could be (and in its essential sense had to be) experienced as already exhausted. The disenchantment *of* the world was simultaneously disenchantment *with* the world and, as the world was now the theatre of self-creation, this implied disenchantment with the self.

The term melancholy, as the condition itself, fell out of fashion at the close of the Baroque age and remained dormant throughout the period of classical modernity; a period for which, re-enchanted by the evident success of reason, science, and the irresistible spread of capitalism and progressive culture, it seemed "natural" for human beings continually to be involved with their world. The notion of melancholy (or melancholia), was revived in the nineteenth century as part of a new psychiatric account of pathological experience. Contrasted with mania it lost its association with cultural accomplishment and its aura of profundity.[12] And, though used in a philosophically sophisticated way by Søren Kierkegaard throughout the 1840s as, later, it was also by Friedrich Nietzsche, for post-Revolutionary Europe melancholy seemed to be out of date. The term re-emerged in a striking manner in the twentieth century as part of a more general post-romantic disillusionment with modernity and, whether under the banner of melancholy or given the new designation of boredom or depression, silently occupied the centre of philosophical reflection.

During the period of high capitalism and the ascendancy of a self-confident bourgeois culture, when the melancholic shadow of modernity was all but invisible under the brilliant halo of a rapidly developing modern urbanity, new forms of unease that were later thematized as boredom, nonetheless became apparent. A connection appeared between, on the one hand, heightened excitement, greater leisure, wealth, and privilege and, on the other, recalcitrant unhappiness. This bewildering interrelation formed the centrepiece of many of Europe's leading literary works, and challenged in different ways the optimistic hope of continuing Enlightenment. After the staggering success of Goethe's *Sorrows of Young Werther* (1774), *ennui* was installed as the dominant feeling-tone of almost all Romantic writing and, persisting throughout the nineteenth century, culminated in thoroughgoing nihilism.

Maria Edgeworth's *Ennui*, for example, written in 1804 and printed 1809, described a well-established condition and already asked rhetorically who "among the higher classes, whether in the wealthy or the fashionable world … is unacquainted with *ennui.*"[13] This aristocratic disdain for the world, indeed, became a fashionable bourgeois disease in the 1830 and 1840s particularly in Paris.[14]

The striking new characteristic of *ennui* has to do with loss of interest in the world resulting from over indulgence and satiety of every appetite; it is a condition of "luxurious indolence."[15] The over provision of every want in the midst of a world that continually pressed more of everything on exhausted and jaded individuals turned pleasure to disgust. The turning away from the world in *ennui* is a protective reaction against its sheer excess.

The term boredom, in fact, seems to be rare in English before the end of the eighteenth century and was established firstly as an equivalent to *ennui*. Certainly, boredom was the more common term by the time of *Bleak House* and was taken to be a quite general, non-pathological condition of the times and of specific situations as much as a description of a particular individual character type or group culture.[16] And in France, Henri-Frédéric Amiel, whose famous *Journal* is one of the key descriptive accounts of boredom in the nineteenth century, also uses the term synonymously with *ennui*. Significantly in the present context Amiel makes a direct link between boredom and disgust: "This afternoon a homicidal languor seized hold upon me—disgust, weariness of life, mortal sadness."[17] This connection, which was developed in a fundamental way in Baudelaire's poetics of *spleen*, was taken up once again in more recent times by E. M. Cioran.[18]

Structure of Detachment

The genealogy of boredom provides a number of clues both to its fascination and its connection to the idea of waste time. To draw together these themes and provide a more substantial account of the characteristic experiences on which they draw, it is helpful to outline a set of key categorical relations presupposed in most modern discussions of the phenomenon. Boredom, that is to say, is constituted both diachronically in a conjectural history of cultural transformation and synchronically as an element in a structure of feeling, action, and ideas.

Boredom is made meaningful in relation to *interest*. Boredom is the loss or absence of interest in life; in boredom we simply cannot take an interest in anything, or find anything of interest; nothing rouses our interest. Significantly, for modern society, interest has an intimate connection with the passage of time. Interest holds our attention, allows us to persist in any activity, and transforms a world of infinite complexity into an ordered field of activity. Above all, interest accumulates; the pecuniary sense of interest is drawn from a more general sense of the continuous flow of time. Value adheres to time "well spent" in productive activity. Time and selfhood are welded together in interest; they grow together in mutual reinforcement. As such, interest is the antithesis of waste; it is value transformed into the continuous and spontaneous flow of life. Interest is seemingly without foundation and remains fundamentally obscure; it is a primary existential orientation to the world. How is interest possible? Given that it is possible, how can it be withdrawn?

The question is deliberately posed in a Kantian form. In his influential analysis of aesthetic judgement that at first seems remote from any consideration of boredom, Kant proposes the notion of "disinterested interest." It is important that the notion of interest emerges in the context of aesthetic judgement and is, therefore, related to the primal experience of pleasure: "The satisfaction that we combine with the representation of the existence of an object is called interest."[19] Interest, therefore, is always connected with the faculty of desire, and connects the human subject with the world through a feeling of want or lack. The possibility of disinterested interest arises, therefore, in calm contemplation or reflection, undisturbed by the urgency of an inner drive directed towards an external object:

> But if the question is whether something is beautiful, one does not want to know whether there is anything that is or that could be at stake, for us or for someone else, in the existence of the thing, but rather how we judge it in mere contemplation (intuition or reflection).[20]

Aesthetic judgement is possible in a pure representation of the object in relation to which there is nothing "at stake" for the self. This is familiarly called *detachment.*

Whether or not this provides the basis for a convincing account of aesthetic judgement, and it is as the distinctive cognitive orientation required of scientific observation rather than the feeling tone of contemplative beauty that this kind of detachment has become normative, Kant's formulation is suggestive, in the current context, of structural relations through which boredom is articulated in modern society.

Detachment, thus, might be contrasted with "interested interest"; that is with a purposive orientation towards the world in terms of goal-directed activity. This is the kind of instrumental relationship that formed the central theme of Max Weber's account of modernity. Here the object is judged in terms of its suitability as a *means* to the achievement of a specific goal with which the self is identified. Modernity, as understood by Weber and others, involves the increasing dominance of this orientation as the ordinary or normal attitude of life. This orientation can be called *purpose.*

Detachment and purpose can also be distinguished, conceptually and existentially, from "interested disinterest"; a possibility that has been almost completely overlooked in the modern period. At first sight, indeed, it seems to be hopelessly contradictory. Interested disinterest, however, is neglected rather than absent as a real experience. The peculiar experience of *patience* might be invoked here. In patience the subject is self-concerned, but not in relation to any immediately present secular object. Patience is simply absorption in the stream of life; working, as it were, with whatever is presented. Patience accepts what "comes along" and is resigned in the face of a world with which it is nonetheless intimate. Patience is not to be concerned over the contingent facts of experience but, rather, assimilates them in such a way as continually to enrich (interest) the self; to find nourishment in every particular. Patience, of course, is out of fashion; it remains, nonetheless the "resting state" and initial mood of every passion that shakes the self from its complacency. Patience is a state of readiness and openness to profound experiences that cannot themselves become the object of instrumental action.[21]

Boredom, in terms of this categorical structure, can be characterized as "disinterested disinterest." Double disinterest is a relation (or non-relation) to the world for which neither self nor object is of vital concern. Boredom simply is; it is the dull immensity of being, without presupposition or position taking. It is for this reason that Walter Benjamin asks pertinently, "What is the dialectical antithesis to boredom," and wonders if it does not rather stand alone as the "in-itself" of the modern world.[22] Boredom is a limiting condition of experience and might better be thought of as non-experience, or the experience of nothing. Yet, while the peculiar quality of boredom remains obscure in this formal analysis, it provides a partial answer to Benjamin's question. Boredom is correctly seen as the antithesis of interest; but it is not a simple negation, or binary contrast. Boredom is situated, rather, in subtle inter-relation of similarity to, and difference from, detachment, purpose, and patience.

These categorical relations, furthermore, can be related, if somewhat imprecisely, to the genealogy outlined above. *Acedia*, if it is to be given a modern meaning at all, can be grasped as the failure of patience; melancholy may be viewed as the negative side of detachment; and *ennui* understood as loss of purpose. Boredom, thus, comprises several prior elements and experiences, drawn together and given a more general meaning in the context of high modernity. This synthesis is accomplished in, and takes the distinctive form of, waste time.

The Sunday of Life

The synthesis of boredom, so far discussed in a historically schematic and abstract way, is given in a more concrete manner through some of the most striking and significant philosophical, literary, and artistic works of the nineteenth and twentieth centuries.[23] A useful point of departure for understanding these varied discourses can be found in some remarks by Hegel. In a discussion of early-modern German and Flemish painting, finding the novelty and significance of this art in its "utterly living absorption in the world and its daily life," he characterizes its celebrated depiction of a "cheerful, roguish, and comic" peasantry in the memorable phrase "the Sunday of life." Here is a world liberated from want, necessity, and labour: "penetrated by a naïve cheerfulness and jollity" in which "heedless boisterousness" "equalizes everything and removes all evil."[24]

Whatever happened to the Sunday of life? The unprecedented productive capacity of modern society, its overwhelming material abundance, practical comfort, and promise of continuous progress has conspicuously failed to realize even this modest, secular vision of happiness. The Land of Cockagyne remains distant; indeed, it has withdrawn farther into the unattainable realm of myth. The "Sunday of life" has become the everyday of everyday life. For modern life the difference between Sunday afternoon and every other afternoon, between any morning and any afternoon, was gradually but relentlessly suppressed. It is not just that the distinction between work-time and leisure-time has been wrenched from the collective chronology of days and weeks but, more significantly, the quality of experience routinized in work and leisure has become indistinguishable; both are equally boring. The tedium of the factory, the office, and the school, their stuffy, enclosed, and oppressive atmosphere, has been rediscovered in the shop, the museum, and the home, in the cinema, theatre, and on television. Leisure experience, which had been grasped as recreative and reviving freedom from work, proved to be no more or less boring than any other experience. When the phrase "the Sunday of life" recurs in the still optimistic writing of Ernst Bloch in 1920s, it is already intended ironically to refer to the utter emptiness of bourgeois life. And the expression reappears in a bitter mode a generation later, as an expression of the utter disillusionment that is the underlying mood of the postwar world: "*The universe transformed into a Sunday afternoon* ... it is the very definition of ennui."[25] Such a view has been confirmed more recently, for instance, in J. G. Ballard's bleak view of contemporary affluence as "the longest afternoon in the world."[26]

The "Sunday of life," in other words, is an expression for the concealed identity of, on the one hand, the empty unreality of boredom and, on the other, the frantic distraction through which we seek to avoid that same unreality: an identity that is nothing other than the essential experience of modern life. Where Burton and Montaigne recommended work to dispel the melancholy clouding the free enjoyment of modern life, and Hegel could find innocent pleasure in temporary release from toil, high modernity revealed the absolute indifference of every activity in its failure adequately to pass the time. Work and distraction were the same because both stood apart from time as a flowing stream and all too literally actualized a peculiarly western longing for eternity.

This identity had, in fact, already been fully exposed by Blaise Pascal, writing in the mid-seventeenth century. For Pascal, people are wretched and their wretchedness consists above all in an obscure awareness of a state of happiness from which they have been separated, and with which they might still be reunited. But, rather than seek happiness, they cannot resist merely diverting themselves from wretchedness and spend their time on,

> some novel and agreeable passion which keeps them busy, like gambling, hunting, some absorbing show, in short by what is called diversion … That is why men are so fond of hustle and bustle … All our life passes in this way: we seek rest by struggling against certain obstacles, and once they are overcome, rest proves intolerable because of the boredom it produces. We must get away from it and crave excitement … Man is so unhappy that he would be bored even if he had no cause for boredom, by the very nature of his temperament, and he is so vain that, though he has a thousand and one basic reasons for being bored, the slightest thing, like pushing a ball with a billiard cue, will be enough to divert him.[27]

Pascal criticized diversion just because it was successful; "diversion passes our time and brings us imperceptibly to our death."[28]

For many twentieth-century cultural critics and novelists (including Walter Benjamin, Ernst Bloch, Fernando Pessoa, and Albert Cohen), however, the hidden identity of boredom and diversion openly declares itself in the banal spectacle of everyday life. For them all time is waste time; the problem is not the fretful tension of having too much to do and there never being enough time to "fit things in"; rather, however busy we become, and however actively we pursue diversions, time remains distant, impassive, and closed against us. Søren Kierkegaard, in the early 1840s, had already said as much:

> Time passes, like a stream, etc., so people say. That is not what I find: time stands still, and so do I. All the plans I project fly straight back at me; when I want to spit, I spit in my own face.[29]

This is just the period, in fact, which Siegfried Kracauer, one of the most sensitive and insightful observers of Weimar culture, identifies as the point at which boredom (*ennui*) and distraction come together in the emergence of luxury, display, fashion, and hyper consumption among a rich, complacent, and no longer ascetic bourgeoisie. For a class that felt itself called to the urgent task of self-indulgence, "the exhaustion of pleasures," rather than the fatigue of work, was an oddly tormenting condition of life from which "the only distraction lay in distraction … It was characteristic of the time that the object of all forms of entertainment was entertainment only. The sole object was to kill time, not give it a meaning."[30]

The failure of distraction any longer even to disguise boredom—far less to act as its antidote—is illustrated in a striking manner in Georges

Seurat's major paintings, *Bathers at Asnières* (1884) and *Sunday Afternoon on the Island of la Grande Jatte* (1884). These works, as prescient as Kierkegaard's writings, anticipate the leading cultural critics of the following century in depicting the extraordinarily democratic spirit of boredom. It is in the experience of boredom that the exhausted working class, slumped in doleful inactivity, and the distracted bourgeoisie, radiant with terminal glamour, share a valid experience of modernity. Certainly, social classes are kept apart, and Seurat includes the island of Grande Jatte, as a distant view for the disconsolate bathers. But, while paying close attention to the stark differences in comfort, wealth, and style between them, he makes of both a "mosaic of boredom" in which "every object seems to fade into watery tepidness."[31] Here is neither leisure time nor free time, but waste time; time, withdrawn from experience, taking with it the compelling sense of a living reality, and transposing the scene into an ontologically puzzling picture of itself. The image of absolute stillness does not, as it does in Vermeer, capture the plenitude of a lived moment, but serves, rather, to reinforce the conviction that eternity has, indeed arrived; and there is no escape!

Being Bored

The essential connection between boredom and modernity, at once evident and obscure, is the particular focus for one of Martin Heidegger's most important philosophical works. The first part of *The Fundamental Concepts of Metaphysics*, which was conceived as the continuation of *Being and Time* and drafted in the late 1920s but not published until 1978, is devoted to an extraordinary discussion of boredom. Heidegger presents boredom as a "fundamental attunement" to reality which, given his previously elaborated analysis of human being as essentially historical (*Dasein*), has become the characteristic *mood* of contemporary life. Boredom is not, for Heidegger, an orientation or disposition that retains a voluntaristic element; it is not an attitude. He asserts boldly that it is the world that is boring, and the experience of boredom has an essential validity for us. It is not that we turn our back on the world; in boredom, rather, the world withdraws from us and leaves us stranded in a void.

Heidegger distinguishes two common forms of boredom. The first he describes as "becoming bored by something." In an obscure way boredom arises "*from out of things themselves*" and is revealed first of all to the extent that we continually seek to escape it.[32] Above all we seek to "pass the time." But the object that is boring *does not concern* us and remains something to which we remain indifferent. Heidegger arrives at an initial, challenging formulation: "Boredom is that which we seek to drive away by passing the time."[33]

He considers the example of waiting for a train. Waiting is not itself boredom; not every waiting is boring—sometimes it is a condition of keen anticipation, suspense, tension, excitement, and so on. As Benjamin also notes, "we are bored when we don't know what we are waiting for."[34] Nor is boredom identical with *impatient* waiting. Rather,

> our passing the time has this peculiar character of a fluttering unease that brings this impatience with it. For what happens in becoming bored is that our unease while having to wait does not allow us to find anything that could grip us, satisfy us or let us be patient.[35]

When bored and seeking to pass the time, "we do not look at anything in particular because nothing in particular offers itself to us. Indeed the inherent predicament of becoming bored is precisely that we cannot find anything in particular."[36] In this relation time is a "dragging" oppressiveness that keeps itself apart from us and "holds us in limbo." Things are withdrawn from us, in other words, and seem to become part of a quite different stream of time in which we have become stuck.

The second type of boredom is being bored *with* (not by) something. For example, an evening out may seem easy, pleasant, and unoppressive. There is no sense here of withdrawnness and its accompanying arousal of an effort to pass the time.[37] Many conventionally pleasant activities, in fact, actually have the character of passing the time as their essence. In the second form nothing in particular bores us. This type of inconspicuous boredom, which is the most general form of distraction, has the character of "letting oneself be swept along by whatever is transpiring there."[38] This "whiling" away the time—casually "going along" with events—in fact leaves time "standing." In "whiling" we close ourselves off from the "unsettling and paralyzing sound" of time ticking away. The sheer transition of the now is extended, sealed off, and eternalized. Nothing actually happens; there is a sense of pure duration without transition. And this standing time, for Heidegger, is a purely formal self-experience stripped of any particular content. Boredom, here, is "self forming emptiness"; it is oneself that is bored and the incidental events become boring as the bored self touches them.[39]

Heidegger's detailed investigation of these two common forms of boredom, insightful as it is, serves primarily as an introduction to his original and striking account of a third form; profound boredom. Here:

> *indifference* yawns at us out of all things, an indifference whose grounds

> we do not know … a profound boredom draws back and forth like a silent fog in the abysses of *Dasein.*[40]

Profound boredom is "boring for one," rather than boring for me, or you, or us. There is a collective experience of boredom, a mood that settles over an entire period in an undisguised way, and does not arise from a particular situation, or from within the self held in suspension. Here emptiness and being held in limbo are overwhelmingly forced upon one. Boredom here is a general indifference; it is a world for which "*everything is of equally great and equally little worth* … it takes us back to the point where all and everything appears indifferent to us."[41] Indifference "enveloping beings *as a whole*" is, in fact, the only remaining form of wholeness and, as distinct from any notion of positively shared values or beliefs, might be viewed as the distinctive mode of negative integration that sustains contemporary society.

This is "a telling refusal on the part of beings as a whole," and Heidegger is uncompromising in his insistence that it characterizes every aspect of contemporary experience:

> All beings withdrawn from us without exception in every respect (*Hinsicht*), everything we look at and the way n which we look at it; everything in retrospect (*Rücksicht*), all beings that we look back upon as having become and as past, and the way we look back at them; all beings in every prospect (*Absicht*), everything we look at prospectively as future, and the way we have thus regarded them prospectively. Everything—in every respect, in retrospect and prospect, beings simultaneously withdrawn.[42]

It is from within the attunement of profound boredom that a "*single* and *unitary universal horizon of time*" comes into view.[43] Being left empty is nothing other than Being's refusal to be drawn into differentiated forms; or, positively, "our *being entranced by the temporal horizon as such.*"[44]

For Heidegger, as previously for Kierkegaard, human being is inherently interested in its world; so the experience of boredom seems fundamentally to contradict the very possibility of *Dasein*. Heidegger's radical solution is to shift the entire burden of Being into a region prior to all particularity. Boredom, like Pascal's "wretchedness," is a trace or intuition of the withdrawn fullness of Being. Boredom is fascinating because its empty eternity, pale and insubstantial as it is, makes human spirit present in a more vivid way than is open to any interested experience. Profound boredom should not be mistaken for abstract, worldless, and indeterminate Being; it is an ordinary world filled with indifferent content.[45]

Heidegger's view of profound boredom revives and radicalizes a genealogy that serves to designate, for modernity, the borderland between humanity and spirit. And Heidegger is not alone. At the same time that he was formulating his thoughts on the withdrawnness of Being, Alberto Moravia's startling first novel with the precisely forensic title *Time of Indifference* (1929) was published. It was a theme to which he returned throughout his writing life. Unsurprising, then, that he should say: "I recall having suffered always from boredom," a vague and indefinite state of mind he characterizes as "lack of contact with external things."[46]

And the Portuguese writer Fernando Pessoa, who, like Heidegger, had learned everything from Kierkegaard, concurs. His *Book of Disquiet* revels in boredom as the most powerful spiritual solvent of modernity. Boredom is not just democratic in the sense that, as depicted by Seurat, it is open to all, but, more significantly, it advances a radical equality in which every distinction is nullified and every object loses its value as a particular thing. Boredom is the truth of modern life:

> The tedium of the forever new, the tedium of discovering—behind the specious differences of things and ideas—the unrelenting sameness of everything … the eternal concordance of life with itself … all of it equally condemned to change.[47]

This is a strangely comforting, tensionless world; for "to suffer without suffering, to want without desire, to think without reason [is] like being possessed by a negative demon, like being bewitched by nothing at all."[48]

Negation

Boredom is the infinitely plastic medium in which everything becomes possible—the empty coherence of the multiple realities of contemporary life. The radical discontinuity in experience that founded modernity, and which is repeated with every novelty it ushers into existence, is assimilated without disturbance to a familiar world of enduring, banal objects through boredom: "Boredom, extinction, is precisely a continuity in nothingness."[49]

Kierkegaard, not simply anticipating, but surpassing Heidegger, finds in boredom a veritable negative principle of modern society. It is more than withdrawnness and indifference; there is, rather, a genuinely negative relation to the world. Proceeding from "the basic principle that all people are boring," Kierkegaard discovers a principle of existential movement; the exact point of his rejection of Hegel was just that it introduced such a principle into a system of philosophical abstraction. But a system is defined by formal relations and rules of transformation that are bound by the conservation of identities; only in living beings can movement occur, because existence is not something identical with itself. Boredom is in some respects like a philosophical system in its detachment of reality, in its eternity, and disinterest in the world; but on closer inspection it is clear that boredom actually repels, it pushes back:

> This basic principle has to the highest degree the repelling force always required in the negative, which is actually the principle of motion. It is not merely repelling but infinitely repulsive … It is very curious that boredom, which itself has such a calm and sedate nature, can have such a capacity to initiate motion.[50]

Kierkegaard links boredom and the culture of the present in the idea of "demonic pantheism." The spirit of modernity, thus, is profoundly negative:

> Pantheism ordinarily implies the qualification of fullness; with boredom it is the reverse: it is built upon emptiness, but for this very reason it is a pantheistic emptiness, but for this very reason it is a pantheistic qualification. Boredom rests upon the nothing that interlaces existence (*Tilvaerelsen*); its dizziness is infinite, like that which comes from looking down into a bottomless abyss … demonically possessed by boredom in an attempt to escape it, one works one's way into it.[51]

Nature of Disgust

Disgust is also conjugated in the negative mood as an involuntary recoil and turning away from the world. Yet, the spasm of rejection lacks conviction and remains incomplete. Where in boredom "You hover above yourself" (Kierkegaard) and remain bound to the pure exteriority of empty time (eternity), so in disgust the world is "held at arm's length," at a distance that can neither be ignored nor overcome, and withdraws into the pure exteriority of filled space (substance).

And modernity is disgusting as well as boring. Melancholic vapour oozes from every pore and seeps relentlessly into the air. Baudelaire discovers in the "sad, lethargic world" (*Parisian Dream*) that "Foul yellow mist had filled the whole of space" (*The Seven Old Men*).[52] The body of society, and the individuated human being, are smeared with the grime of modernity. Sooty grime fills the air, clings to every surface, and shrouds the world in gloomy indistinction. Fog, the dense miasma of productive life, thickens the air with melancholic substance. The clarity with which objects had appeared and occupied the foreground of life, dissolve into a background grown heavy with disgusting residues. Being withdraws into remote impenetrability; locked in to the sluggish effluence of time and space. The air, the breath of life and medium of speech, turns fetid and rank. The moods of boredom and disgust appear, first of all, in the heavy, thick atmosphere of the industrial city.[53] The subject is overwhelmed and, finally shaken from dreamy torpor, recoils in disgust.

Modernity, that is to say, holds the subject in suspension, at once poised between spirit and nature and crushed between boredom and disgust. The world of modernity is constituted not only through the pure dimensionality of time and space but also through their choked and viscous waste forms; which is to say, through the modalities of boredom and disgust.

Offence

In a provocative gesture Søren Kierkegaard described Christianity as an offence to modern values of universal reason and morality. Authentic Christianity is the experience of faith that is ultimately without foundation in historical evidence or scientific understanding of nature and, worse, is scandalously lacking ethical justification. When it comes to the ultimate possibility of salvation Christianity fails humanly to distinguish between the righteous and the sinner. In another sense the gloomy indifference and stupefying triviality of contemporary secular culture is also offensive; it is disgusting. In Kierkegaard's work boredom and disgust come together in the banal culture of the "present age."

Disgust is "taking offence" at something. In the contemporary experience of life, offence is taken at waste, especially at waste space; that is, at space filled and made unusable by vile mess. Offence is also taken at religious ideas and practices, usually other people's (Kierkegaard is exceptional in being offended by his own, corrupted religion), other cultures, other languages, other food, jokes, clothing, conventions of everyday life, and so on. Disgust, in other words, is excited by a vast range of circumstances that have little in common apart from the contemptuous manner of their rejection by individuals and groups complacent in their self-regarding superiority.

The associative power of disgust is formidable, but what might rather too readily be termed its primal form is invariably connected with bodily discharges and decaying life. In a pioneering paper, Andras Angyal provides a valuable psychological account of disgust that correctly identifies offence as its key defining criteria. Disgust is a reaction to specific objects related primarily to wastes, products defined broadly as "coming out of the body." Importantly, Angyal points out that the disgust reaction is not related to an evident threat:

> There is no clear notion present of a damage or harm which the disgusting object would inflict upon one's person. Disgust is directed rather against close contact with certain objects, implying mainly the fear of being soiled ... the nucleus of the disgust reaction, the main threat against which disgust is directed, is the oral incorporation of certain substances.[54]

Disgust is recoil from contact and the possibility of incorporation, and is characterized by prohibition on touching and rituals of cleansing: "Wastes, to our minds, are something base, and contact with them is experienced as debasing, degrading rather than harmful."[55] In this regard disgust is related to the sacred, with which, in fact, it is often implicated.[56]

Angyal's analysis has been influential in the more recent, wide-ranging studies of Rozin.[57] According to these researches, disgust is roused primarily by the threat of "psychic incorporation" linked to body-image and selfhood rather than to any simple notion of corporeality. These observational and laboratory studies support the important and still neglected work of Silvan Tomkins, for whom disgust is "recruited to defend the self against psychic incorporation or any increase in intimacy with a repellent object."[58]

Notwithstanding the insights gained in these and related psychological studies, the overwhelming physical aspect of disgust, the spasm of rejection, remains puzzling. Paul Ricoeur, thus, pertinently remarks:

> What resists reflection is the idea of a quasi-material something that infects us as a sort of filth, that harms by invisible properties, and that nevertheless works in the manner of a force in the field of our undividedly psychic and corporeal existence. We no longer understand what the substance-force of evil, the efficacy of a something that makes purity itself an exemption from defilement and purification, an annulment of defilement, could be.[59]

Ricoeur, without resolving this issue, elaborates on the significance of defilement for the development of a symbolism of evil. Ricoeur sees an essential link between the experience of the impure (disgust) and the idea of human fault (defilement). The experience of disgust is the primal encounter with, and thematic constitution of, evil; that is to say, disgust is elemental rejection of moral-physical waste.

However, by viewing defilement as a starting point for the emergence, in turn, of experiences of sin, guilt, and responsibility, Ricoeur makes the problem yet more intractable. In this view, the contagious potency of disgusting matter, whatever its origin, should conspicuously be reduced, if not entirely eliminated, in modern society. Yet, the acquisition of spiritually developed experiences of human fault coexists with the persistence of disgust, and, equally, disgust is "developed" also into contempt for, and degradation of, innumerable "out groups," unorthodox ideas, transgressive conduct, and so on. Older anthropological accounts of disgust, going back to nineteenth-century ideas of "primitive thought," sympathetic magic, and empiricist associative psychology, which are invoked by Rozin and continue to play a significant implicit role in many contemporary discussions, also leave obscure the strange power of abjection that is seemingly untouched by modernity.

Disgust is disturbing rather than painful or harmful; it is an experience involving debasement, contamination, impurity and defilement. Both the nature and origin of disgust, however, remain obscure.

Logic of Space

The peculiar character of disgust may be clarified by an approach that considers its essential connection to the logic of *space*. Disgust, a "literal pulling away from the object," is, first of all, an experience of space.[60] But the logic of space is complex and disgust must be considered in relation to a number of distinct spatial relations and qualities.

Categories: In celebrated texts, Mary Douglas provides a suggestive starting point.[61] She argues that our idea of dirt is compounded of two things: care for hygiene and respect for conventions.[62] Dirt "is the by-product of a systematic ordering and classification of matter"; it is *matter out of place*, and pollution is a consequence of disorderliness, the result of disruption and nothing in itself.[63] In this perspective disgust emerges as a mechanism for boundary maintenance; which is to say, for keeping things in their "proper" place. The aversive feeling of disgust is directed towards misplaced objects, motivating actions directed at restoring order.

Additionally, and even more significantly, powerful feelings of disgust are roused by ambivalent and marginal objects. The ultimate arbitrariness of any system of classification is, to varying degrees, obscured by the "naturalness" of the disgust reaction to marginalia that threaten its integrity. This is pre-eminently the case with respect to the boundaries of the human body, which constitute a "natural symbol" for any classificatory unity. The creation and maintenance of bodily unity and integrity, thus, play a pre-eminent role in the ordering of human experience generally and, above all, in organizing views of the world as a structure of differentiated and individuated objects. We take care, therefore, to protect the body from "foreign" matter invading or adhering to it and, equally, distance ourselves from substances issuing from the body:

> We should expect the orifices of the body to symbolize its especially vulnerable points. Matter issuing from them is marginal stuff of the most obvious kind. Spittle, blood, milk, urine, faeces or tears by simply issuing forth have traversed the boundary of the body. So also have bodily parings, skin, nail, hair clippings and sweat.[64]

Corporeal secretions and wastes have to be reclassified as other than, and unambiguously outside, experiential bodily unity. Corporeal coherence and social integration, that is to say, are *both* conventional orders that depend upon continuous boundary maintenance and, ultimately, both are precarious unities constituted through and protected by disgust reactions. Consequently, both bodily unity and the categories of everyday culture that appear to us as "natural" unities have to be learned. The extent to which this is the case is evident from the absence of disgust reactions in babies to their own waste and any other "filth," and, in a different but related perspective, the special attachment young children have for soft toys, blankets, and other "transitional objects" not yet symbolically detached from their bodies.[65]

The recoil of disgust throws us back onto (that is, into) ourselves, and restores the world to its accustomed order. In this context the "body politic" is no metaphorical extension of a wholly natural object. And social hygiene (one thinks, of course, of "ethnic cleansing") follows "logically" from exclusive identification with a social group or cultural tradition.

Substance: William Miller, alternatively, provides a view of disgust in terms of substance; it is reaction to a particular kind of stuff rather than the property of a distinctive relation. Awareness of being defiled is rooted in bodily experiences, but for Miller it is a particular kind of body that is at issue:

> When our inside is understood as soul the orifices of the body become highly vulnerable areas that risk admitting the defiling from the outside. But when our inside is understood as vile jelly, viscous ooze, or a storage area for excrement the orifices become dangerous as points of emission of polluting matter, dangerous both to us and to others.[66]

Disgusting material issues from inside us, and is disgusting even while it is inside us. It is not excretion that makes shit disgusting; it *is* disgusting. And other things are disgusting to the extent that they are like interior matter:

> It is not that animal bodies decay, excrete, suppurate, and die that makes these processes sources of disgust to us: it is that ours do. The animals that

> disgust us do not disgust us as animals but because they have characteristics that are disgusting: sliminess, slitheriness, teemingness ... What the animals remind us of, the ones that disgust us—insects, slugs, worms, rats, bats, newts, centipedes—is life, oozy, slimy, viscous, messy, uncanny life.[67]

Life, that is to say, that ceaselessly transforms, gives birth, decays, and putrefies. Disgusting matter lacks form; it is at once *hyletic* substance from which spring all living beings, and the nonbeing to which they return.

Form: Disgust also bears importantly on the notion of form in a more particular aesthetic sense. In a remarkable book Winfried Menninghaus highlights the almost wholly unacknowledged role of disgust in the development of modern aesthetics.[68] Disgust is central to the emergence of the aesthetic in the eighteenth century, particularly for Moses Mendelssohn's pioneering and influential *Letters Regarding Literature* and remains fundamental throughout its development. For Mendelssohn, as for Kant, disgust is accorded a certain categorical supremacy. Disgust registers the absolutely real, which is not only exempt from the doubt infecting all our sense based knowledge of the world, it is also immune to the "aesthetic illusion" through which that world is beautified. Irrespective of its mode of presentation, disgust is *always* disgusting; a picture or memory, for example, of something disgusting *is* disgusting. Disgust is unconditional; it is a pure, elemental experience rather than a form synthesized from simpler or more elementary elements. Thus, while disgust is reactive, it gives immediate access to a reality that aesthetics can approach only indirectly.

Disgust emerges within the structure of appetites and drives, and like boredom is not defined in terms of a simple binary opposite. Thus, while disgust might be thought as the negation of appetite, it is more accurately characterized in terms of a double relation. It is tempting at this point to elaborate a structural space related to Freud's analysis of drives and corresponding to the pseudo-Kantian structure of "disinterested interest." Such a project, however, lies beyond the scope of this essay, so the formulation of disgust as "repulsive repulsion" is advanced simply on intuitive grounds. *Disgust* is being repelled by something repellent. This seems obvious, but it is also possible to be attracted by something repellent (starving people might eat vermin, or human flesh); this might be termed *perversity*. It is also possible to be repelled by something attractive (an extravagant display of rich food right after eating); this is *satiation*. While the normative condition of *desire*, corresponding to "interested interest," is to be attracted by something attractive.

Yet there is something that binds us even to repulsive repulsion, just as there is something that binds us to the double negative of boredom. What is the nature of this boundedness? This is the still unresolved issue at the root of our fascination with waste.

Phenomenon: Some revisions to the pseudo-Kantian approach to an understanding of waste space are suggested by the neglected work of Aurel Kolnai.[69] Writing in the modern phenomenological tradition established by Edmund Husserl his essay, first published in 1929, remains unequalled in terms of subtlety and rigour. Kolnai's painstaking description of the actual experience of disgust is exemplary in its precision and careful differentiation of related states with which it is often confused. Thus, for example:

> In connection with disgust one thinks of the two opposite poles of contempt and nausea. Hate and even anger are less bound up with bodily phenomena than is disgust, in spite of the more violent physical phenomena attendant upon anger. This is because the sensuous impressions involved in disgust and the suggestion of physical reaction of vomiting play here a more essential role in a way that is much more specific and concrete than the raging, kicking, and throwing which may arise through anger. Every kind of disgust—even moral disgust—is, even if not more physical, still somehow more physiological than anger. On the other hand disgust should not be confounded with nausea as such, nor with the tactile sensations connected with the latter (especially with the sticky, the clammy, or the lukewarm). However there is a sense in which fear is even more bound to states of the body than disgust: for once any feeling of fear, in contrast to disgust, has been related to something physical, it thereby becomes also directed to the body itself and to its "intactness."[70]

Kolnai draws attention to the spatial aspect of disgust, particularly the sense that space is constricted and filled; disgusting stuff is distressingly close. The *proximity* of the disgusting means that, even without direct physical contact, it "feels" slimy and warm: "disgust is distinguished also by the way in which it adheres to the object which is its cause."[71] Kolnai's philosophical interest in disgust lies particularly in the peculiar way in which it binds together the subject and object of ordinary experience in the modern world. In disgust:

> the location of the subject and the disgust-arousing object merge together, as it were, to constitute what might, somewhat crudely, be called a harmonious unity … one particular aspect of proximity constitutes … the character of disgust. This is its will to be near, its non-self-containedness, or, as I would rather put it, its shameless and unrestrained forcing itself upon us. The disgusting object grins and smirks and stinks menacingly at us.[72]

In fact, "everything disgusting somehow 'adheres' to the subject, embraces it with its proximity, with its miasma."[73] In this perspective dirt is not just matter out of place, or the insides turned outwards; it is a cloying residue of life. "This mushy mixture of particles without qualities attaches itself, as it comes into being, to the skin of human beings as they go about their activities."[74] At the same time disgust involves:

> a turning-outwards and a certain "seizing" of the object … There is without doubt a certain invitation hidden in disgust as a partial element. I might say, a certain macabre allure … not only is an aversion to its object characteristic of disgust, but also a superimposed attractedness of the subject towards the object.[75]

Kolnai's insightful account suggests that a precise dialectical opposite to disgust is to be found in the special character of "craving." Disgust arises from overindulgence and superabundance which is connected in a particular way with "the disgusting character of bodily secretions … (which) carry the motif of 'an indecent surplus of life'."[76]

Boundaries: A somewhat different approach to the experience of space links disgust to the emergence and development of modernity as a project of human self-determination. This is connected with the transformation of the world view from a "closed world to an infinite universe."[77] Space and time are here conceptualized as infinitely extended, uniform, and indifferent modalities within which objects appear, are differentiated, located, and interact. A corresponding process of infinitization also works itself out in the cultural history of modernity. The infinitization of the human, however, cannot be conceptualized as an empty medium and *requires* concrete images of freedom and the unbounded. Disgust is one such concrete image; it involves the transgressive abolition of a limit, rather than simply crossing over a territorial marker. The passage from inside to outside overwhelms the boundary and gives rise to a new experience of unbound space.

The peculiar dialectic of boundaries is revealed in the ambivalence of disgust. It is not just a zone of ambiguity; but a relation of bound to *unbound* space. Boredom, similarly, marks a boundary between bound and unbound time. The boundary is not a line of demarcation, as if drawn on a page, between two ontologically equivalence regions of space and time. The non-transitive character of the boundary follows from the incommensurable character of the realities constituted within its distinctive regions; rather than marking a simple a simple division within uniform space, the boundary becomes an unapproachable limit. From within either zone, that is to say, the possibility of the other ceases to exist and, consequently, the world appears monistic.

The space of disgust is unbound, and filled with a continuous, mucous substance. The fierce contagion of disgust is a result of the unbound quality of its space. This is why we cannot pull ourselves away from disgusting substance, we cannot free ourselves from it because it is space filling and, once we become disgusted, we occupy and are bound to its space. A small stain on our clothing, if it disgusts, is transformed into an absolute; defiling the entire experiential world. Just as, for the bored, all existence is boring.

Conventionality

Disgust is the force behind the mechanism of repression that enables conventions to be established and society to take form. Freud provides fundamental psychological insight into this mechanism which, in turn, Norbert Elias has relied upon for the development of a sociological understanding of disgust. Yet disgust, for Elias, remains curiously unexamined. It is assimilated to "shame" and taken uncritically as a "given" of modern life; it is "the emotion of civilization, and socialization."[78] Thus, while offering a detailed historical account of the transformation of manners, the effective mechanism of socialization to new conventions of bodily self-control and public interaction, is obscure.

The role of disgust in founding behavioural conventions is equally obscure, but seized in a provocative and subversive way, in the heterodox writings of Mikhail Bakhtin, Georges Bataille, Michel Foucault, and others associated with surrealism.[79] For them, transgression, which is signalled by disgust, is an open invitation to plunge wholly into a world beyond: beyond shame, disgust, and the self-limiting conventionality of modern bourgeois society. Disgust is a barrier, not a boundary, and one we need not fear to cross.

Julia Kristeva, whose wide ranging work includes separate full length studies of both boredom and disgust, builds on these insights and suggests an important reformulation. The abject (disgust) is not the repressed self, nor an *other* self, but other *than* the self. The abject is not repressed, though it furnishes the motive force for repression; it does not require repression because it cannot enter directly or fully into ordinary experience. Ordinary experience is, after all, founded in and through modalities of extended, empty, and uniform time and space; whereas in disgust, as in boredom, these modalities are radically transformed and with them the possibility of experience itself. As substance the abject is disgusting, formless, engulfing, and overwhelming; it is radical exteriority. The abject as a feeling does not correspond to an object:

> What is abject is not my correlative, which, providing me with someone or something else as support, would allow me to be more or less detached and autonomous. The abject has only one quality of the object—that of being opposed to *I*.[80]

The abject is, literally, the excluded and, thus, "the place where meaning collapses."[81] It is a relation to the superego not the ego, thus: "to each ego its object, to each superego its abject."[82] There is here, Kristeva writes:

> A massive and sudden emergence of uncanniness … radically separate, loathsome. Not me. Not that. But not nothing, either. A "something" that I do not recognize as a thing. A weight of meaninglessness, about which there is nothing insignificant, and which crushes me.[83]

There is nothing here of the logic of desire; the "I" is expelled. The abject is not a sign. The abject is not the ambiguous borderland; it is outside, but outside the world for which that border has a meaning or significance; it is another world, not another and dangerous region of this world. This is the connection between abjection and the sacred.

Spirit, Nature, Waste

Boredom is the borderland of spirit realized as waste time; disgust is the borderland of nature actualized as waste space. Boredom and disgust are the moods (modes/modalities) of wasted time and space. Boredom is the experience of eternal indifference; disgust is the experience of primordial actuality. Boredom and disgust are the margins of humanity: the boundaries between the bound and the unbound. In disgust, as in boredom, though constituted as cultural unities related in essential ways to the culture of the present, we confront something absolute. The subject encounters itself as pure exteriority as other than a self.

Boredom is not time but eternity; time that piles up and cannot be "got through." The subject cannot close with, and identify itself with, time as boredom; yet it is bound to it and cannot avoid it, cannot remove itself wilfully, or distract itself from the eternal, which forces itself on the subject, turning viscous and opaque the sheer fluid transparency of time. Boredom is the unbound to which the subject is bound; undoing the subject's identity, undoing its capacity for selfhood, ultimately transforming the subject into pure, indifferent objectivity. Disgust is the experience of waste space; space completely filled with amorphous, impenetrable substance. Waste space is infinitely dense, so that, in disgust, as in boredom, the subject is immobilized and helpless.

Heidegger's work, resting on Kierkegaard's insights, provides a philosophically subtle account of the *spirit* of boredom as, to paraphrase Marx, "the spirit of a spiritless age." Time is spirit, and waste time is spiritless spirit. In a less developed form, and resting on Freud's insights, Max Scheler suggests that disgust (shame) arises when the person is overwhelmed by an upsurge of animal life.[84] Shame is a reminder of our creaturely existence, which we can never wholly transcend or master. And more recently Martha Nussbaum, in a different perspective, describes disgust in similar terms: "The core objects of disgust are reminders of mortality and animality, seen as pollutants to the human."[85] Disgust, thus, clings particularity to phenomena connected with mortality and decay.[86]

Boredom and disgust are human experiences of the non-human; symptomatic residues of Spirit and Nature that modernity, positing itself as self-experiencing autonomous humanity, constituted as exterior, transcendental, and impenetrable realities. Waste, thus, poses a real existential challenge; it is an inescapable encounter with the inexperiencable. Or, to put it otherwise, for modernity waste is thematized as the *debris* of premodernity; and the premodern, fallen into oblivion, is ideal, invisible waste. Garbage, that is to say, cannot be acknowledged in the contemporary world. It should not exist, not because we prefer a sanitized and continuously engaging environment, but because such an environment is an essential element of our self-understanding of the present. In fact, we deny that they do exist. The overwhelming abundance of real garbage, like the stubborn persistence of boredom, is surprisingly easy to ignore. We literally turn a blind eye, not from wilful negligence, but because, for the world of modernity, emerging from the depths of empty space and time as a wholly and uniquely human experience, such realities, literally, cannot exist. In so far as we demand of ourselves and our world the coherence of systematic modernity, we do not have to ignore garbage; for us, it simply does not exist. But inasmuch as we fall into lethargic indifference, or recoil from the mess of the world in involuntary repugnance, we confound the myth of modernity and, albeit it in degraded forms, once again participate directly in the life of Spirit and Nature.

Modern humanity creates itself by drawing a boundary around itself, but cannot wholly succeed in laying hold of the region of self-determination thus opened up. Contemporary culture, in fact, indicates the extent to which this hard won region has been surrendered to the ambivalence of phenomena that now press in on both sides and fill the world of human experience with nothing but "absolute banality."[87] The genius of the present is to merge boredom and disgust into the absolute waste that has become the new measure of our humanity. Kitsch, glamour, celebrity, and the virtual no longer inhabit the shadowy margins of modernity; they have taken centre stage. And this is simply because, to produce ourselves as absolutely free and self-sufficient human beings, we have created a society that produces the absolute, and self-sufficient: that is, waste. The product of contemporary society is waste. The inconsumable is not its image or emblem; it is its truth and ultimate reality. Waste cannot be consumed, cannot be used, it stands outside us just as we now stand apart from God and Nature.

We are fascinated by waste because it shows us *what* we are, not *who* we are, or would like to be, or would like *to be like*. Waste reveals to us in the most direct way the reality and fate of a being who demands autonomy, and heroically refuses to accept anything less. Boredom and disgust are not metaphors and do not rely on mental associations. To the contrary, they are variant forms of the *longed-for* of western metaphysics: which is to say, absolute being, the indubitable. But the veritable "thing-in-itself" is encountered as radically other than ourselves, other *than* a self. The radical alterity of waste is encountered as sheer exteriority—as *alien*. The overwhelming "givenness" of boredom and waste manifest neither self-presence nor the "saturated" qualities of intensified phenomena. Their irresistible obscurity is the token, rather, of a borderline personality: the dark plenitude of humanity in its dreadful vulnerability.[88]

Notes

[1] Ernst Bloch, *Traces*, trans. Anthony A. Nassar (Stanford, CA: Stanford University Press, 2006), p. 87.
[2] Julia Kristeva, *Powers of Horror: An Essay on Abjection*, trans. Leon S. Roudiez (New York: Columbia University Press, 1982), p. 2.
[3] Siegfried Wenzel, *The Sin of Sloth: Acedia in Medieval Thought and Literature* (Chapel Hill, NC: University of North Carolina Press, 1967).
[4] *Ibid.*, p. 5.
[5] Anne Larue, *L'autre mélancolie;* acedia, *ou les chambers de l'esprit* (Paris: Hermann, 2001); Michael L. Raposa, *Boredom and the Religious Imagination* (Charlottesville, VA: University of Virginia Press, 1999).
[6] Jean-Louis Chrétien, *De la fatigue* (Paris: Editions de Minuit, 1996).
[7] Larue, *L'autre mélancolie.*
[8] Raymond Klibansky, Erwin Panofsky and Friz Saxl, *Saturn and Melancholy: Studies in the History of Natural Philosophy, Religion, and Art* (London: Nelson,

1964).
[9] Juliana Schiesari, *The Gendering of Melancholia* (Ithaca, NY: Cornell University Press, 1992).
[10] Reinhart Koselleck, *Futures Past: On the Semantics of Historical Time,* trans. Keith Tribe (Cambridge, MA: MIT Press, 1985); Pierre Manent, *The City of Man,* trans, Marc A. LePain (Princeton, NJ: Princeton University Press, 2000); F. R. Ankersmit, *Sublime Historical Experience* (Stanford, CA: Stanford University Press, 2005).
[11] Wolf Lepenies, *Melancholy and Society,* trans. Jeremy Gaines and Doris Jones (Cambridge, MA: Harvard University Press, 1992).
[12] Michel Foucault, *Madness and Civilization,* trans. Richard Howard (London: Tavistock, 1967).
[13] Maria Edgeworth, *Castle Rackrent and Ennui* (Oxford: Oxford University Press, 1992), p. 144.
[14] Siegfried Kracauer, *Jacques Offenbach and the Paris of His Time,* trans. Gwenda David and Eric Mosbacher (New York: Zone Books, 2002).
[15] Edgeworth, *Castle Rackrent and Ennui*, p. 144.
[16] Orrin E. Klapp, *Overload and Boredom* (Westport, CT: Greenwood, 1986).
[17] Henri-Frédéric Amiel, *Amiel's Journal,* trans. Mrs Humphrey Ward (London: Macmillan, 1889).
[18] E. M. Cioran, *A Short History of Decay* (New York: Arcade Publishing, 1998).
[19] Immanuel Kant, *Critique of the Power of Judgment,* trans. Paul Guyer (Cambridge: Cambridge University Press, 2001), p. 90.
[20] Kant, *Critique of the Power of Judgment*, p. 90.
[21] Harvie Ferguson, "Patience: The Critique of Pure Naïveté," in Robert Perkins, ed. *International Kierkegaard Commentary*: *Eighteen Upbuilding Discourses* (Macon, GA: Mercer University Press, 2003), pp. 265-88.
[22] Walter Benjamin, *The Arcades Project,* trans. Howard Eiland and Kevin McLaughlin (Cambridge MA: Harvard University Press, 1999), p. 105.
[23] Elizabeth S. Goodstein, *Experience Without Qualities: Boredom and Modernity* (Stanford, CA: Stanford University Press, 2005).
[24] G.W.F. Hegel, *Aesthetics: Lectures on Fine Art,* vol. II, trans. T. M. Knox (Oxford: Clarendon Press, 1998), pp. 886-87.
[25] Cioran, *A Short History of Decay*, p. 22.
[26] J. G. Ballard, *Cocaine Nights* (London: Flamingo, 1997), p. 218.
[27] Blaise Pascal, *Pensée,* trans. A. J. Krailsheimer (London: Penguin, 1966), p. 414.
[28] *Ibid.*, p. 414.
[29] Søren Kierkegaard, *Either/Or,* ed. and trans. Howard V. Hong and Edna H. Hong, vol.1 (Princeton, NJ: Princeton University Press, 1971), p. 26.
[30] Kracauer, *Jacques Offenbach and the Paris of His Time,* p. 122.
[31] Ernst Bloch, *The Principle of Hope,* trans. Neville Plaice, Stephen Plaice and Paul Knight, vol. II (Oxford: Blackwell, 1986), p. 873.
[32] Martin Heidegger, *The Fundamental Concepts of Metaphysics,* trans. W. H. McNeill and N. Walker (Bloomington, IN: Indiana University Press, 1996), p. 83.

[33] Heidegger, *The Fundamental Concepts of Metaphysics,* pp. 92-93.
[34] Benjamin, *The Arcades Project,* p. 105.
[35] Heidegger, *The Fundamental Concepts of Metaphysics,* p. 94.
[36] *Ibid.*, p. 99.
[37] *Ibid.,* p. 111.
[38] *Ibid.,* p. 117.
[39] *Ibid.,* p. 127.
[40] *Ibid.,* p. 77.
[41] *Ibid.,* p. 137.
[42] *Ibid.,* p. 145.
[43] *Ibid.*
[44] *Ibid.,* p. 147.
[45] Rudi Visker, *The Inhuman Condition: Looking for Difference after Levinas and Heidegger* (Dordrecht: Kluwer, 1995)
[46] Alberto Moravia, *Boredom,* trans. Angus Davidson (New York: NYRB, 1999), pp. 5, 16
[47] Fernando Pessoa, *The Book of Disquiet,* trans. Richard Zenith (London and New York: Penguin, 2002), p. 110.
[48] *Ibid.*, p. 208.
[49] Søren Kierkegaard, *The Concept of Anxiety,* trans. Reidar Thomte (Princeton, NJ: Princeton University Press, 1980), p. 133.
[50] Kierkegaard, *Either/Or,* p. 285.
[51] *Ibid.*, p. 291.
[52] Charles Baudelaire, *The Flowers of Evil,* eds, Marthiel and Jackson Mathews (New York: New Directions, 1989).
[53] Alain Corbin, *The Foul And The Fragrant: Odour And The French Social Imagination* (London: Picador, 1994); William A. Cohen and Ryan Johnson, eds, *Filth: Dirt, Disgust, and Modern Life* (Minneapolis, MN: University of Minnesota Press, 2005).
[54] Andras Angyal, "Disgust and Related Aversions," in *Journal of Abnormal and Social Psychology,* 36 (1941), p. 394.
[55] *Ibid.*, p. 397.
[56] Franz Steiner, *Taboo* (London: Cohen and West, 1956).
[57] Paul Rozin, J. Haidt, and C. R. McCauley, "Disgust: the Body and Soul Emotions," in *Handbook of Cognition and Emotion,* ed. Tim Dalgleish and Mick Power (New York: Wiley Blackwell, 1999); Paul Rozin, J. Haidt, and C. R. McCauley, "Disgust," in *Handbook of Emotions,* ed. M. Lewis and J. Haviland (New York: Guilford, 2000).
[58] Silvan S. Tomkins, *Affect, Imagery, Consciousness Volume II: The Negative Effects* (New York: Springer, 1963), p. 233.
[59] Paul Ricoeur, *The Symbolism of Evil,* trans. Emerson Buchanan (Boston, MA: Beacon Press, 1967), pp. 25-26.
[60] Tomkins, *Affect, Imagery, Consciousness Volume II*, p. 128.
[61] Mary Douglas, *Purity and Danger* (London: Routledge, 1966); Mary Douglas, *Natural Symbols* (New York: Pantheon, 1970).

[62] Douglas, *Purity and Danger,* p. 17.
[63] *Ibid.*, p. 48.
[64] *Ibid.*, p. 145.
[65] D. W. Winnicott, *Playing and Reality* (London: Routledge, 1971).
[66] William Ian Miller, *The Anatomy of Disgust* (Cambridge, MA: Harvard University Press, 1997), p. 2.
[67] *Ibid.*, p. 50.
[68] Winfried Menninghaus, *Disgust: Theory and History of a Strong Sensation,* trans. Howard Eiland and Joel Golb (Albany, NY: SUNY Press, 2005).
[69] Aurel Kolnai, *On Disgust,* ed. Barry Smith and Carolyn Korsmeyer (Chicago and La Salle, IL: Open Court, 2004).
[70] *Ibid.,* p. 32.
[71] *Ibid.,* p. 39.
[72] *Ibid.,* p. 41.
[73] *Ibid.,* p. 55.
[74] *Ibid.,* pp. 55-56.
[75] *Ibid.,* p. 42.
[76] *Ibid.,* pp. 54-55.
[77] Alexandre Koyré, *From the Closed World to The Infinite Universe* (Baltimore, MD: The Johns Hopkins University Press, 1957).
[78] Norbert Elias, *The Civilizing Process* (Oxford: Blackwell, 1999), p. 434.
[79] See, for example, Mikhail Bakhtin, *Rabelais and His World,* trans. Hélène Iswolsky (Bloomington, IN: Indiana University Press, 1984); Georges Bataille, *Eroticism,* trans. Mary Dalwood (London and New York: Marion Boyars, 1987); Michel Foucault, *The History of Sexuality,* vol. 1, trans. Robert Hurley (London: Allen Lane, 1979).
[80] Julia Kristeva, *Powers of Horror: An Essay on Abjection,* trans. S. Roudiez (New York: Columbia University Press, 1982), p. 1.
[81] *Ibid.,* p. 2.
[82] *Ibid.*
[83] *Ibid.*
[84] Max Scheler, *Person and Self-Value,* trans. M. S. Frings (Dordrecht: Nijhoff, 1987).
[85] Martha C. Nussbaum, *Hiding from Humanity: Disgust, Shame, and the Law* (Princeton, NJ: Princeton University Press, 2006), p. 99.
[86] *Ibid.,* p. 90.
[87] On "absolute banality", see Vladimir Jankélévitch, *l'Aventure, l'Ennui, le Sérieux* (Paris: Editions Montaigne, 1963), reprinted in Vladimir Jankélévitch, *Philosophie morale* (Paris: Flammarion, 1998).
[88] Jean-Luc Marion, *In Excess: Studies of Saturated Phenomena,* trans. Robyn Horner and Vincent Berraud (New York: Fordham University Press, 2004).

Chapter Fourteen

The "Litter Nuisance" in Twentieth-century Britain

Timothy Cooper

The Litter Nuisance is not a trivial thing ... there is no room in this island for untidiness, and we must behave accordingly
—SCAPA, "The Litter Nuisance," 1929

In recent years, academics have discovered an interest in the social, cultural and political ramifications of waste and pollution. Much of this interest has focused on the role of waste and disposal in the making of modernity.[1] Among historians, Chris Otter has recently studied the relationship between waste and modernity in the context of the Foucauldian idea of "governmentality."[2] Otter has argued that behind modern techniques for controlling the visual and olfactory environments, particularly of urban spaces, lay the "liberal bourgeois" desire to restructure the moral and political behaviour of the working class. Otter describes how nineteenth-century utilitarian liberals saw or smelt in urban refuse and rubbish evidence that the poor "could not be trusted with social and political freedom."[3] Before political enfranchisement would be possible, therefore, it was first necessary to cleanse the slums and produce a hygienic, self-governing bourgeois mentality within the poor. This characterisation of filth and waste as formative concepts in the formation of characteristically "modern" modes of governance has not found universal acceptance. Mark Jenner has introduced objections to this "modernization" interpretation, instead arguing that it is first necessary to historicize adequately the role of filth and waste within early modern society.[4] Indeed, Mary Douglas's classic study, *Purity and Danger*, confirms that ideas of pollution and taboo are anthropologically universal, and therefore suggests that their ideological operation is far from confined to the modern era.[5]

In this essay, however, I wish to stress a slightly different objection to Otter's approach. Employing the concept of "governmentality" to environmental debates in a manner that fails to reflect the contested meanings applied to the concepts of filth and waste, and which assumes the success of such methods of social control, is flawed both theoretically and empirically. Historical uses of "governmentality," despite their interpretive utility, often over-emphasize the oppressive content and power of such intellectual formulations of waste and disposal, and neglect their hidden liberatory potential as symbolic challenges to the social order. Waste, in other words, can operate not only as a trope within the positive rationality of the bourgeois order, but also as a negative idea that undermines it. It is waste's bi-polar character as a point of contest between the claims of instrumental reason and its own often contradictory effects that make it such an crucial point of study for those interested in the origins of environmental politics. Otter largely ignores the way in which waste has been employed to challenge hegemonic ideas of hygiene and efficiency. This essay attempts to probe these questions in the context of the making of the "Litter Nuisance" in the twentieth century. On one level, the litter debate certainly involved attempts to invoke public compliance with a bourgeois vision of appropriate behaviour within both urban and rural environments. However, the results of self-education and self-discipline in creating the desired behaviour were ultimately perceived to be inadequate. The "litter-lout" proved to be a recalcitrant and a repeat offender, not to be eliminated simply through bouts of "environmental education." Ultimately, in the absence self-discipline, it was necessary to resort, much against initial liberal instincts, to criminalisation.

The character of evolving environmental politics in twentieth-century Britain has to be understood in the context of the simultaneous development of a democratic polity and mass consumerism. Increasing geographical mobility and affluence presented a challenge to the dominant aesthetic conceptions of the environment established in the latter half of Victoria's reign. Litter, the flotsam and jetsam of the new affluent culture, threatened new ways of disordering and defiling either preserved natural landscapes or newly cleansed townscapes. To preservationists litter represented the material manifestation of human behaviour that was both anti-natural and anti-social, and demanded a response in defence of an idealized version of landscape. The debate on what shape this response ought to take was influenced by the contemporaneous development of "behavioural" approaches to human psychology.[6] Early anti-litter campaigners adopted two visions of human nature that were in creative tension with one another. Both positions took the view that littering was a bad habit, the result of human conditioning in urban environments where concern for the welfare of nature was unable to develop fully. However, the first variant, which might be broadly termed the "liberal" view, emphasized the malleability of human character and believed that environmental education was the most practicable solution to the litter problem. The second response might be termed "pessimist," and was more authoritarian in tone; it emphasized the irredeemable nature of recidivist, urban litter-louts, and advocated legislative compulsion as the only realistic means of compelling submission to bourgeois ideals of cleanliness and order. These two views were, of course, never completely divergent. Their basic understanding of the origins of the litter problem was the same. However, it was the growing influence of the "pessimist" viewpoint that ultimately propelled both preservationists and parliament toward adoption of legislative compulsion in the Litter Act of 1958. The Litter Act provided the first nationally applicable statutory punishments for "litter-louts", replacing the range of local bylaws that had traditionally dealt with the problem. Alongside the Clean Air Act of 1956, the Litter Act represented the logical culmination of the trajectory taken by preservationist sentiment in Britain since its inception.[7] The litter debate contributed to the formation of many of the characteristic elements of contemporary environmental discourse. It led to a behavioural paradigm that dominated early environmental politics, in which the role of state and voluntary organisation was to mould a self-willed, habitually anti-environmental "public" to manifestly good ends. This ensured that responses to environmental problems were framed as issues of either education or of discipline, a fact convenient to the growing retail and packaging industries.

The emergence of the "Litter Nuisance" can be dated with precision to the years immediately following the Great War. The earliest reference, discovered by the present author to litter as carelessly discarded street refuse, was made in 1901 by William Weaver in response to a paper on "The State of London Streets," given to the *Royal Society for the Promotion of Health*. Weaver complained that "Since the multiplication of newspapers, opening of stores, and the vast importation of foreign goods wrapped in paper, the litter that accumulated in the streets of London was enormous."[8] Although Weaver referred to the problem as the other "great nuisance on the roads" after house refuse, this kind of complaint does not appear to have been extensive before 1914. The cleansing departments of local authorities were still very much concerned with removing large quantities of mud, dust and manure from the roads, rather than post-consumer discards.[9] It was not until the immediate post-war period, therefore, that litter became a topic of widespread concern in the letters pages of the press and among social commentators.

Nonetheless, many of the intellectual roots of the litter problem lay within the mainstream of nineteenth-century preservationist argument.[10] Historians and literary critics have outlined the importance of the picturesque in developing Victorian perceptions of the desirable landscape, and the subsequent emergence of an argument that rural landscapes required protection from the forces of urbanisation and industrialisation.[11] Ruskin's defence of Thirlmere, and Victorian campaigns for preservation of the commons, were illustrative of the political consequences of this consciousness. As Harriet Ritvo has recently pointed out, preservationist claims that "a nation should have some say in the disposition of significant landscapes" had radical implications.[12] However, during the early twentieth century, the utilitarian impulses within Victorian preservationism gradually came to the fore. Environmental explanations of poverty, social decay and disease had long been promoted by preservationists in order to indicate the ways in which the health, labour-efficiency, and self-discipline of the working class would be aided by access to green spaces.[13] The Commons Preservation Society, Metropolitan Public Gardens Association and the National Trust all emphasized that the working class would benefit physically and morally from exposure to open spaces.[14] Before the First World War, they encouraged and legitimated working-class exploration and use of "nature," but did so in the context of an ideological commitment to bourgeois standards of behaviour.[15] Consequently, while organisations like the Commons Preservation Society maintained a position on the "land question" that was arguably sympathetic to the ideas of radicals, they nonetheless also preserved a hierarchical view of who should police behaviour within "common" spaces, and what the nature of that behaviour should be. Following the Great War, the open spaces movement moved away from concern with issues of health and access to the land, and emphasized a much more authoritarian view of the need for surveillance of public spaces.

Of course, complaints against urban litter were also common, and they provided an important point of coalescence between the concerns of rural conservationists and the defenders of urban amenity.[16] In the city, the litter nuisance occurred among concentrated populations and in a political context in which the cost of street cleaning was an important factor in local public spending. After 1875, refuse disposal had been progressively municipalized and removed from the *ad hoc* responsibility of private contractors and scavengers, becoming a professionalized and specialized activity within towns across Britain. Street cleansing also became increasingly technologically sophisticated and, along with the decline of the horse as a major source of motive power, ensured that pollution was increasingly displaced from the ground to the air. Despite the continued existence of slums, poverty, unemployment and ill-health, urban spaces were, in purely visual terms, increasingly clean places. Consequently, the physical presence of litter in urban spaces became more palpable at the same time as the increasing purchasing power of working class and white-collar workers, and changes in the nature of consumption and salesmanship were increasingly the sources of litter.[17] Tinned, pre-packed foods and bottled drinks proved popular; cigarettes, with accompanying wrapping, were more readily available; newspapers became cheaper. In the competition to attract buyers, packaging became an increasingly important means of persuading shoppers to make choices from a growing superfluity of goods. As *The Times* despairingly observed in 1925:

> whatever means are taken … it will be as much as all of them together will be able to do to suppress litter; for the simple reason that the materials for it multiply every year. Newspapers are printed and read in ever increasing numbers; a newspaper is now almost the natural concomitant of every human being daily, Sundays included, and Sunday is a great day for excursions. Likewise, many more cigarettes are in packets are sold than formerly; they and chocolates, also in packets, are carried on the persons of many more than formerly. These three materials, together with paper bags for wrapping food, are almost ubiquitous in the summer months, and for a long time they resist the forces of disintegration.[18]

Matthew Hilton has argued that the early twentieth-century debate on the morals and ethics of consumerism articulated growing middle-class concern with the "breaking down of external indicators of social hierarchy."[19] The "litter nuisance" should be added to the list of middle-class concerns with status hierarchy.[20] David Matless has shown that the conservatism of post-war English preservationism needs to be understood in the context of the increasing purchasing power and spatial mobility of the working-class. Widespread "leisure" and engagement with consumer culture partially dissolved the traditional symbols of social hierarchy. The "Litter Habit", as *The Times* termed it in 1935, grew alongside the phenomenon of public holidays: "People who never thought of getting away a few years ago now go off as a regular thing. And the multiplication of transport facilities feeds the practice indefinitely."[21] Consequently, as Matless argues, the ways in which opportunities for leisure were used became increasingly important as signs of individual merit, status and citizenship: in short, they were means of reproducing hierarchy.[22] The knowledge of how to use leisure, and of how to interact with nature, became symbolic of social and civic responsibility.[23] This was the lesson of works such as W. Boyd's, *The Challenge of Leisure* (1936) and H. Durant's, *The Problem of Leisure* (1938), both of which emphasized the need for newly empowered social groups to learn how to employ their countryside leisure in a spiritually and intellectually meaningful manner.[24] There remained however widespread middle class suspicion that the newly empowered urban working class was either too ill-versed in the ways of the country or too uncivilized to appreciate properly the opportunities they had been granted. Developments such as the motor-car and motor-bus made urban workers "free of the country," but such a sudden liberation meant that they could not necessarily "be expected to know how to behave there."[25] This idea famously found its most extreme expression in the somewhat jaundiced views of C.E.M. Joad. In *The Untutored Townsman's Invasion of the Country* (1945), Joad looked back on changes in the inter-war countryside and among the sources of despoliation he found litter: "for democracy, when it goes to call upon the country, insists on leaving its visiting card behind it." Joad's explanation for all this littering was explicitly behaviourist: urban dwellers had acquired bad habits as a consequence of being cut-off from the country. How were those bred up in the "wilderness" of the slum or suburb expected to possess the taste required to appreciate nature without the impulse to "deluge it with litter."[26]

The question arises why litter became such a significant issue. Part of the answer lies in the symbolic power of waste as a challenge to restricted property ownership. While popular desire for access to natural amenities was increasingly seen as legitimate, there was an accompanying perception among property owners that it would be dangerous to leave the public with the idea that granting access meant they had also obtained rights of possession. The "litter bug" was one who was sometimes seen to be asserting such a right:

> there is apparently a perverted feeling, which occasionally finds a public expression, that a place dedicated to the people has not really passed into the possession and usufruct of the people unless they are allowed to do exactly what they please in it, to trample freely and unrestrainedly upon any verdure that may be growing there, and to leave behind any litter or rubbish which they are too lazy to conceal or to take home with them.[27]

When activities such as rambling continued a long tradition of asserting use rights that implied limits to the rights of property, it was far from irrational for property owners to fear that the public might actually begin to use the countryside as if it were their own.[28] G. Parker and N. Ravencroft have convincingly argued that during the twentieth century landowners conceded the right of access to the public only in return for the acceptance of their right to determine "proper" land-use.[29] Preservationist organisations co-operated in emphasizing this contractual notion, that conceding access to the land was not to be interpreted as granting the freedom to behave as one wished there. Richardson Evans, Chairman of Scapa Society, clearly expressed the politics of the case:

> A large percentage of holiday visitors of every class enjoy and respect the charms of nature. But, for most, the fascination lies in the sense that they are free to go where they like and do as they please. They have not been disciplined into the knowledge that licence is not liberty, and so they leave behind them the *debris* of picnics, the discarded news-sheet, the emptied cigarette box.[30]

One way in which the "Litter Nuisance" contributed to framing this control was through encouraging the formation of *ad hoc* local environmental police forces in the shape of the "countryside wardens." A number of areas experimented with countryside wardens between the wars, including Cheshire and Derbyshire, where voluntary wardens patrolled lands to which landowners had granted access, "checking the litter nuisance, and damage to wild plants, walks, trees and so on."[31]

The relationship between the "Litter Nuisance" and property is highlighted on the rare occasions that we encounter the voice of the "litter lout." *Litter in the Royal Parks* stated that one of the most common responses to park keepers challenging litter-droppers was, "It's our own Park and we can do what we like in it."[32] In the debate on the Litter Bill in 1958, the somewhat blimpish character of Sir Henry Studholme, MP for Tavistock, recounted an encounter with the "litter lout":

> Not many years ago, I found a family picnicking on my drive. I was glad that they should, but they were about to leave behind them a mountain of mess. As politely as I could, I asked whether they would mind picking it up. The father replied, "if you can afford a place like this, you can afford to have somebody pick up the mess". That may not be a very prevalent attitude but it is a deplorable one.[33]

While Studholme may have invented this encounter, it is nonetheless indicative of the fact that littering should not necessarily be accepted as the mindless vandalism that its opponents presented it as. The role of litter and the activity of littering may have much to do with contesting the rights of private property and government to regulate behaviour, either in the countryside or urban spaces. Significantly, one of the aims of the 1958 Litter

Act was to extend protection to private landowners, who were not previously covered by local government by-laws. By the 1970s the majority of prosecutions under the Act actually came from private individuals.[34]

The early anti-litter movement certainly had an elite, propertied character. The beginning of formal anti-litter campaigning can be dated to 25 April 1924. On that date, a Mansion House protest meeting was convened under the auspices of the "Scenery Preservation Committee" to promote an anti-litter campaign against:

> the ever increasing nuisance and injury to public amenities caused by the growing habit of littering public thoroughfares, parks, commons, and the countryside generally with waste paper, cigarette-boxes, empty tins, picnic refuse, and other rubbish.[35]

There was a strongly anti-socialist political atmosphere to the meeting. Among those in attendance were the art historian, mountaineer, and Conservative MP, Sir Martin Conway; the anti-socialist Liberal, Hilton Young; and the free-trade Liberal, Henry Lunn. Lord Curzon, Sir Baden-Powell, and Conservative MP, Sir William Joynson-Hicks also attended. The composition of delegates emphasizes the relationship between property, authority and anti-litter sentiment. These connections were further displayed in July 1925, when King George V weighed in with an attack on littering during the public dedication of the recently purchased Ken Wood in Hampstead.[36] The precise achievements of the Mansions House meeting remain unclear, but the Scenery Preservation Committee claimed that its own "crusade" against litter aroused "great interest in all parts of the country", and its main effects certainly appear to have been the promotion of local anti-litter initiatives.[37]

Probably the most significant expressions of anti-litter activism in this period were the local anti-litter leagues, which emerged in a number of areas between the wars. Of these, the Surrey Anti-Litter League was the largest and best organized. The Surrey League, which was founded by Shere Women's Institute in July 1928, grew rapidly, incorporating members from neighbouring branches of the WI.[38] By 1930, it claimed an astonishing membership of 24,000 people. The League operated in a self-disciplinary fashion, similar to that of Victorian temperance societies, asking members to sign a "pledge" never to leave any litter.[39] It also distributed pamphlets, emptied litter bins and brought prosecutions against "litter louts" under County Council by-laws. The Surrey League took a particular interest in a place known as Newlands Corner, an area of wood and grassland near Guildford that was popular with picnickers and walkers.[40] They also established "watchers" whose role was to police the area and to "prosecute those who leave litter about"; in 1929 Godfrey Wyndham was one unhappy victim of these "watchers" and he pleaded guilty to what he called "a technical offence," and a local magistrate fined him one pound for dropping orange peel.[41] This rural anti-litter activism found an urban counterpart in the London Anti-Litter League, which became part of the London Society in 1930.[42] The London League expressed similar concerns with litter, but emphasized neighbourly duty and orderliness rather the "despoilation" of natural scenery: "the majority of people in London have no conception of their duty towards their neighbours in the matter of the cleanliness and tidiness of public thoroughfares," the London League proclaimed.[43]

The anti-litter leagues, and similar *ad hoc* organisations, relied heavily on public educational events such as exhibitions and demonstrations. The London League adopted educational tactics such as anti-litter exhibitions, which were exported to areas considered in special need of such education, such as the new Becontree housing estate where, "they suffer a great deal from this particular nuisance."[44] Demonstrations were not restricted to the organisational entrepreneurship of the leagues however. On 6 September 1930 Lady Montagu of Beaulieu led an anti-litter demonstration in the New Forest. This involved the collection and ritual public burning of litter. The litter, collected in the New Forest by an assorted group of volunteers, was ceremonially incinerated to the sound of the local Girl Guides singing the ditty:

> There are lots of people everywhere
> When they take their meals in the open air
> Who either don't think or else don't care,
> And leave all their rubbish behind them

On occasion such public burnings included an effigy of the "litter lout," as occurred at an anti-litter demonstration at Newby Moor in the West Riding on 27 September 1930. After the Moor had been cleared of rubbish, there was a ceremonial burning of the "Litter Lout," who, it must be said, had the last laugh:

> A fearsome creature who had arrived in a bath chair oozing the most horrible kinds of litter from each of his sack clothed limbs. Having been condemned to a fiery death, he was duly placed upon the flames. But even at the end the litter lout dies hard, and his final affront to the moor took the form of a disconcerting blast of charred paper, caught in a sudden blast of wind, which it took the combined efforts of the whole party some considerable time to recover.[45]

Similar "burnings" took place at various times elsewhere.[46] The mimicry of Fifth of November celebrations is suggestive of the subversive role attributed to the "litter-lout," who now found himself in the company of Guy Fawkes. These sources also demonstrate the importance of women, especially elite women, in anti-litter demonstrations. The language of anti-litter activism stressed a civic role for women that reinforced gendered concepts of domestic order and cleanliness, demonstrating, as Caitríona Beaumont indicates, that equal suffrage and citizenship could co-exist with the traditional female virtues of domesticity and the rural way of life.[47]

There was a strong tendency, particularly in rural anti-litter activism, to assume that the urban environment encouraged the development of false forms of behaviour when translated into a rural context. Urban dwellers, by which was implied the urban working class, were believed to have been habituated to litter through "public indifference to seemliness and order," which was "the greatest impediment to establishing a national habit of tidiness."[48] H. E. Cookson thought there were three types of litter lout: recidivists who simply enjoyed littering from what Cookson termed "moral perversity"; those whose environment mitigated against a proper appreciation of nature, "who having been nursed in the slums and starved of natural beauty sin involuntarily … as they are acclimatized to ugliness"; and, thirdly, there were the lazy and selfish who "violate the beauty of nature, which they see but do not feel, though they flock in hundreds to her shrines."[49] In each case, it was assumed that "education" or, perhaps more accurately, habituation to the correct mode of behaviour was the cure. This concern with the acquired environmental habits of the urban working class in rural spaces had a counterpart in discussion of the urban litter problem, which tended to emphasize civic duty, neighbourliness, citizenship and good order. Throughout, a behaviourist language of "habituation" can be discerned:

> nowadays we all live very close together, and the man who will not consider other people's convenience is a nuisance. Civilization starts when we begin to consider other people's rights and happiness, and goes from better to better as that habit becomes more general. There is nothing better than to cultivate the mind of the good citizen. And thoughtlessness is the chief enemy.[50]

It was partly because of this focus on habituation, education and self-discipline, that the anti-litter movement initially eschewed the idea of pursuing parliamentary legislation as illiberal and ineffectual. Indeed, where local by-laws were in place, and had been enforced, the experience had often proved less than rewarding. As *The Times* observed in 1925, "Prosecutions taken under the by-laws which have been obtained by some local authorities are looked upon as a cumbrous and not necessarily successful method of handling the problem."[51] "Legislation is of little avail," argued Sir Martin Conway, "A sound public opinion must be created with the assistance of educational authorities, Boy Scouts, Girl Guides, village institutes and every association that can be drawn in to help."[52] He repeated this view at the Mansion House meeting of 1925:

> What they had to do was create a public opinion against the practice [of littering]; to cause it to be considered a thing that was "not done"—make it something to be ashamed of.[53]

Established organisations like the Scapa Society took a similar behaviourist view of the malleability of the human character, arguing that schools should be involved in teaching children tidiness: "The habits of older people are formed already, but those of children are still in the making. [Education] is therefore probably the most effective method of all."[54] The propensity to focus on the education of children reflected the assumption that adults were largely irredeemable because the urban environment had already negatively formed their behaviour. "Nothing will teach these strangers to the country the decency which the people who live in or love the country are scrupulous to observe."[55] The social reformer and environmental activist, Mary Higgs, whose books included the psychological study *Evolution of the Child's Mind*, emphasized the achievements of her Beautiful Oldham Society in establishing a concern for environmental aesthetics among children in local Oldham schools. She believed that something like the temperance pledge could be employed in schools to develop what she believed were innate tendencies to admire natural beauty that were otherwise stunted under urban conditions.[56] In Oldham's schools, she claimed,

> generations of children have been trained to preserve and not destroy flowers and trees, given a chance in the schools to grow bulbs admitted to the annual spring flower show, and encouraged to exhibit nature drawings.[57]

For Higgs "training" people to be averse to litter was part of nurturing the nature habit in general.

Calls for a programme of public education certainly had an influence on official circles. The institutional epitome of bourgeois fantasies about the reforming influence of education, the British Broadcasting Corporation, broadcast regular appeals to bank holiday-makers not to drop litter.[58] Yet, whether the anti-litter movement's programme of public indoctrination had any actual effect on behaviour is difficult to assess. By the early 1930s, the Scapa Society felt able to report that progress in habituating the public in aversion to litter was being made "and that people are beginning to feel self conscious when they throw away their paper bags."[59] In 1934, a census of litter undertaken in Oxford appeared to show that the amount of litter left had declined.[60] But in accounting for this decline, three causes were elaborated: anti-litter propaganda; new anti-litter by-laws, introduced by Berkshire County Council in 1929; and the increased provision of litter bins.

However, while the anti-litter movement was largely content to focus on education during the inter-war years, and could arguably point to some progress, there were always plenty of pessimists who doubted the idea of a malleable human psychology, and who propounded a more aggressive approach. Ivor Brown, writing for the *Manchester Guardian*, best summarized the pessimists' view:

> The truth of the matter is that all our campaigns of persuasion about litter have failed. By royalty and notables, by the leaders of democracy, in the newspapers, on the wireless, in the schools, on noticeboards, the appeal to decency has been constantly made for years past, but we have preached to the converted and the stubborn ears have not been reached. It is time that the community began to protect itself instead of making a sweetly reasonable supplication to the oafs and louts.[61]

A particular difficulty was perceived to be the local variations in enforcement caused by reliance on local by-laws. If the right habits were to be absorbed, the lessons taught had themselves to be uniform. Howard Hazell asked whether national legislation should not deal with the problem of certain spaces being covered by local by-laws while others were not: "a uniform law for the whole country would inculcate habits of tidiness, particularly when they are enforced by fines."[62] This was education not by reason and the acquisition of good habits, but by aversion therapy. By 1929 the idea of legislation, "applicable to the whole country instead of only to isolated districts," was gathering support from both *The Times* and the Chairman of the Scapa Society, Lawrence Chubb, a veteran of the late Victorian preservation movement.[63] Some even went so far as to argue that nature herself had made man a litter bug, and that, as a consequence, the liberal educative effort was a waste of effort.[64] Growing scepticism with the capacity of persuasion to remould behaviour encouraged the post-war campaign for legislative intervention.

There were a number of reasons for the revival of interest in the litter nuisance in the post-war period. One of the most important changes was the greater direct involvement of the state in environmental issues following the National Parks and Access to the Countryside Act 1949.[65] The new National Parks Commission received powers to make by-laws covering rural areas under its control and sponsored a new "Country Code" that gave "Leave no Litter" a place among ten regulations on right use of the country. Although it was not a penal code, the Commons Preservations Society recognized its moral force, and suggested that those who failed to follow it "failed in citizenship."[66] Through direct involvement in countryside preservation, the state absorbed many of the basic assumptions of earlier preservationist organisations, including the idea that litter-louts exhibited the behavioural patterns of urbanites possessed of an inadequate knowledge of the country: "frequently families out for an afternoon or day excursion who may have little knowledge of, or respect for, the countryside."[67]

Concern over the litter problem was further focused by the Coronation of 1953, which drew a great deal of unfavourable comment on the amount of litter left by the viewing crowds. That the nation had, at a critical public moment, so badly failed to represent itself to the outside world with the necessary ritual decorum was seen by government as a cause for serious concern. Litter and public discipline became drawn into the discussion of national decline. A report by the Ministry of Works' Departmental Committee on Litter in Royal Parks observed the damage done to a nation's reputation by its own insufficiently orderly inhabitants:

> We have only to travel to certain foreign cities to realize that litter is no problem where high standards of living are accompanied by a parallel development of civilized habits and where public opinion makes an offender realize that to drop litter in public places is a social offence. Much of a nation's reputation is built up of impressions of a way of life of its capital. Litter harms this reputation and is often at its worst where it can harm most. London, and particularly its central open spaces, carries the responsibility of presenting British life to a growing number of visitors from overseas.[68]

While recognizing that "waste material increases with the development of higher standards of living and hygiene," the Departmental Committee also somewhat incongruously concluded that that the packaging industry itself had "no moral responsibility for the litter scatterer."[69] Overall, the Departmental Committee maintained a "liberal" stance, and remained against either the introduction of summary fines or a campaign of prosecutions, imploring instead the need for an "emphasis on constructive ideas:

on the preservation of the beauty of the parks, and the dignity and majesty of ceremonial."[70]

Education remained the main plank of preservationist approaches to litter in the immediate post-war years; systematic critique of the causes of litter in an affluent society remained largely absent. This was particularly remarkable in the context of the revolution in retailing practices occurring in the 1950s. The dramatic changes in packaging methods and uses associated with this in turn transformed the nature of the post-consumer waste stream. Yet instead of challenging the packaging, retailing and advertising industries which were constructing the "throwaway society," litter campaigners remained transfixed by the myth of an irresponsible mass of litter-louts corrupted by an urban environment that bred contempt of both human society and nature. This perhaps explains the apparently incongruous willingness of some anti-litter campaigners to forge an alliance with the consumer society's apparently successful methods of persuasion. If one could not reason with the litter-lout, maybe advertising could subconsciously instil the requisite degree of self-control. It had, after all, apparently borrowed from the lessons of behavioural psychology with immense success:

> It is a well-tried maxim, as any advertiser will tell you, that if you wish to plant an idea in the minds of the public so firmly that it will take root and produce results in action, you must bring it to their attention so often that consciously or subconsciously, or both, it "registers" in their brain cells and influences their behaviour in the desired direction.[71]

This kind of bald assertion that environmental education ought to ally itself with the methods of the advertising industry was unusual, but it represented a logical further development of preservationist ideas.

Although education would retain its importance in anti-litter campaigning, the post-war era also saw increasing political influence accrue to the pessimistic view that legislative compulsion was necessary. Increasingly there was, what the Royal Commission on Common Lands in 1958 called, a "scatter-gun" identification of litter-louts among property owners who were "tempted to blame all alike."[72] The development of an indiscriminate pessimism about the behaviour of an increasingly classless public composed of irresponsible consumers apparently raised new concerns. Little more than a decade after the war, the first legislation imposing a uniform penalty on those casting litter was passed. Although a number of Acts, including the Police Act (1847), the Local Government Act (1933), and the National Parks Act (1949), enabled a range of local and national authorities already enforcing penalties on litterers, the application of by-laws was patchy and their powers did not apply to private land. Where such by-laws were in force, they often only became effective when civic organizations were willing to bring prosecutions. The Peak District National Park relied on voluntary wardens drawn from the Ramblers Association to police its land, asking members of the public to take away their litter.[73] Pressure for legislation partly came from these civic organizations. The Women's Institute led the foundation of the Keep Britain Tidy Group in 1954, following a London conference.[74] This new national campaign group led a range of organizations in successfully arguing for a statutory penalty.[75]

The first attempt to introduce an Act of Parliament was made in 1956 by the Conservative MP, J.E.B Hill. His private member's Bill would have made littering an offence throughout England and, importantly, it included private land. The bill provided for a £10 fine to be imposed by magistrates. In arguing for the Bill, Hill raised the familiar arguments about the results of affluence and indiscipline on human behaviour. Motor car travel "means more and more litter," he claimed; increased wrapping and the concern with hygiene meant "much greater litter potential."[76] His central contention was, however, that education had failed. Despite all the efforts, the "litter lout" was still a threat against which the "ultimate pressure . . . is some form of control or deterrent." At this stage there was still strong opposition by the Conservative government to the idea of national statutory intervention in the litter problem, partly because of the Conservative Party's post-austerity commitment to decontrol. This opposition ensured that Hill's Bill did not become law; a similar Bill, proposed by a Conservative colleague, Rupert Speir, failed in April 1957, when the Government raised objections to the duplication of existing by-laws. The idea of compulsion, it was claimed, clashed with Conservative notions of individual freedom. Responding to the bill in debate, the Parliamentary Secretary to the Minister of Housing and Local Government, R. Bevins, argued that a Bill that "makes it a criminal offence to leave a piece of paper in somebody's back garden or somebody's field and which gives numerous persons the right to demand names and addresses under penalty," infringed Conservative ideals of English liberties and the right to privacy, it established "a principle which is totally unacceptable to the government."[77] The government's position was reversed however in the period preceding May 1958, when it supported a third Bill proposed by Speir. This reversal of position was a result of a strenuous press campaign, led by groups such as Keep Britain Tidy and the Campaign for the Preservation of Rural England, and a growing consensus among MPs in favour of the legislation.[78] The Litter Act 1958 introduced a maximum tend pounds fine, enforceable by magistrates, for littering on lands subject to public access. The Act brought very tangible benefits to the Keep Britain Tidy Group, which received a £1,500 per annum grant from the government to assist in publicising the Act's contents.[79] By 1971, the Tidy Britain Group operated a network of 6,000 school committees that offered a range of environmental education to children.[80] By the 1980s, while still retaining its charitable status, it also received official recognition as the "national agency for litter abatement," and attracted an annual income of £500,000 from government and private industry.[81]

Obviously, what the Litter Act did not succeed in doing was to eliminate litter. In 1967, the same year as Keep Britain Tidy's first National Anti-Litter Week, Alan Major could still complain of Britain being seen as "the untidiest filthiest country in Europe."[82] In 1971, the Countryside Commission issued a report on *Litter in the Countryside* in which all the old concerns were rehearsed. Some changes were apparent however. The most important of these was the emergence of packaging as a concern, which increasingly tended to redirect attention from the litter-lout to the packaging industry. This reflected the changing nature of post-war environmental politics, especially the emergence of a radicalized environmentalism. New environmental organizations such as Friends of the Earth placed far more emphasis on the entropic and polluting effects of waste on ecological systems than had previously been the case. They departed significantly from traditional concerns with the visual impact of litter. This ecological and critical emphasis placed far more blame on the manufacturer than the litter-lout. As Andrea La Moury observed:

> The way to avoid packaging becoming litter, many manufacturers believe, is to educate the public and punish them heavily for offences, but amenity groups and local authority associations seem to think that if the amount of non returnable packaging is substantially reduced, the volume of litter will automatically fall.[83]

In the wake of the Oil Crisis, as finite resources became a central concern of both environmentalists and policy-makers, the packaging industry found blaming the litter bug an increasingly inadequate response to environmental challenges now based on concerns over the sustainability of high levels of consumption. New environmental lobby-groups were relatively sanguine about litter: unpleasant it may be, but it did not reflect the essential problem of too few consuming too much:

> The problem of litter, while significant, must not be allowed to distort the overall perspective when considering packaging and its problems. The packaging industry is fond of focusing attention exclusively on litter as

> "the number one environmental problem," to quote one advertisement. It is, of course, nothing of the kind. Litter is unaesthetic, a nuisance and from time to time locally dangerous. But the use of resources in packaging is a much more serious problem.[84]

For the first time, the packaging industry was faced with a situation in which they were seen to be one of the sources of the litter problem. In response, the industry moved quickly to reinforce existing behavioural explanations of littering. In May 1972, at a conference on "Packaging and Litter," hosted by the Secretary of State for the Environment, Peter Walker, the packaging industry's representative, John Gordon, presented a paper entitled, "Education, example and enforcement."[85] Arguing for "analysis, not emotion" and that "packaging is essential," Gordon asserted the utility of plastics packaging in providing hygiene, convenience and value. Non-returnable packaging was not only essential but unavoidable, he argued, and recent concerns with recycling were misplaced. Rather than making a scapegoat out of the packaging industry, he put the blame on "an anti-social behaviour pattern, in which people improperly discard unwanted material."[86]

> [Litter] is a problem caused not by packaging—although it must be admitted that packaging features largely in its total mix—but by people. It is caused by the thoughtless, selfish, inconsiderate actions of individuals whose behaviour in public is often far removed from their habits at home.[87]

The Keep Britain Tidy Group supported an analysis that fitted their traditional explanation of littering: "without the human link packaging would not become litter."[88] Litter was "refuse in the wrong place" and Keep Britain Tidy's role was to "to inculcate a positive social message into the minds of people, both young and old."[89] The Conservative MP, Eldon Griffiths, attending a conference on packaging organized by the Plastics Institute in 1974, similarly defended the uses of plastic packaging. "Packaging is often the scapegoat," Griffiths suggested, "but the real culprits are people"; waste simply needed to "be satisfactorily disposed of."[90] In the "Age of Environment," behavioural explanations of litter found a new use as a defence of packaging from the systemic critiques of environmentalists.

This short essay has attempted to provide a brief outline of how the debate on the litter nuisance developed during the twentieth century. It rejects viewing litter, and other environmental concerns, simply through the lens of "governmentality." If anything, the history of litter suggests the enduring difficulty of constructing the self-disciplining subject. Despair over the achievement of education, of which so much was initially expected, forced a turn to openly disciplinary practices of government, invoking law and fines and negating liberal bourgeois ideas of individual civic responsibility. This history of litter reflects the inherent political and ideological possibilities of waste. Litter was always potentially oppositional; it could be invoked (or misinterpreted) as an act of rebellion against the rights of property, or as evidence of the environmental failures of capitalist industry. To exercise control over the structure of ideas surrounding litter was undoubtedly crucial: to ensure that litter represented individual moral failure, rather than systemic crisis, was crucial to ensuring its validity to the instrumental rationality of preservationist organizations, government and capitalist industry. Redefined as "environmental crime," litter and graffiti still offer presentiments of disorder and social decay. As the House of Commons Environmental Audit Committee noted in 2004:

> While increasing levels of litter and graffiti lead at first, through complacency, to more of the same and to augmented levels of general anti-social behaviour and low-level crime, it seems well understood that continuing local degradation of the environment will lead ineluctably to more serious crime. There does indeed appear to be a link between litter and graffiti on the one hand and drug-peddling, muggings and burglary on the other.[91]

Again the authority of law is invoked as solution, now in terms of "zero-tolerance" policies:

> Of course enforcement of the law is key. Much of the fight against litter and graffiti can be carried out at a low level. Casual litterers can be humiliated and shamed into better behaviour and, if necessary, fined—but to do this requires those willing and able to do this difficult duty.[92]

Yet, attempts to persuade or enforce cleanliness have always been less than successful, and the idea of what constitutes "litter" is itself unstable. The story of why individuals littered is clearly an important one here, but unfortunately it is difficult to access through the sources. It is clear that littering had the potential to be an act of resistance to authority and its politics should be the subject of further study. Similarly, environmentalism introduced new ways of conceptualizing waste that challenged the primacy of litter within environmental discourse, and demanded a response. Litter—along with waste and hygiene more generally—are not just weapons in the armoury of liberal-bourgeois "governmentality," but also offer possibilities to those who exist outside, or oppose these structures.

NOTES

[1] For some examples, see E. Cockayne, *Hubbub, Filth, Noise and Stench in England, 1600-1770* (London: Yale University Press, 2007); W. A. Cohen and R. Johnson, *Filth: Dirt, Disgust and Modern Life* (Minneapolis, MN: Minnesota, 2005); M. Gandy, "The Paris Sewers and the Rationalization of Urban Space," *Transactions of the Institute of British Geographers*, 24 (1999), pp. 24-44; C. Hamlin, *Public Health and Social Justice in the Age of Chadwick: Britain 1800-1854* (Cambridge: Cambridge University Press, 1998); D. Laporte, *History of Shit* (Cambridge, MA: MIT Press, 2002); J. Scanlan, *On Garbage* (London: Reaktion, 2005); S. Strasser, *Waste and Want: A Social History of Trash* (New York: Henry Holt, 2000).

[2] C. Otter, "Making Liberalism Durable; Vision and Civility in the late-Victorian City," *Journal of Social History*, 27, 1 (2002), pp. 1-15; "Cleansing and Clarifying: Technology and Perception in Nineteenth-Century London," *Journal of British Studies*, 43, 1 (2004), pp. 40-64. For a discussion of the relationship between environmentalism and "governmentality" see T. Luke, "Environmentality as Green Governmentality," in *Discourses of the Environment,* ed. E. Darnier (1999), pp. 121-151.

[3] Otter, "Making Liberalism Durable," p. 3.

[4] M.S.R. Jenner, "Civilisation and Deodorisation: Smell in Early Modern English Culture," in *Civil Histories: Essays Presented to Sir Keith Thomas,* ed. P. Burke, B. Harrison and P. Slack (Cambridge: Cambridge University Press, 2001), pp. 127-144.

[5] Mary Douglas, *Purity and Danger: An Analysis of the Concepts of Pollution and Taboo* (London: Routledge, 1995).

[6] D. J. Murray, *A History of Western Psychology* (Englewood Cliffs, NJ: Prentice Hall, 1983), pp. 258-266.

[7] The Clean Air Act was presented by some opponents as an oppressive force, a "Gestapo of the Grates." See P. Thorsheim, *Inventing Pollution: Coal Smoke and Culture in Britain Since 1800* (Athens, OH: Ohio University Press, 2006), p. 182.

[8] See the discussion by W. Weaver of T. Blashill, "The State of London Streets," *Journal of the Royal Society for the Promotion of Health*, 22, 6 (1901). Accessed online 24/02/2008, DOI: 10.1177/146642400102200102.

[9] Horse transport grew rapidly in the second half of the nineteenth century with obvious consequences for city streets. See R. Turvey, "Horse traction in Victorian London," *Journal of Transport History*, 26, 2 (2005), pp. 38-59.

[10] The debate sprang from Wiener's controversial account of the origins of British industrial decline *English Culture and the Decline of the Industrial Spirit*, published in 1981, in which he argued that the emergence of preservationism was underpinned by development of a rustic "Englishness" among England's elite. See M. J. Weiner, *English Culture and the Decline of the Industrial Spirit, 1850-1980* (Cambridge: Cambridge University Press). This interpretation has received trenchant criticism, not least from Peter Mandler in his classic essay "Against 'English-

ness': English Culture and the Limits to Rural Nostalgia, 1850-1940," *Transactions of the Royal Historical Society*, 6th ser, 7 (1997), pp. 155-75.
[11] Jonathan Bate, *The Song of the Earth* (Cambridge, MA: Harvard University Press, 2000).
[12] Harriet Ritvo, "Manchester v Thirlmere and the Construction of the Victorian Environment," *Victorian Studies*, 49, 3 (2007), pp. 457-482, 463.
[13] D. Porter, "'Enemies of the Race': Biologism, Environmentalism, and Public Health in Edwardian England," *Victorian Studies*, 34, 2 (1991), pp. 159-178.
[14] See, H. L. Malchow, "Public Gardens and Social Action in Late-Victorian London," *Victorian Studies*, 29, 1 (1985), pp. 97-124; J. Ranlett, "'Checking Nature's Desecration': Late-Victorian Environmental Organization," *Victorian Studies*, 26, 2 (1983), pp. 197-222.
[15] Peter Bailey, *Popular Culture and Performance in the Victorian City* (Cambridge: Cambridge University Press, 1998).
[16] On landscape conservation and an argument for its inadequacy with regard to urban environmental needs see P. Mandler, "Politics and the English Landscape since the First World War," *Huntingdon Library Quarterly*, 55, 3 (1992), pp. 459-476.
[17] P. Johnson, "Conspicuous Consumption and Working-class Culture in Late-Victorian and Edwardian Britain," *Transactions of the Royal Historical Society*, 5th ser. 38 (1988), pp. 27-42; Matthew Hilton, *Consumerism in Twentieth-century Britain* (Cambridge: Cambridge University Press, 2003).
[18] *The Times*, 22 July, 1925, p. 15d.
[19] Matthew Hilton, "The Fable of the Sheep, or, Private Virtues, Public Vices: The Consumer Revolution of the Twentieth Century," *Past and Present*, 174, (2002), p. 227.
[20] Ross McKibbin, *Classes and Cultures, 1918-1951* (Oxford: Oxford University Press, 1998), pp. 85-105.
[21] "The Litter Habit," *The Times*, 29 August, 1935, p. 11d.
[22] David Matless, *Landscape and Englishness* (London: Reaktion, 1999), pp. 67-70.
[23] Matless has traced this development of littering as an indicator of "cultural trespass" in which the imagery of the "Cockney" is invoked as a stereotyped "anti-citizen," ill-educated, "vulgar" and undeserving of the beauties of the countryside. See Matless, *Landscape and Englishness*, pp. 67-70; "'The Art of Right Living': Landscape and Citizenship, 1918-1939," in *Mapping the Subject: Geographies of Cultural Transformation,* ed. Steve Pile and Nigel Thrift (London: Routledge, 1996), pp. 94-96.
[24] William Boyd, *The Challenge of Leisure* (London: New Education Fellowship, 1936); Henry Durant, *The Problem of Leisure* (London: Routledge & Kegan Paul, 1938).
[25] "Litter," *The Times*, 7 August, 1926, p. 11d.
[26] C.E.M. Joad, *The Untutored Townsman's Invasion of the Country* (London: Faber and Faber, 1946).
[27] *The Times*, 22 July, 1925, p. 15d.

[28] M. Tebbutt, "Rambling and Manly Identity in Derbyshire's Dark Peak, 1880s-1920s," *Historical Journal*, 49, 4 (2006), pp. 297-313; B. Harker, "'The Manchester Rambler': Ewan MacColl and the 1932 Mass Trespass," *History Workshop Journal,* 59, (2005), pp. 219-228.
[29] G. Parker and N. Ravenscroft, "Benevolence, Nationalism and Hegemony: Fifty Years of the National Parks and Access to the Countryside Act 1949," *Legal Studies*, 18 (1999), pp. 297-313.
[30] Letter from R. Evans, *The Times*, 28 June, 1916, p. 12f.
[31] *Manchester Guardian*, June 4, 1934, p. 10; "Country Wardens for Cheshire," *Manchester Guardian*, March 10, 1939.
[32] Departmental Committee on *Litter in the Royal Parks* (1953), p. 5
[33] *Hansard*, 5th ser., 587, 2 May 1958, p. 801.
[34] Countryside Commission, *Litter in the Countryside* (1970), p. 15.
[35] "Disfigurement of Scenery," *The Times*, 26 April, 1924, p. 15d. See also, P. A. Readman, "Landscape Preservation, 'Advertising Disfigurement' and English National Identity, *c.* 1890-1914," *Rural History*, 12, 1 (2001), pp. 61-83.
[36] "The Curse of Litter," *The Times,* 20 July, 1925, p. 15e.
[37] "Danger of Glass Bottles," *The Times*, 25 August, 1924, p. 6e.
[38] "No more Litter," *The Times,* 12 July, 1928, p. 10d.
[39] *Scapa Society Quarterly Papers,* New Series, 2 (1930).
[40] "Anti-Litter League," *The Times*, 9 July, 1928, p. 11b.
[41] "Litter at Newland's Corner," *The Times*, 25 May, 1929, p. 14d.
[42] "Notes of the Month," *Journal of the London Society*, 147, May (1930), p. 67.
[43] "Litter," *Journal of the London Society*, 145, March (1930), p. 40.
[44] "Litter," *Journal of the London Society*, 149, July (1930), p. 100.
[45] "The 'Litter Lout'," *The Times*, 29 September, 1930, p. 9f.
[46] See the burning of the "litter lout" on the Yorkshire Moors in 1937, *Manchester Guardian*, 12 August, 1937.
[47] C. Beaumont, "Citizens not Feminists: the Boundary Negotiated between Citizenship and Feminism by Mainstream Women's Organisations in England, 1928-39," *Women's History Review*, 9, 2 (2000), pp. 411-429.
[48] *Manchester Guardian*, 19 April, 1930, p. 16.
[49] "Litter," *The Times*, 13 September, 1934, p. 6c.
[50] "Thoughtlessness," *Manchester Guardian*, 13 November, 1935, p. 9.
[51] "Litter in Public Parks: Educating the Minority," *The Times*, 21 July, 1925, p. 16c.
[52] "Litter in the Parks," *The Times*, 21 April, 1924.
[53] "Disfigurement of Scenery," *The Times*, 26 April, 1924, p. 15d.
[54] *Scapa Society Pamphlets*, 7 (1929).
[55] *The Times*, 7 August, 1926, p. 11d.
[56] M. Higgs, "The Litter Nuisance," *Manchester Guardian*, 20 November, 1928, p. 22.
[57] "The Litter Nuisance," *Manchester Guardian*, 20 November, 1928, p. 22.
[58] One such broadcast went as follows: "With every general holiday there comes the same story of field, moorland, footpath, and road bestrewn with litter, sandwich

papers, chocolate cartons, photographic film covers, tobacco covers, orange peel, match boxes and other forms of refuse, which are to be found wherever the holiday maker has penetrated. One hears of walls pulled down, farm gates left open, flowers torn up, and of wanton damage in many forms. The litter is chiefly caused by thoughtlessness. Will all holiday makers please try to remember that besides offending the eye this litter is sometimes a danger, and that the carelessly discarded match and cigarette are possible causes of serious moorland, heathland and woodland fires? We appeal to all listeners to do their utmost to prevent litter being left about or damage done to the countryside."

[59] "Rural Amenities," *The Times*, 4 April, 1931, p. 7c.

[60] "Census of Litter," *The Times,* 10 July, 1934, p. 11c.

[61] Ivor Brown, "Flowers and Filth," *Manchester Guardian*, 2 June, 1934, p. 11.

[62] "Litter in Public Places," *The Times*, 21 July, 1925, p. 10c; "Litter," *The Times*, 10 August, 1926, p. 13e.

[63] "London Litter," *The Times,* 3 June, 1929, p. 17d; "Litter," *The Times,* 5 June, 1929, p. 10c.

[64] "Litter and Man's Nature," *The Times,* 5 Sept, 1935, p. 8b.

[65] J. Sheail, "The Concept of National Parks in Great Britain 1900-1950," *Transactions of the Institute of British Geographers*, 66 (1975), pp. 41-56.

[66] *Journal of the Commons Open Spaces and Footpaths Preservation Society*, x, 3 July (1951), p. 91.

[67] *Litter in the Countryside*, p. 2.

[68] *Litter in the Royal Parks*, p. 4.

[69] *Litter in the Royal Parks*, p. 4.

[70] *Litter in the Royal Parks*, p. 5.

[71] "Litter," *The Journal of the Commons, Open Spaces and Footpaths Preservation Society*, x, 7 August (1952), pp. 208-9.

[72] Quoted in *Journal of the Commons, Open Spaces and Footpaths Preservation Society*, xiii, 7 July (1958), p. 252.

[73] "Litter by the Ton," *The Times*, 14 November, 1957, p. 13f.

[74] R. Westaway, "Women's Institutes and Nature Conservation," in Northants Naturalists Trust, *European Conservation Year 1970* (Northampton, 1970), p. 26.

[75] National Federation of Women's Institutes, *Annual Report* (1958), p. 8.

[76] *Hansard*, 5th ser. 557 (1955-56), pp. 1753-60.

[77] *Hansard*, 5th ser. 568 (1956-57), pp. 1474-1480.

[78] "Minister on Litter Campaign," *The Times,* 19 October, 1957, p. 4f.

[79] National Federation of Women's Institutes, *Annual Report*, 1958, p. 8.

[80] *Litter in the Countryside*, p. 11.

[81] "A New System to Make Britain Tidy," *Heritage Outlook*, July-Aug (1981), p. 99.

[82] *Municipal Review*, 453, (September, 1967), pp. 484-5.

[83] A. La Moury, "Litter and Packaging," *Municipal Review*, 509, May (1972), p. 138.

[84] Friends of the Earth, *Packaging in Britain* (1972), p. 55.

[85] "Conference on Packaging and Litter," *Public Cleansing*, May (1972), p. 240-293.
[86] "Conference on Packaging and Litter," p. 247, 249.
[87] "Education, example and enforcement," *Public Cleansing*, May (1972), p. 249.
[88] "Conference on packaging and litter," p. 243.
[89] "Conference on packaging and litter," p. 243.
[90] E. Griffiths address to the conference on "Plastics Packaging and our Society," *Solid Wastes Management*, Jan (1974), p. 34.
[91] House of Commons Environmental Audit Committee, *Environmental Crime: Fly-tipping, Fly-posting, Litter, Graffiti and Noise* (2003-4), HC 445, p. 22.
[92] House of Commons Environmental Audit Committee, *Environmental Crime,* p. 22.

Chapter Fifteen

Shipping and Receiving: Circuits of Disposal and the Social Death of Electronics

Jennifer Gabrys

Figuratively, we "use up" places and dispose of them in much the same way that we dispose of Kleenex or beer cans
—Alvin Toffler, *Future Shock*[1]

The "Social Death" of Electronics

In *Invisible Cities*, Italo Calvino describes a metropolis, Leonia, which refreshes itself by discarding all its objects on a daily basis. So persistent is the process of using up and expelling goods that this becomes Leonia's defining attribute, its apparent source of pleasure. The city's constant stream of refuse is transported by anonymous workers to unknown places located on the urban periphery. Yet the practice of expulsion grows to such epic proportions that an increasing quantity of debris accumulates and threatens to unleash in a cataclysmic landslide. In the process of disposing of its remains, Leonia unwittingly constructs orders and spaces of enduring and even menacing materiality. At the same time, the city establishes circuits of disposal that become defining routes of renewed consumption, duration, and value. These circuits are invisible, overlooked; yet the remainders that move through these spaces of disposal give Leonia its "definitive form."[2]

Similar daily rituals of consuming and wasting emerge in even greater relief in Cornucopia City, an imagined geography that postwar cultural commentator Vance Packard describes as an example of the furthest extreme of overproduction. In this metropolis, temporary buildings are constructed from papier-mâché, and the factories produce a heap of products that are trucked directly to the dump before they are even able to inundate the consumer market. Through his concocted city, Packard expresses a perceived "crisis of production," a crisis that threatens to saturate markets to such an extent that it overwhelms the possibility for consumption to keep pace.[3] In these cities, disposal, invisible and abundant, is continual and essential to the renewal of production and consumption. Yet there is more to the process and geography of disposal than this loop between production and consumption. As abandoned goods make their outward journeys, they undergo transformations and deformations; they accumulate in peripheral spaces and define well-traveled circuits of disposal. These circuits and spaces of disposal are often hidden, but as Leonia and Cornucopia City imply, they are indispensable to everyday material practices.

This chapter focuses on electronic waste in the form of discarded devices in order to describe the circuits and spaces of disposal through which abandoned electronics travel. Disposal is not just about garbage trafficked to waste sites, and it involves much more than simply throwing unwanted items in the rubbish bin. Disposal, as it turns out, involves the holding patterns, stockpiling, recycling, and salvaging of materials before they further dissolve or enter another stage of waste. Electronic waste moves not just out of centers of production but also through marginal storage spaces and into recycling depots and, via shipping containers, toward developing countries. In this sense, disposal requires complex infrastructures, practices, and relationships in order to shift devalued objects into spaces for potential revaluation. Such circulations more fully describe the material geographies and practices of disposal, since there is no simple periphery to which objects can be jettisoned. The imagining of the periphery, furthermore, constitutes a topic of investigation: where is there an "outside" to which wastes can travel?

Electronics is a rapidly growing industry, with increasing rates of consumption and obsolescence, and for this reason, its waste stream has increased as well. While the exact delineation of what constitutes electronic waste varies, "consumer electronics" of all sorts are scrapped in numbers that are now reaching numbers of units in the billions.[4] Although electronic waste is growing at a rapid rate, the circuits and practices of disposal are not clearly delineated, often because this is a relatively new form of waste. Even with the obvious growth in the number of electronics bought, sold, and discarded, it is actually quite difficult to determine how many of these devices enter the waste stream at any given time, because owners often store and stockpile them for several years beyond their useful life. To further add to the confusion, many countries that export or import electronic waste do not use a specific code to track its delivery, so the trail of disposed devices becomes further obscured in the process of shipping and receiving.[5] The processes and spaces of disposal are not singular but open into expanded geographies. Similar to Leonia and Cornucopia City, the peripheral routes for the disposal and displacement of electronic waste accumulate and congeal into a "definitive," if makeshift, form. This form emerges through disposal practices that are relatively obscured but that are essential in maintaining the apparent immateriality of electronics, even while they are enduring and toxic.

As I have discussed in *Digital Rubbish: A Natural History of Electronics* in a more extended study of electronic waste, modes of electronic waste extend to wastes generated from electronic production and transmission to consumption and disposal. The focus on consumption here specifically considers how it is continuous with disposal and how consumption patterns can even inform the ability of materials to be "used up." This chapter examines another aspect of digital technology and "use"—not necessarily to concentrate on patterns of interaction between "users" and technology, but to consider instead the more extensive material networks that enable relatively transient forms of "use." But the relationship between consumption and disposal is often neglected. Some studies on consumption suggest that we trace the "social life of things" in order to understand the "trajectories" of commodities.[6] Yet there is a certain difficulty in following "things" in a study on electronic waste. These electronic commodities rapidly expire, have numerous hidden inputs and fallout, and are stockpiled or enter dubious routes of disposal upon their expiration. On many levels, electronic disposal, then, offers a more complete account of electronic consumption, since these technologies have been designed and developed within material cultures of disposability.

Disposal is formative in the making and unmaking of the materiality of electronics. The practices of disposal involve multiple modes of material disassembly and depend on interconnected geographies for the circulation and recuperation of discarded devices. Disposal—in the form of use and using up—is a complexly situated process of materialization. To study these material processes specific to electronics, it is useful to account for the multiple "hidden flows" that enable their formation and deformation. Waste is a significant part of the flows of materials that are present not as consumer goods, but as the fallout from production and disposal. Indeed, at any one time, the majority of global material flows are made up of some form of waste. As estimated by the World Resources Institute, these material flows are typically comprised of the by-products and resources that are necessary for the formation of commodities.[7] But these estimates of material flows typically account for the waste generated from production processes and further assume that every item produced will eventually migrate toward consumption and then disposal. Consumption and disposal are protracted spaces and practices that do not necessarily involve a unit-by-unit correspondence. There are vague spaces and processes of expenditure that take place between consumption and disposal. Indeed, a "unit" of consumption does not automatically translate into a unit disposed; rather, consuming, using up, and disposing generate extended spaces of delay, deformation, and demattering. To map these spaces and movements, I take up anthropologist Rudi Colloredo-Mansfeld's suggestion that we should go beyond the social life of things to consider the "social death" of things.[8] This attention to social death can bring to light the extended processes, practices, and places that emerge with the disposal of objects.

In this chapter, I discuss a natural history of electronics in relation to the transience and migration of electronics as they pass through and are suspended in circuits of disposal, which cross local and global environments, depend on formal and informal labor economies, and at times require material movements much slower and heavier than the dematerialized networks of electronic markets. To describe these circuits and spaces of disposal, it is also necessary to describe how electronics became so disposable in the first place. As they shift around the globe, disposed electronics sediment as residues from the processes that have contributed to the "throwaway society." This chapter explores how electronics developed within a culture of disposability and how advances in automation, together with new material developments, actually intensified processes of disposability. In particular, the development of plastics played an important role as an ephemeral and disposable material, as well as a material that might be valued for its performance and functionality rather than its durability and solidity.[9] The ease of disposability, the material transformations of electronics, their consumption and disposal, along with the storing, shipping, and stripping of these technologies—these material practices and spaces together form this account of how electronics turn into waste.

Appliance Theory

During the spring of 2005, in London, a "humanoid" sculpture of electronic proportions loomed seven meters above the river Thames. Composed of refrigerators and computer mice, mobile phones and microwave ovens, computer monitors and washing machines, the three-ton structure represented the amount of electronic waste a typical Briton would generate in his or her lifetime. Five hundred and fifty-three electronic devices in total contributed to the architecture of this sculpture. Yet the number of electronics is as striking as the diversity of devices that now constitute electrical and electronic waste.[10] The pervasiveness of the microchip manifests in an equally pervasive array of electronics and appliances, including everything from irons to vending machines. Many of these devices are more or less "electronic," because microchips and printed circuit boards that channel the flow of electrical currents and information power them. But these microchips are also encased in a skeletal body of plastic and copper, glass and lead. The extended material infrastructures required to house and enable microchips are evident in this motley assortment of plastic appliances. Microchips and plastic assemble into simultaneously pervasive and disposable devices. Leftover electronic devices are primarily composed of plastic and thus appear to be disposable.

The microchip, that miniature conductor and amplifier of electricity, is now neatly sealed in the contours of the everyday. Under the influence of the chip, appliances of all sorts have acquired new "functionalities" and speeds. The ways in which electronics have led to the transformation of objects, materials, and environments may be described as what the now-obscure packaging designer Vernon Fladager has called a "new machine economy." In every such economy," Fladager suggests, and with "every increase in machine speed," new materials, designs, and packages emerge. In fact, "the perfect package material of today can go out the window tomorrow because a new machine economy may make an alternate material a better choice."[11] The electronic package of microchips and plastic is bound up with processes of materialization that can be described through the quickening of matter, proliferation, and increased disposability. This is a machine economy that not only describes altered rates and scales of production but also establishes a temporal mechanism for the disposal of existing materials and designs. Electronics even appear to be programmed for their own elimination, as though an expected part of electronic processing has to do with eventual disposal and erasure.

Electronics, it seems, are prime operators in this transient machine economy. In many respects, electronics are situated within a larger culture of disposability that significantly expanded with the advent of automation after World War II. With automation, there was a general explosion of many consumer goods, which were typically produced to the point of market saturation. New practices of consumption and wasting arose in relation to automated production. In a similar way, practices of electronics consumption and disposal have emerged to facilitate these particular machine economies. Disposability may even constitute an "inventive" use of electronics and peripherals. DVDs have been developed that would expire upon 48 hours after their packages were opened,[12] and certain varieties of mobile phones have been designed to last for only a few days of use.[13] The duration of electronics has dwindled from at least a decade to, in some cases, a matter of hours. Devices appear disposable because they are at once freely available, constantly updated, bound to cycles of fashion, and often increasingly miniature in size. These machine economies encompass more than microchips simply acting on matter. Instead, they evidence the changing material arrangements and practices that sediment within particular technological and material forms. Automation, altered consumption patterns, material developments in the form of plastics, packaging, and shipping technologies and economic geographies have all informed electronic processes of materialization and disposal.

The Throwaway Revolution

The term throwaway revolution is used by Packard to describe—if not denounce—the postwar rise in automation and disposability in the United States, when objects with short life spans or limited use increasingly appeared on the market. Technological advancements in automation led to lower production costs, which led, in turn, to a flood of cheap goods on the market, the rise of disposability, and the decline of repair. This was a moment when, as is typically the case now, it became much cheaper to dispose of and replace objects than to repair them. Commenting on the rise of the throwaway revolution within his time, Packard suggests that automation led to an explosion in the number and type of disposable goods available. "Paper plates, cups, bottles, containers have long been disposable," he writes, and "these are now joined, according to a recent report, by 'everything from bikinis to men's blazers, nightwear to student's gowns, curtains to bathmats.'"[14] In Packard's popular critique, economic progress seems to require even more elaborate forms of waste making. If new and improved goods were to be made available and if the economy were to continue to grow, new strategies of consumption and disposal were necessary. Cornucopia City was simply the most ideal—if perverse—installment of this logic: wasting, in the end, stimulates growth.

With automated mass production, a greater store of goods was made available, which enacted changes not just in patterns of consumption but also in patterns of disposal. These changes extended to the availability of a greater variety and volume of disposable goods; yet they also included, as waste theorist Gay Hawkins suggests, "the fundamental logic of the commodity form, seriality."[15] The repetitive production of goods meant they could be easily replaced, old things disposed for new, without any relative concern for where the disposed objects went. Far from constituting a continuation of existing patterns of disposability, the postwar orders of disposability that emerged marked a fundamental shift, not just in the form of commodities, but also in the dynamics whereby they were valued or devalued. Technological advancements that allowed for more rapid product manufacture contributed to the sense that objects were less enduring and more replaceable. Transience and substitution became motivating factors in consumption. This is another aspect of the way in which waste is a generative dynamic, a necessary movement of goods out of consumption-bound circuits and into other circuits of disposal and removal. Practices of consumption become inseparable from practices of disposal.[16]

Disposability is evident not just in the materiality and consumption of goods but also in the growth of the automated production process. With the rise of automation in the mid-twentieth century, changes occurred not just with the gadgets and products available but also to the processes of manufacture and to what it meant to be "automatic." When goods became "electric," they became fluid, moving just as easily from warehouses to markets, homes, and rubbish bins. Matter is programmed—as much for fluidity as for disposability. This stage of automation not only made available a greater abundance of goods but also contributed to the transformation of matter. The accelerated movement of goods was concomitant with a greater sense of dematerialization, plasticity, and disposability. Plastic objects in particular appear to be inscribed with their inevitable movement toward rubbish.[17] These objects tip toward disposal and waste more readily not just because they are more abundant or made of more ephemeral materials but also because they are produced through technologies that enable speed and transience.

The postwar history of technology is a legacy of successively intensifying attempts to electrify objects. Things quicken under the influence of electricity. Once inert objects transform and are animated by the quiver of electricity. These permutations of matter and electricity corresponded to goods that became more and more transient. The prefix e- now potentially can precede even more than markets. The electronic conjoins and augments material and transactions from electronic mail to electronic money and electronic waste. Phones and ovens, cameras and books, leaf blowers and teakettles all submit to the same hazy law of the electronic. Every appliance presents an electrical mutation of an object that once stood still. Matter is charged, but what does it generate? In this general economy of electrification, matter does not just levitate, emanate, conduct, and mobilize; it also circulates, leaks, dematters, disappears, and wastes. The boundaries of objects break down at the same time as they receive an intensifying jolt. Just as the early pioneers from Texas Instruments and Intel anticipated, electronics are now so pervasive that nearly everything is informed in some way by electronic processing. But this pervasiveness is now part of the dilemma, where electronics have proliferated to such a degree that their volume and transience constitutes a material-handling problem.[18] In what might be called a "revolution operating on matter,"[19] electronic objects are produced and designed with increasingly shorter life spans. The effects of increased production and shortening life spans become most evident through the accumulation of electronic waste.

With the rise of automation and electronicization, materials become increasingly indistinguishable from their performance. Materials such as plastic are defined in relation to their functions, as designer Ezio Manzini suggests, from "mechanical function" to "surface quality" to "special electric properties" and even integrating "information input and output systems" into materials.[20] Materials are assessed for their performativity; they are engineered for efficiency, functionality, and, on a certain level, elimination. Objects become smaller, and extraneous components are removed. Function and flow stand in for matter—qualities that are ultimately symptomatic of the electric and the electronic. Matter performs as a package, a surface, a plastic medium for the delivery of function. These operations are also processes of materialization. Electronic technologies enable the capacities for acceleration, proliferation, and destabilization. Yet these same dynamics contribute to the transformation of material and its exchange, as well as the generation of waste and remainder. Electronics and electronicization have as much to do with material developments as with innovations in technology and manufacturing.

Packaging Electronics

It may be the case that electronics owe as much of their development and evolution to the history of plastics as they do to the history of silicon and transistors. It may also be the case that the plastic and the electronic—and, by extension, the plastic and the virtual—have more in common than previously imagined. Plastic is the material that enabled the profusion of disposable packages; it is abundant and pervasive, malleable, and suitable for an infinite variety of uses. But plastics and silicon are also functional materials; they perform operations, so they do more than provide the "raw" material for technologies and objects. These materials in fact inform the possibility of emerging technologies. As "informed material[s],"[21] they exist within processes of materialization and not simply as inert matter.

Informed materials, as discussed by Bernadette Bensaude-Vincent and Isabelle Stengers, operate as more than raw materials, but in fact contribute to the possibility for new technologies and functionalities to emerge.

Electronics are comprised of informed materials: silicon enables the flow of electricity and the apparent dematerialization of matter; plastic is inscribed with the capacity for disposability and mass production that now characterizes electronics. Plastic, as a functional material, could be produced in relatively unlimited quantities; it was inexpensive, easily replaced; it could embody the instantly disposable and the imminently possible all at once. As commentators in the *Modern Packaging Journal* opined, "The biggest thing that's ever happened in molded plastics so far as packaging is concerned is the acceptance of the idea that packages are made to be thrown away."[22] Plastic packaging came to embody all the defining traits of disposability: cheap, abundant, and expendable after a single use. The transience of packaging ultimately contributed to increases in production volumes, where millions of packages eventually grew to billions of packages discarded annually.[23] The single-use purpose of packaging easily extended to all objects made of plastic. Suddenly, not just the casing but entire goods were subject to the logic of abundant, single use.

Spectacular examples of multi-million unit uses of expendable molded plastics in containers for razor blades, ice cream and other foods, in tomato trays and berry baskets, are demonstrating that a plastic package, while it may be a thing of beauty, need not and should not be a joy forever. Consumers are learning to throw these containers in the trash can as nonchalantly as they would a paper cup—and in that psychology lies the future of molded plastic packaging.[24]

Plastic took the place of paper as the ultimate disposable material, and by doing so, it redefined the material sense of disposability. The rise of plastic packaging was, at one level, part of an effort to minimize the weight of goods previously packaged in glass. The use of plastic in order to minimize associated material, energy, and transport costs was related, then, to a certain drive toward dematerialization. The drive toward dematerialization became continuous with elimination, where goods and packages became more expendable as they required fewer material inputs.

Modernized packaging not only extends to the cellophane and molded polyethylene surrounding tomatoes and soap but also includes the skin around the increasingly transient technological "guts" of machines.[25] In this sense, packaging became a model for disposability that began to inform a whole range of goods, including electrical appliances. Electronics, as with the force of electricity that preceded it, depend on the design of these packages and fluid materialities. Designed packages in the form of electric appliances may enable a sense of efficiency, futurity, and disposability.[26] With electrical appliances and electronics, increasing consumption depended as much on the disposability offered by the package as on the promise the futuristic package presented in the form of technological fashion. Electronics perform in relation to imagined futures; they are packaged in a forward and instantaneous passing of time. Electronics of all sorts have been packaged in ephemeral plastic containers, disposable shells for the conveyance of information.

Plastic, as Roland Barthes writes, "is in essence the stuff of alchemy," because it enables "the transmutation of matter." So thorough is this transmutation that plastic appears to dematerialize completely in the production process, where it moves from "raw telluric matter" to the "finished object."[27] In this dematerializing movement, plastic acquires infinite possibilities for transformation. Any number of objects appears in plastic shells, molds, and packages. Plastic, similar to electronics, mobilizes matter toward apparent invisibility, lending a sense of dematerialization through miniaturization and through accelerating rates of circulation. Plastic is in many ways continuous with the changes enacted by the microchip: these are materials and technologies that emerge as programmed matter, engineered to express malleability, invisibility, and disposability. It was the proliferation of plastics that gave concrete—if immaterial—form to this sense of dematerialization. Plastics are in fact also the material carriers of many seemingly immaterial information and communication media.[28] Just when plastic became so pervasive that it even became the common carrier for electronic technologies, it receded from view. For this reason, plastic partly enabled the sense of virtuality, the sense that digital media somehow operate free from materiality.[29] As discussed earlier, in many ways, immateriality has less to do with the actual removal of matter and more to do with the alteration and "destabilization" of materials.

Indeed, the microchip is a kind of plastic, a reverse packaging that renders malleable the electronics and appliances that it powers. But in fact, these devices take on another level of materiality through electronicization. Objects that were once inert, durable, and relatively benign are now plastic, toxic, disposable, and yet enduring. Electronics do not dematerialize as much as they rematerialize through such programming of matter. Plastic, metals, and glass are the primary materials that make up electronics. As the icon of disposability, plastic is part of a group of material composites that often fade from view. These plastic composites constitute what Manzini calls "a world of nameless materials." No longer are objects made of materials that are readily identifiable, such as wood or clay; instead, they are typically composed of a highly engineered and mysterious mix of substances. Computers are assessed less for their material integrity and more for their performance; materially, they may appear at most to be "plasticky" and disposable.[30] What we see with these opaque materials is the operation and image of the devices. Material becomes synonymous with its function and appearance and effaces its own substance. This shift was inevitably aided in large part by plastics. "Plastics have played a fundamental role," Manzini notes, "in triggering the technical, economic, and cultural dynamics that led to the current new scenario of materials."[31] Advances in plastics led not only to the "unrecognizability of materials" but also to the constant redesign of products with materials that promised better performance, with "less matter, less energy, more information."[32] These are the new and nameless materials that dematerialize through the force of information. But when they resurface, they are increasingly difficult to salvage and recycle. Because of the wide variety of plastic composites used in electronics, it is often difficult to sort and recycle these materials for additional use.[33] They are also increasingly troublesome as pollutants and objects that linger indefinitely.

While the electronic industry has speed and turnover in mind, it typically employs materials that last for decades. Here are copper and plastic, mercury and lead, substances thicker and more enduring than any transcription of ones and zeros. Yet for all their endurance, these substances have been essential to the emergence of new orders of ephemerality. Plastic is nearly synonymous with disposability; yet it is also the enduring discardable. Packaging carries with it this deeply ambivalent relation to materiality. Inside the plastic shell that constitutes the predominant material for most electronics are also beryllium, cadmium, and brominated flame retardants.[34] Materials are caught in a tension between the quick and the slow. Ephemerality can only hold at one level; it instead reveals new spaces of permanence. Throw away plastic to discover it lasts for an ice age. The balance of time shifts. The instant plastic package creates new geologies. We now have mountains of congealed carbon polymers. Entirely new landscapes are built up around the fallout from the momentary and the disposable. So this is not just a story about the vaporization of "all that is solid"; rather, it suggests that new forms of solidity—new types of "hardware"—emerge with the program of disposability. Disposability is, then, about more than just overproduction; it also includes conditions of material transience and pliability. Electronic technology may have ephemerality as its guiding agenda, but it unwittingly produces new orders of permanence and new spaces and artifacts of indeterminable duration. The remainders that move through the circuits of disposal, in contrast to the accelerated networks of production and consumption, are drawn into these extended orders of duration and material solidity.

Circuits of Disposal

Disposal and disposability distinctly inform processes of materialization and dematerialization. Disposal and disposability correspond to spaces of removal that stretch beyond singular disposable objects. These are the hidden flows of disposal, involving not just the wasted materials that are used in the manufacture of goods but also the murky spaces where abandoned electronics are dismantled, trafficked, and repurposed. These circuits of disposal reveal how and where these technologies dissolve. The plastic package that encases most electronics has a life beyond its immediate disposal. Indeed, the plastic packaging surrounding electronics enables disposability, a relative sense of immateriality, and mobility.

"The distinction between disposability and mobility," as cultural commentator Alvin Toffler notes, "is, from the point of view of the duration of relationships, a thin one."[35] "Mobile technologies" acquire an expanded meaning, for the most mobile of technologies are, no doubt, often the most disposable. The discards that are mobilized, packaged, and shipped across watery networks give rise to new places and new formations. But where are these circuits and places of disposal?

When we trace through the circuits of disposal, we move closer to what might be Leonia's nebulous boundary between garbage mounds and city. Dirt is supposedly "outside the system."[36] But disposal is not just about attempted elimination but also arranging and ordering, putting aside or situating in relation to networks of exchange.[37] While many studies on waste suggest that garbage is a relationship between "matter in place and matter displaced,"[38] the very process of displacement can, in fact, give rise to places. These places emerge as the residue from attempting to relocate dirt toward an outside. There are many stages and places within disposal, which may extend to sites of storage, reuse, and recycling; transfer stations; and incinerators and landfills. The remainder of this chapter addresses those sites of disposal that are prior to and in transition to the salvage yard and dump, before electronics have reached terminal waste sites.

Disposal does not necessarily involve an absolute expelling of unwanted material but, rather, reveals attempts to recuperate or delay the demise of objects in order to postpone their decline of value. Yet the margins where trash is shifted or held are not necessarily sharply delineated but overlap and intersect. Electronics are left on curbsides and in skips, packaged in closets, bundled up in warehouses. These peripheral sites are often actually central but invisible. Part of the process of disposal and displacement involves a willful overlooking of the electronic material debris that surrounds us. Debris lingers in places and often compels us to contend with its dissipated value. A disposed object has, in addition to mobility, a sort of "motility" or stickiness, as geographer Kevin Hetherington notes: objects appear to vanish "only to return again unexpectedly and perhaps in a different place or in a different form."[39] When waste returns and resurfaces, it becomes clear that disposal is about more than matter out of place. Instead, disposal involves a set of practices for dealing with waste (even if this means overlooking it). When we dispose of something, we create places and relations out of the residue of this displacement.

In an attempt to map out these extended spaces of electronic waste disposal, I took a friend's aged personal computer to the nearest recycling facility (at the time, in Montreal). Like many devices of its kind, this PC had sat in a closet gathering dust. Outdated, with a DOS operating system, the petrified machine was a bulky object that one felt should be put to good use but that was no longer functional. As mentioned previously, as much as 75 percent of obsolete electronics are currently stockpiled in the United States.[40] If all the devices that had been stowed away entered the waste stream suddenly, en masse, they would completely overload most waste-handling systems. But there is a good reason why these devices do not unilaterally go in such a direction and why they continue to linger past the point of optimum performance. Not only are the circuits for electronic disposal undefined, but electronics are caught in a set of holding patterns that typifies disposal. The spaces of stockpiling and delay involve sites where "uncertain value" can be assessed.[41] The pause before a more terminal disposal in the dump or before packaging in shipping containers bound for the shores of China and India, is necessary in order to assess the lapsed value of the item. Disposal involves strategies of deferring the moment when objects become rubbish. Electronics initially undergo just such a holding pattern. No doubt, electronics stick around because of the relatively high price paid for them in proportion to the shortness of their useful life. What was at one time a device at the cutting edge of performativity has become an inert black (or beige) box, a device awaiting its final dispatch but remaining in the dim margins.

In my electronics disposal experiment, I located the nearest certified electronics recycler—situated, inevitably, well outside the city center, so that I had to drive the device to its proper waste-handling home. Following this path of disposal, I drove to the near edge of the airport, to a row of nameless light-industrial structures. Numbered loading docks edged up against a continuous plane of corrugated steel architecture, which was interrupted only by the company logo and front entrance. Carting the PC from the car trunk to the front lobby, I noticed that I was the only person in sight, and silent parking lots stretched into the distance. Inside the waiting room, it was clear that this act of singular recycling was unusual, even absurd. I met with the recycler and asked for verification of how the machine would be recycled and if the hard drive could be "wiped" of data (evidence of the success of this process was later sent to me in an e-mail with thirteen lines of zeros, indicating no data found).[42] With the recycler's assurances, I handed over the ancient machine, which transferred to the shop floor for disassembly and recycling.[43]

Businesses, institutions, and manufacturers are the primary recyclers of electronics. These groups are often prohibited from sending their electronics to landfills, so they are bound by law to find a recycling option for their machines.[44] While it is not yet illegal in many places for consumers to place their electronics in the trash for eventual shipment to the landfill, more policies now require that electronics are not interred in landfills, as many of the components in these devices are hazardous and present the possibility for environmental damage upon their breakdown and decay.[45] Increasing pressure has also been placed on governments to mandate an "extended producer responsibility," or EPR, that would require electronics manufacturers to take back the devices that they produce, for disposal and treatment.[46] EPR is often seen as a more ideal solution than a mandate that would only require the recycling of electronics, as the latter does not address the fact that the vast majority of electronics collected for recycling are eventually sent, in varying states, to developing countries, where they are processed and handled in relatively unsafe and environmentally unsound conditions.

When we follow electronics beyond their initial disposal, we find that even the apparently final forms of disposal are not nearly so complete and that value is never quite fully exhausted. If we unfold the stages of electronic disposal, we begin to see that there are multiple possible stages of removal, depending on the route that electronics follow. From Montreal to the Bronx and from Pennsylvania to New Jersey, I have visited electronic waste recyclers who have detailed the process of electronics disposal and recycling. Typically, electronics are first collected by recyclers in North America or Europe, who salvage high-grade machines for resale and extract valuable metal from devices for scrap or who alternately bundle defunct machines in shipping containers. In either case, at some stage down the line of processing, the electronics are usually sent to developing countries for scrap and salvaging of components, copper, gold, iron, plastic, nonferrous metals, cables, cathode-ray tubes, printed boards, and more. Raw materials markets thrive on and reincorporate these materials.

The disposal of electronics, then, follows a trajectory between developed and developing countries, where devices migrate from technology rich regions to those places with an abundance of cheap labor and a high demand for raw materials. While countries such as China are currently regulating against the importation of electronic waste, shipments continue to make their way to Asia, Africa, and other developing countries for recycling and disposal.[47] Using GPS to track the fate of a television recycled in the United Kingdom, Greenpeace activists have mapped how this legitimately recycled electronic device was eventually retrieved in a secondhand market in Nigeria. But there were many stages to locating and recovering the television as it moved across the ocean, from recycler to port, and from port to market.[48] At the same time, many used computers and electronics are sent to developing countries as donations. These devices are meant to contribute to overcoming the "digital divide" by supplying electronics to people who might not otherwise have access to them. Yet the donation of obsolete electronics does not contend with the dilemma that these machines will eventually become waste and will linger in places that often lack the infrastructure for handling these wastes properly.[49]

Indeed, this geographical relation between waste and raw materials is critical to the formation of the "third world."[50] Even when electronics are collected by recyclers in developed countries, the cost of recycling materials and the geography of markets for raw materials make developing countries a more "viable" place for disposed electronics to be sent in the end. But the cycle of production, consumption, disposal, and recycling is not a machine in perpetual motion, and as the recent collapse in the global market for recyclables suggests, the geographic relationship of manufacturing and waste is not fixed. When developed countries experience slower rates of growth and consumption, the developing countries that supply the products and remove the wastes similarly experience a slackening of activity. During recessions, piles of recyclables stack up in developed countries, as the usual routes for shipping and reusing these materials freeze up. Prices for raw materials can move with the same volatility as apparently abstract indices within electronic markets.[51] Recyclables may even begin to move in new circuits, shifting the relationship between manufacturing and raw materials from more disparate trajectories to nearer geographies; or materials are repurposed not for production but for incineration.[52]

Not only is it often cheaper to send electronic waste across the ocean than to process it locally in places such as North America, but because so much manufacturing takes place in China, the enormous demand for raw materials means the movements of electronic commodity and electronic waste nearly collide with one another, as electronic waste often makes the loop back to the site of its manufacturing.[53] In an account that is reminiscent of Packard's Cornucopia City, journalist Heather Rogers describes how "some shipping companies that bring consumer goods into the United States have taken up rubbish handling. Instead of returning with empty vessels, they fill their cargo containers with U.S. wastes, which they then sell to recycling and disposal operations in their home countries."[54] Shipping containers become part of a veritable conveyor belt, where the movement of goods back and forth across the ocean operates as some well-oiled machinery. Commodity and rubbish anticipate each other. The ease with which these goods move, the lack of distinction between goods for market and goods for disposal, increasingly functions as an abstract system of exchange, as the shipping and receiving of goods now takes place through the automated movement of sea containers.[55] The jumble, reek, and materiality of shipped goods are neatly sealed in containers that do not reveal the contents within. The same containers that ship electronic goods to market could just as likely contain electronic waste: the specificity of these materials has been eclipsed within a standardized container and mechanism of movement.

The majority of electronic waste, then, moves from developed to developing country by ship, which constitutes yet another space of delayed disposal. Electronics that have benefited from advances in plastics, packaging, and automation are then shuttled across the ocean by virtue of this other advance in "packaging." Shipping containers advanced as a maritime technology at the same time that automation and packaging emerged. Shipping containers enabled a new and automated ease of movement, which had a particular influence on the global transfer of cargo. The automated, containerized, and efficient movement of goods by ship resembles those other material, economic, spatial, and temporal changes that were taking place, from plastics to electronics. A technical innovation and newly fluid network of containerized shipping emerged to facilitate the distribution of goods and wastes.[56] These containers move in a liquid and global organization that shifts in relation to cheap labor. Newly discovered peripheries can then become sites for the mobilization and shipment of waste.

Yet within these watery circuits of transport and communication are spaces of material delay. Even at its most routinized, shipping constitutes an extended temporality that undergirds the instantaneous time of electronics. The age of information is more approximate to what artist Alan Sekula calls the "third industrial revolution," which crucially depends not just on electronic technologies but also on these technologies and networks of shipping. While the instant and virtual transport that occurs in digital space often holds sway over our sense of mobility—global, material, or otherwise—in fact, the "forgotten space" of the sea actually enables the movement of most materials, including electronics. So binding are these material flows that they serve as a significant counterpoint to the dematerialized flows of "cyberspace." Sekula writes,

> Large-scale material flows remain intractable. Acceleration is not absolute: the hydrodynamics of large-capacity hulls and the power output of diesel engines set a limit to the speed of cargo ships not far beyond that of the first quarter of this century. It still takes about eight days to cross the Atlantic and about twelve to cross the Pacific. A society of accelerated flows is also in certain key aspects a society of deliberately slow movement.[57]

Electronics and electronic waste trail through these spun-out liquid networks. The suddenness of disposal is drawn out again into orders of material time that are neither plastic nor virtual but, rather, extend into the indeterminable durations of delivery, disassembly, and decay. Just as we position ourselves in the "information revolution," we find that in many ways we are still entrenched in the measured material networks of the Industrial Revolution. In the paused space of shipping, all that had apparently dematerialized rematerializes. Electronics pass through, collect, and sediment in the delay between material registers and in the delay between continents.

Container ships loaded with electronic waste are primarily sent from North America to China, although other circuits of electronic disposal may be traced from Europe to Africa and from Singapore to India. In its report Exporting Harm, the Basel Action Network estimates that as much as 50 to 80 percent of electronic waste that is collected in recycling centers in the United States is eventually shipped to locations in developing countries. Guangdong, Lagos, and Delhi receive and distribute used electronics, which move from harbors inland to scrap yards, recycling sites, and resale markets. While electronic waste may have been displaced from one location, it resurfaces in these sites as material for potential reuse and recycling. The question of which "system" is displacing its wastes and how these wastes are configured looms large with the issue of electronic waste. While electronics may have reached the end of their useful life after 18 months in developed countries, becoming "inessential," these same devices are incorporated into other systems where value and use is recuperated and where waste becomes scrap and commodity. These disposed materials are further delayed from complete rubbishing, as they are processed and repurposed in locations often distant from their use and consumption.[58]

What makes electronic waste of particular concern is not just its volume and the fact that it now constitutes the fastest-growing waste stream in developed countries but also that its components are potentially hazardous upon disassembly and decay.[59] The practice of recycling may reinforce a sense that electronics are prepared and processed in a responsible way. But in developing countries, the recycling of electronics occurs through often crude and unsafe methods, including "open burning, acid baths and toxic dumping," which pollute the environment and endanger the workers and local population.[60] Residents in developed countries are relieved of responsibility for these materials, and residents in developing countries process materials and waste that often they did not generate. Perhaps for this reason, sociologist Zygmunt Bauman has spoken of how workers that sort through wastes and recycle materials seem, in the global economy, to also be "disposable people,"[61] expendable and made to deal with wastes from the wealthiest parts of the world. The murky but inevitable relationship between disposability and accountability materializes in concrete form with electronic waste. Circuits for the disposal of electronic waste do not enable its complete elimination; instead, they mobilize these materials toward other sites, forms of labor, and salvage practices.

Dirt, Displacement, Dematterinng

Recycling may potentially have the effect of increasing or encouraging disposability.[62] Materials may be just as rapidly thrown away, but the sorting, delay, and reintegration of these materials suggests that any problems arising from disposability can be addressed through this reuse. The distinction between recycled matter and rubbish is important in understanding the dynamic of electronic waste and rubbish in general. Recycling is another space of delay within disposal. It draws out materials for sorting, the recuperation of value, and reintegration by transforming rubbish into new commodities. Recycled material can even reenter spaces of exchange and renewed production. In many ways, this transformation takes place through the almost complete devaluation of goods and return to raw material, so that recycled materials move in and out of the economy; they are transformed from commodity to waste and raw material and from raw material into commodity again. But this process involves not just the abstract transformation of materials and values but also the formation of places where material rejection and devaluation takes place. Wire villages, canals flush with broken monitor glass, and alleys full of chemical barrels, which are the typical sites for recycling electronic waste in developing countries, are the actual sites in which these transformations occur. Far from the dematerialized specter of cyberspace, these practices of disposal continually provide evidence of just how material—if dispersed—electronic technologies are.

When we recycle, we repeat the process of delaying the inevitable return to rubbish. Electronic waste may be discarded in one location but then surfaces in another to be processed as goods with marginal scrap value. Yet when that scrap is processed into new electronic components, for instance, it reenters a value system that will mobilize again toward rubbish. Waste becomes an operative dynamic within the constitution and reconstitution of value and things. A thing may be reconstituted—as the preceding discussion on plastic reminds—in an infinite number of ways. It may pass into states of disposal and then enter several stages of delay, recuperation, and reentry. When electronics pass through disposal, they undergo such transformations. The displacement of this electronic "dirt" further gives rise to places, social relations, and environmental effects.

It is useful at the end of this chapter to return to the earlier discussion on the relevance of approaching consumption through disposal, of understanding the role that consumption plays in using up and dissolving goods and how these practices are guided by the dynamic of dirt. Addressing the interdependent relationship between production and consumption, Marx articulates that "a product becomes a real product only by being consumed" and that "only by decomposing the product does consumption give the product the finishing touch." In this sense, "consumption creates the need for new production."[63] Marx's schema creates a loop between production and consumption and focuses on production as the condition to which economies return, where consumption provides the necessary dissolution of products in order to spur new production—hence his phrase "Consumptive production. Productive consumption."[64] While Marx crucially draws attention to the dissolution that characterizes consumption, his analysis does not draw out the spaces and processes of dissolution and does not consider that dissolution may, in fact, be a condition guiding economic exchange. Waste, in this respect, is typically unaccounted for within discussions of production and consumption. Yet waste is a dynamic that influences all phases of economic exchange, providing the basis for the rise and fall of value and the formation of new commodities. Indeed, Marx says as much when he argues, "Consumption accomplishes the act of production only in completing the product as product by dissolving it."[65] While Marx goes on, in the same passage, to the renewed need for production, a slight interpretive realignment indicates that what is guiding these economic exchanges most of all is the inevitable dissolution of products. Here, products are complete only when wasted. This is a dissolution that occurs not only in consumption but also, by extension, in disposal and the recuperative spaces of recycling.

When we focus on these spaces and processes of dissolution, we can reconsider consumption not only as a process of acquisition but, equally, as a matter of how and where we rid ourselves of objects that are typically manufactured for disposal. Consumption is continuous with using up, and disposal is a critical part of the use of electronics, even if these devices are not in direct control of users. Consumption emerges not just as a process of dissolution that spurs new production, but also as a drawn-out process of "dispossession" and "demattering" that critically calls attention to how we get rid of things, how they circulate, where those things go, what residues they leave behind, and what political economies and ecologies they bind together. Disposal provides a way to focus on consumption without eliding this act of using up and without seeing disposal as the simple discarding of matter. Instead, disposal brings into relief those practices, spaces, temporalities, and performances that emerge through the removal and dermattering of goods in general and of electronics specifically. Colloredo-Mansfeld argues that "what it actually means to consume an object remains curiously unexamined" and that, in fact, this aspect of consumption as using up is not only a necessary area of study but also reveals how consumption can articulate social relations that "act as generative moments" through expenditure.[66] Consumption and dissolution do not return exclusively to production in this analysis but open up into other spaces that are shaped through the practices and materialities of disposal.

The circuits of disposal discussed in this chapter reveal the locations—often not so officially designated—where the demattering of electronics takes place. As these practices demonstrate, such demattering is too multi-layered and multilocated to occur in any single designated place. If we return to Leonia and Cornucopia City, we arrive this time with a much different sense of the circuits of disposal in these places. Cornucopia City trucks its goods from the production line to the dump: it does not account for the necessary role of consumption in using up goods and extending disposal into multiple places. Leonia simply shifts its continually discarded goods outward, to an unnamed margin, which could just as well be some electronic waste dump in Guangdong. We not only need places of demattering; we already have them. They just tend not to register as places of regard. But these places of disposal continue to exceed their boundaries, forcing us to reconcile ourselves to the effects of our wastes—electronic and otherwise.

NOTES

[1] Alvin Toffler, *Future Shock* (London: The Bodley Head, 1970), p. 69.
[2] Italo Calvino, *Invisible Cities*, trans. William Weaver (New York: Harcourt Brace Jovanovich, 1974), 91.
[3] Vance Packard, *The Waste Makers* (New York: David McKay, 1960), pp. 4-5.
[4] International Association of Electronics Recyclers, *IAER Electronics Recycling Industry Report* (Albany: International Association of Electronics Recyclers, 2003, revised 2004), p. 7.
[5] Basel Action Network, *The Digital Dump: Exporting Re-use and Abuse to Africa* (Seattle: Basel Action Network, 2005), p. 12.
[6] Arjun Appadurai, "Introduction: Commodities and the Politics of Value," in *The Social Life of Things: Commodities in Cultural Perspective*, ed. Arjun Appadurai (Cambridge: Cambridge University Press, 1986), p. 5.
[7] Allen Hammond et al., *Resource Flows: The Material Basis of Industrial Economies* (Washington, DC: World Resources Institute, 1997); and Robin Murray, *Zero Waste* (London: Greenpeace Environmental Trust, 2002).
[8] Rudi Colloredo-Mansfeld, "Introduction: Matter Unbound," *Journal of Material Culture,* Vol. 8: 3 (2003), p. 246.
[9] For a more extended and interdisciplinary study on plastics, see Jennifer Gabrys, Gay Hawkins and Mike Michael, eds. *Accumulation: The Material Politics of Plastic* (London: Routledge, 2013).
[10] While early studies on electronic waste focused on PCs and mobile phones, the full delineation has now been extended to include this range of electrical and electronic devices. For a discussion on the early estimates of electronic waste volumes, see H. Scott Matthews and Deanna Matthews, "Information Technology Products and the Environment," in Ruediger Kuehr and Eric Williams, eds. *Computers and the Environment: Understanding and Managing Their Impacts* (Dordrecht: Kluwer Academic, 2003), pp. 17-40.
[11] Vernon L. Fladager, *The Selling Power of Packaging* (New York: McGraw-Hill, 1956), p. 94.
[12] Katie Dean, "Disposable DVDs at Crossroads," *Wired News*, 7 February 2005 (available at: www.wired.com).
[13] Amanda Onion, "Buy, Use, Dispose: A Spike in Disposable Products Has Environmentalists Worried," *ABC News*, December 4, 2002.
[14] Packard, *Waste Makers*, p. 54.
[15] Gay Hawkins, "Plastic Bags: Living with Rubbish," *International Journal of Cultural Studies* 4, no. 1 (2001): 9.
[16] Kevin Hetherington, "Secondhandedness: Consumption, Disposal, and Absent Presence," *Environment and Planning D: Society and Space,* Vol. 22: 1 (2004), p. 158.
[17] Hawkins, "Plastic Bags," p. 9.
[18] With new levels of convergence and pervasive computing with wireless technologies, electronic waste may increase even further. For an extended discussion of this dilemma, see Andreas Köhler and Lorenz Erdmann, "Expected Environmental

Impacts of Pervasive Computing," *Human and Ecological Risk Assessment,* Vol. 10: 5 (2004), p. 831–52.

[19] Michel Serres, *Hermes: Literature, Science, Philosophy*, ed. Josue V. Harari and David F. Bell (Baltimore: Johns Hopkins University Press, 1982), p. 56.

[20] Ezio Manzini, *The Material of Invention* (Milan: Arcadia Edizioni, 1986), p. 29.

[21] Bernadette Bensaude-Vincent and Isabelle Stengers, *A History of Chemistry*, trans. Deborah van Dam (Cambridge: Harvard University Press, 1996), pp. 205–6.

[22] "Molded Plastic Containers," *Modern Packaging Journal,* Vol. 31 (1957), p. 120.

[23] According to William Rathje and Cullen Murphy, however, packaging accounts for a comparatively moderate proportion of municipal solid waste in the United States, and they suggest that packaging has actually helped to reduce some types of waste, including food waste. See William Rathje and Cullen Murphy, *Rubbish! The Archaeology of Garbage* (New York: Harper Collins, 1992), pp. 216-20.

[24] "Molded Plastic Containers," p. 120.

[25] Ellen Lupton and J. Abbott Miller, *The Bathroom, the Kitchen, and the Aesthetics of Waste: A Process of Elimination* (New York: Princeton Architectural Press, 1992), pp. 65-66.

[26] Adrian Forty, *Objects of Desire: Design and Society*, 1750-1980 (London: Thames and Hudson, 1986), pp. 190-93.

[27] Roland Barthes, "Plastic," in *Mythologies*, trans. Annette Lavers (New York: Farrar, Straus and Giroux, 1972), p. 97.

[28] Jeffrey L. Meikle, *American Plastic: A Cultural History* (New Brunswick, NJ: Rutgers University Press, 1995), p. 299.

[29] *Ibid.*

[30] Manzini, *Material of Invention*, p. 31.

[31] *Ibid.*, p. 32.

[32] *Ibid.*, p. 39.

[33] Even though they increasingly appear to be disposable, electronics are not typically designed for disassembly. Electronics recyclers often point out the difficulty of salvaging machines that are not designed with standardized disassembly in mind, which means that most machines must be stripped by hand. Electronic waste then constitutes a formidable waste problem, not least because there is no established or systematic infrastructure to handle this type and volume of waste.

[34] The "Electronic Waste Guide" suggests that the hazards of electronic waste can be minimized if properly handled: "The formation or discharge of hazardous emissions during the recycling of electrical and electronic equipment depends highly on the handling of electronic waste. Hence hazardous substances contained in computers and televisions don't lead automatically to a risk for the environment and the human health" (available at: http://ewasteguide.info).

[35] Alvin Toffler, *Future Shock* (New York: Random House, 1970), p. 55.

[36] Mary Douglas, *Purity and Danger: An Analysis of Concepts of Pollution and Taboo* (London: Routledge, 1995).

[37] Hetherington, "Secondhandedness," p. 160.

[38] Michael Shanks, David Platt, and William L. Rathje, "The Perfume of Garbage: Modernity and the Archaeological," *Modernism/Modernity,* Vol. 11: 1 (2004), p. 80.
[39] Hetherington, "Secondhandedness," p. 162.
[40] Environmental Protection Agency, "Waste Wise Update: Electronics Reuse and Recycling," EPA 530-N-00-007, October 2000 (available at: http://nepis.epa.gov).
[41] Hetherington, "Secondhandedness," p. 169.
[42] The process of destroying data on hard drives is notoriously difficult, and many electronics recyclers first set up business to deal not with the reuse of machines but, rather, with the destruction of sensitive data on computers. Typically, recyclers will offer a certification that ensures the elimination of data from hard drives. Many computers, however, are "recycled" without this guarantee, and as a result, hard drives have been scoured for personal data. The Basel Action Network documents leftover sensitive data on hard drives sent from the United States to Africa in their report *The Digital Dump* and in media related to this project on their web site, www.ban.org.
[43] This recycling experiment was conducted at a time when relatively few channels existed for the recycling of consumer electronics. These programs have since become better established, although they are not without their difficulties.
[44] Basel Action Network and Silicon Valley Toxics Coalition, *Exporting Harm: The High-Tech Trashing of Asia* (Seattle and San Jose, 2002), p. 6.
[45] The WEEE Directive is the first of such systematic attempts to prevent the flow of electronics to landfills. Initiated by the European Union, the WEEE Directive has recently required that manufacturers and producers take responsibility (extended producer responsibility, or EPR) for electronics at end of life by providing options for take-back and recycling of their machines. The program has had difficulties with enforcement, however. As a result, some commentators suggest the WEEE and RoHS regulations have become prime examples of "greenwashing." See Fred Pearce, "Greenwash: WEEE Directive is a Dreadful Missed Opportunity to Clean up E-waste," *Guardian*, June 25, 2009.
[46] See Bette K. Fishbein et al., *Extended Producer Responsibility: A Materials Policy for the 21st Century* (New York: Inform, 2000).
[47] Up to 80 percent of electronics shipped to Asia for recycling have their point of origin with raw material brokers in the United States. See Basel Action Network and Silicon Valley Toxics Coalition, *Exporting Harm*, pp. 1-2. See also Jennifer Clapp, *Toxic Exports* (Ithaca: Cornell University Press, 2001), for an in-depth analysis of the geopolitical implications of the movements of hazardous waste.
[48] "Greenpeace Deploys GPS to Track Illegal Electronic Waste," Environment Blog, *The Guardian*, 18 February 2009 (available at: www.guardian.co.uk); "Undercover Operation Exposes Illegal Dumping of E-waste in Nigeria," *Greenpeace News*, 18 February 2009 (available at: www.greenpeace.org).
[49] See Basel Action Network, *Digital Dump*; and Bette K. Fishbein, *Waste in a Wireless World: The Challenge of Cell Phones* (New York: Inform, 2002).
[50] David Harvey, *Spaces of Capital: Towards a Critical Geography* (London: Routledge, 2001), pp. 252-53.

[51] Dan Glaister, "US recycling: 'I Don't Even Think We Have an Industry,'" *Guardian*, 9 January 2009.
[52] Tania Branigan, "From East to West, a Chain Collapses," *Guardian*, January 9, 2009; Christine Oliver, "Recycling in the Credit Crunch," *Guardian*, January 9, 2009; Leo Hickman, "The Truth about Recycling," *Guardian*, February 26, 2009.
[53] "UN Programme Aims at Environmentally Sound Disposal of Electronic Waste," *UN News Centre*, 25 November 2005 (available at: www.un.org).
[54] Heather Rogers, *Gone Tomorrow: The Hidden Life of Garbage* (New York: New Press, 2005), 201. Increasingly, the production of electronics has been offshored to India, China, and Taiwan—the same places where electronic waste returns after it has been shipped and consumed in the United States and Europe.
[55] Allan Sekula, *Fish Story* (Düsseldorf: Richter Verlag, 1995), 12. See also Marc Levinson, *The Box: How the Shipping Container Made the World Smaller and the World Economy Bigger* (Princeton: Princeton University Press, 2006).
[56] Sekula, *Fish Story*, p. 49.
[57] *Ibid.*, 50. Not only is shipping a more protracted space of material movement, but it also requires extensive material resources while contributing significantly to greenhouse gas emissions. See National Oceanic and Atmospheric Administration, "Maritime Shipping Makes Hefty Contribution to Harmful Air Pollution," 26 February 2009 (available at: www.noaanews.noaa.gov).
[58] While North America is currently the largest consumer of electronics, China and India are now experiencing the greatest growth in electronics consumption. In this case, the countries that have traditionally been the recipients of electronic waste from other parts of the world must now deal with their own internal electronic waste problem. See the "Electronic Waste Guide," which notes, "China had the highest growth in number of computer users per capita in the period 1993-2000. It grew a massive 1052%, compared to a world average of 181%" (available at: http://ewasteguide.info).
[59] Basel Action Network and Silicon Valley Toxics Coalition, *Exporting Harm*, p. 5.
[60] *Ibid.*, p. 18.
[61] Zygmunt Bauman, *Wasted Lives: Modernity and Its Outcasts* (Cambridge: Polity, 2004), 60. Bauman notes that electronics are so easily outdated that not even charities will accept them as donations. To process the waste, instead, there emerges "human-waste producing plants. In Guiyu, there are 100,000 of them—men, women and children working for the equivalent of 94p a day."
[62] Gavin Lucas, "Disposability and Dispossession in the Twentieth Century," *Journal of Material Culture,* Vol. 7: 1 (2002), p.15.
[63] Karl Marx, *Grundrisse: Foundations of the Critique of Political Economy* (Rough Draft), trans. Martin Nicolaus (London: Pelican, 1973), p. 91.
[64] *Ibid.*, p. 93.
[65] *Ibid.*
[66] Rudi Colloredo-Mansfeld, "Consuming Andean Televisions," *Journal of Material Culture,* Vol. 8: 3 (2003), pp. 275, 283.

CHAPTER SIXTEEN

MANAGING HOUSEHOLD REFUSE: TOWARDS A HISTORY OF THE MUNICIPAL WASTE HANDLING BUSINESS[1]

RAY STOKES AND STEPHEN SAMBROOK

During the last third of the twentieth century, the collection, processing, and disposal of municipal solid wastes—the residue of daily domestic life—emerged as not just a pressing political, social and environmental issue, but also as a major economic one. In fact, waste management came to constitute big business in wealthy and highly industrialised countries. By the turn of the twenty-first century, the industry in Germany was turning over some €10.9 billion annually.[2] The top seven European firms in waste management in 2005 all had sales in excess of €1 billion, and many of them operated internationally.[3] In Britain, the value of the sector passed the £5.5 billion mark the same year.[4]

It was not always so. In the United Kingdom, for example, waste collection and disposal immediately after World War II was not seen as a field for private enterprise, but rather was carried out at the local level by the public sector in a continuation of pre-war practice rooted in public health legislation. Where private companies were involved they generally operated on a small scale in niche markets, such as collecting waste paper and scrap metals for salvage, an area in which there were some opportunities for profits without large-scale capital investment.[5] Some indication of the contemporary status afforded to waste handling is suggested by its virtual absence from published official reporting at the time: there was no discrete entry for waste collection and disposal in national industrial or economic statistics, something which only began much later. How then, and with what consequences, did waste collection and disposal come to be seen as potentially profitable business, moving from the public into the private sphere and becoming "waste management" in the process? What was the extent of the metamorphosis? Why did this process differ from country to country, and indeed from locality to locality?

These are the central questions we have sought to answer in the context of a research project funded by the United Kingdom's Economic and Social Research Council, which examines the causes, course and consequences of the emergence of this industry in two highly industrialised European countries of similar population size and wealth, the United Kingdom and West Germany, between the end of World War II and the early 1990s.[6] Our purpose in this chapter is to examine briefly the context of existing scholarly literature, within which the project is situated, to describe the project and its aims in some detail, and to report on some of the initial results of our research. Our initial findings confirm that the project and its approach offer a rare opportunity for sustained attention to the interconnections between business, technology, economy, politics, and society as they changed through time.

The Scholarly Context

Literature on the history of waste management tends to be technically specialised or essentially general, with a dearth of historically based social science approaches to the topic.[7] The interdisciplinary and largely empirical field of environmental history has only recently begun sustained consideration of issues of waste and disposal, while business history and history of technology approaches are even less well developed.

American environmental historians, such as Martin Melosi and Joel A. Tarr, tackled this topic first, focusing initially on industrial waste and / or urban refuse; this was extended to a broader "social history of trash" by Susan Strasser, while Zimring examined a key niche of the industry's private sector, metal scrap, over time. All provide valuable insights and use sources imaginatively, but focus primarily on the United States.[8]

In this area, as in so many others, however, the experience of the United States was unusual in several dimensions. The "consumer society" arrived much earlier there than elsewhere. The peculiarities of the U.S. federal system also affected the industry's development, as did relatively low population density (even in its urban areas) and, related to this, a large hinterland. Americans also produced on average much more rubbish per capita than other affluent nationalities, just as they used on average far more energy per capita.[9] One key strand of the project will be the extension, adaptation, testing, and revision of the findings of American environmental historians through attention to the other important national contexts, which form the focus of this study.

The literature on these other countries is again, apart from the purely technical, much more limited than for the United States, with very little of it comparative. Matthew Gandy provides an important exception, with consideration of urban waste—especially recycling—in three cities, New York, Hamburg and London, although he focuses on politics at the local level and on the 1980s and early 1990s.[10] Research associated with the Institute for Environmental History at the University of St Andrews has dealt with waste in Britain, with projects on recycling, household waste in the nineteenth century and theoretical aspects of the subject which complement this project.[11] For Germany, there are a handful of studies, including those by Fuchsloch and Park.[12] Stokes also addresses waste management in East Germany in the context of technological history.[13] Each is typical in its own way of this small field within German environmental / technological history: those that do exist are tightly focused on specialised aspects of waste disposal and / or on individual localities. The only overview in German is very general.[14]

This new study will thus not only complement and extend, but also help define a relatively new scholarly field in at least five distinctive ways. It will use a thoroughgoing comparative approach; it will focus on business and economic historical aspects of the development of the industry; it will explore national, regional and local agendas and interactions; it will pay close attention to shifts in the balance between the public and private sectors; and finally it will consider development over several decades. To show how it will do so, we first provide some additional information on the details of the project before turning to some preliminary results of our research thus far.

A Framework for Examining the History of the Waste Management Industry after World War II

During the period after 1945, the United Kingdom and the Federal Republic of Germany shared extensive similarities in terms of their affluence, manufacturing and technological capabilities, and population size and density. They also shared what appear to be the most important drivers in the emergence of a distinctly identifiable waste management industry in the post-1945 period, including increasing affluence, spiralling consumerism, growing sophistication of materials used in manufacturing, limited space for landfill, growing public concern about landfills and incineration, and (perhaps most importantly) changing cultural attitudes and policies, not just towards waste, but also towards the role of the private vs. the public sector.[15] In both countries, central and local government responded to these drivers, helping to shape the industry's development not only through legislation, but also through shaping the conditions within which local government became both a provider of and client for waste management services and facilities. The two countries entered the post-war period with legislation and practices in the industry, which had been established over a long period through to the late 1930s, and they shared problems of shortage and reconstruction in the immediate aftermath of the Second World War, which had significant impacts on the industry. The two also witnessed dramatic changes in the late 1960s and early 1970s—in particular in relation to the rapid growth of the private sector; both in absolute terms and in market share—owing to growing consumption, new materials, and environmental awareness and related legislation. Both were increasingly affected by European legislation and regulation. Finally, the two consolidated new approaches to, and concepts of, the industry in legislation and practice in the early 1990s. The common starting and ending points in the emergence of the modern waste management industry in Britain and West Germany between the immediate post-World War II period and the early 1990s determine the time frame considered in the project.

Despite these commonalities, the two countries evolved very different systems of waste management. The West Germans embraced sorting and recycling regimes much earlier and more thoroughly than the British. In the German case, local authorities had competence over direct levying of fees for domestic waste collection, a practice explicitly proscribed in the United Kingdom and reiterated in new legislation as recently as 1990. The West Germans also pioneered in devising of strategies for reuse of materials which sometimes relied on high technology, as well as developing product packaging legislation and an environmentally oriented industrial policy.[16]

How did the extensive shared drivers in the creation of a modern waste management industry lead to such different outcomes in the two countries with regard to the activities, size, shape, and significance of their respective waste handling industries? One possible answer would be to point towards deep-seated and long-standing social, political, and cultural differences between the two, and examination of such differences (as well as some similarities) through time will be part of our approach. After all, it is impossible to understand the industry's development in either case without attention to the emerging environmental movement: initial efforts to develop recycling schemes and technologies were stimulated there rather than in private industry or the state sector in both countries, although there were also extensive differences in timing and approach between them.[17] What is more, the industry sits at the juncture between economic and innovation policymaking and practice. Some government policymakers have thus come to see the waste management industry as not just vital to sustainable economic development, but also as one of a suite of environmental knowledge-based industries to be fostered by innovation and other policies owing to its perceived potential for economic growth and foreign trade.[18] Again, though, although this tendency is to be observed in both countries, different political and legal traditions led at times to different outcomes. Finally, the waste management industry was the place where the consequences of, and tensions between, industrial production, growing world trade, and increased consumption met and required some sort of resolution. In the process, relationships between a number of crucial areas have been highlighted, confronted, and changed through time, e.g. between: human beings and their material culture; the personal and the collective; the state and the citizen; the public sector and the private sector; and the local, the national, and the international levels. Once more, there were a large number of differences—and some similarities—between the United Kingdom and West Germany in the nature of the response to these challenges.

Our contention, however, is that much of the story of the sometimes parallel, sometimes divergent, development of a modern waste management industry in these two countries must be told not only in terms of politics, society, and culture, but also in terms of the established themes and concerns of business history and business studies more generally. Markets, entrepreneurship, and the role of the state differed in many ways within each of the two countries, and the particular constellation of the political economy in each case was clearly extremely important in the emergence of this new industry (as it was in the case of other new industries) and in the development of companies within it.[19] In particular, evolving markets, regulatory regimes, and attitudes towards the public/private divide in the economy[20] provided incentives and opportunities in this industry (again, as in others) for the private sector to move from being a niche player (albeit an important one) to a major force in all aspects of it. Once again, there were similarities between the two countries in this area, but also many differences. The story also involves issues such as innovation and the firm, strategies and processes of competition, diversification and concentration, and the strategy, structure, organisation and performance of firms in each national context.[21]

The project thus focuses on seven principal research questions:

- How did major local authorities (LAs) in the two countries organise and finance waste collection and disposal in the immediate post-1945 period, and what political, economic, cultural pressures, and / or technological developments caused changes in these practices?
- What factors have shaped markets for "waste" materials in Germany and the UK compared to America, and how have markets been shaped in turn by the industry and its technologies? In particular, to what extent have volatility in commodity prices and perceptions of political risk and/or liability shaped the extent of and changes in private-sector involvement in sub-sectors of the industry?
- To what extent have attempts in each country been successful in transforming "waste" into valuable commodities through recycling and re-use, and why?
- What has been the role of research and development, technological change, and innovation in this process of redefinition?
- How did individual firms and / or industry interest organisations respond to (or create) emerging business opportunities in this area?
- What role did government play in the emergence and development of the waste management business, as participant, customer, and / or regulator?
- To what degree, by what means, and to what effect did actors in business and government in each of the two countries learn from one another in this area?

Some Initial Results

We have begun to explore some of the questions mentioned above by focusing initially on the United Kingdom and by examining not just sources produced at the national level, but also municipal case studies. In the British case, even in the early post-war period, large local authorities performed an extensive and central role in waste handling, involving complex organisational structures, large work forces, and clearly defined managerial hierarchies which operated with substantial budgets. And it was not only in these respects that the waste-handling operations of large local authorities had a considerable amount in common with large private business enterprises. Certainly, the larger municipal bodies treated collection and disposal of refuse as a complex techno-economic system, involving notions of efficiency, effective resource allocation, and carefully costed capital investment. One hypothesis to be tested, therefore, is that emergence of the "waste management" industry actually predated the coining of the phrase "waste management." The phrase itself emerged in the late twentieth century in the context of changes in environmental awareness beginning in the 1970s and of a politically inspired shift towards the privatisation of formerly public service provisions starting in the 1980s. But it is quite possible that the effective emergence of the "waste management industry" lies much further back in the post-war period and has hitherto remained shrouded in the anonymity of local government administration. If this hypothesis is confirmed, of course, an appropriate further question is whether the eventual shift from public service to private enterprise was responsible for any substantial changes in the philosophy and effectiveness of waste management practices, or whether there was a continuation of the pre-existing *status quo* even as ownership and governance structures in the industry changed.

We started to test this hypothesis by trying to reconstruct longitudinal data series in relation to costs, income, quantities, types of waste materials handled, and so on. The fact that published national statistics routinely omitted such information until very recently is a strong indication of the relatively low level of importance attached to waste handling by the nation-state for much of the post-1945 period, and we were concerned at first that such data may not have ever been collected. Fortunately, our concerns proved unfounded in at least this respect: a great deal of data pertaining to waste were actually gathered by local authorities and subsequently collated to some extent by central government, even if they only rarely appeared in publicly available documents. On the other hand, the data were collected for different reasons—and therefore with different implications—than some of the statistics produced more recently. At least until the latter part of the 1960s, domestic waste *per se* was seen as having little importance in the context of either the national economy or the natural environment; the records that were created were thus not intended to foster knowledge of waste itself or its impacts on society, but rather as part of accounting and auditing practices that were rooted in the legislative structure that governed the handling of refuse by local authorities.

In the United Kingdom during the late nineteenth and early twentieth centuries, these authorities, the corporate bodies that were responsible for the organisation of life in both urban and rural communities, became the *de facto* trustees of refuse handling on behalf of the public they served. The collection and disposal of household wastes were progressively taken up and organised within the established framework of local government so that, irrespective of the philosophy driving the service, it was inevitable that it would be regulated by the same processes that applied to all other municipal activities, particularly regarding their financial management. Regulation was largely achieved through the recording of data for operating and capital costs, which provided a tightly defined means to evaluate how efficient a service was. The aim was to develop a quantified expression of activity, which was intended to demonstrate sound stewardship of the public purse. Such oversight applied as much to the handling of household waste as to any other municipal activity

What we now term waste management was originally called "public cleansing," an indication of the philosophy lurking behind the origins of the organised collection and disposal of municipal refuse in Britain. Although the problems of refuse in organised communities had long been recognised, efforts to tackle them were, until the second half of the nineteenth century, haphazard and generally far from successful.[22] A growing interest in the health and welfare of society led to the Public Health Act of 1868 which, among other things, empowered Local Authorities for the first time to recover the costs of removing domestic waste through the local rating levy rather than by means of individual charges. These new "rate fund" costs were therefore borne not directly by the individuals or households actually creating refuse, but indirectly by the ratepayers who made up the local electorate. The expenses of the service fell under local government budget and audit procedures and so, to satisfy those and to demonstrate financial efficiency, a system of monitoring costs eventually appeared based on the prices of handling weights of waste and servicing each dwelling in the local authority.[23] It is these financial records leading up to the 1980s which—almost accidentally—allow insight into how much waste had to be dealt with, thus permitting us to form an overall picture of how relatively important refuse handling was in the local and national economies.

The 1868 Act was supplemented by further public health legislation, leading to the consolidating Public Health Act of 1936, which formed the basis for municipal waste practice through to the 1960s. None of this imposed a statutory duty on local authorities to collect and dispose of refuse, although they had the legal right to do if they wished.[24] It was a right that was taken up voluntarily by local authorities after 1868 as part of what became increasingly perceived as a civic duty, which could only be effectively carried out through public administration. This was very much in contrast to practice in the United States, where there was no similar unanimity of opinion and municipal bodies frequently entrusted the task to third parties in the private sector with varying degrees of efficiency and economy.[25]

Essentially, then, the British model involved the devolution of actual organisation of waste handling to local level as a public health issue. In part as a result of this, national collation of data was approached in a somewhat haphazard and incomplete way. Public health matters, and hence "cleansing," fell within the domain of the Ministry of Health, and during the 1930s the Ministry began to request from English and Welsh (but not Scottish) authorities details of costs incurred in handling domestic refuse. The purpose of this was principally to promote increased financial efficiency amongst bodies of all sizes, rather than to analyse the nature of refuse or assess its impact on the society creating it. Published *Public Cleansing Costing Returns*, whose generally low-ranking priority for the national government was perhaps indicated by their suspension from 1937 until 1952 (when they reappeared under the aegis of the Ministry of Housing and Local Government), provide a useful, if incomplete, picture of the amount of waste being dealt with and the sums of money spent doing so. The returns expressed costs in terms of tonnages handled and numbers of houses serviced, along with information about the income from what were then known as "salvage" (recycling) activities. Unfortunately for historians, there was no legal requirement or compulsion to provide the information, let alone to guarantee its accuracy, so that some authorities (LAs) failed to provide any data at all and others only estimated very roughly the weights they handled, often based on samples as small as one or two percent.[26] The returns from 1952 to the mid 1960s covered only two thirds of the population and were sometimes based on figures that were often scarcely more than guesses. Nevertheless, the data are the nearest thing to a national overview and despite their limitations they do give an indication of the contemporary scale and cost of municipal waste handling.

The amounts of material collected and disposed of, and the money spent doing so, were considerable. The first post-war survey, for the financial year 1952-53, covered approximately 70 percent of the population of England and Wales and showed that almost £17.1 million had been spent managing over nine million tons of refuse gathered from domestic premises.[27] That sum represented some 2.3% of all LA expenditure for England and Wales, with the five largest cities—Birmingham, Liverpool, Manchester, Leeds and Sheffield—being responsible for one seventh of total reported expenditure on refuse collection and disposal.[28] Collection costs greatly exceeded those for disposal; nationally the ratio was 4:1, a figure which may imply that the gathering up of refuse was of greater concern and expense than its disposal, but also may have to do with income from disposal offsetting some of the costs and / or with the relative labour intensity of each activity. We will return to the point about offsetting of costs later. By the mid-1960s the returns were still covering a similar proportion of a larger population, showing an increase of almost ten percent in tonnages handled. In the financial year ending March 1965, expenditure had more than doubled to £36.16 million within a broadly similar distribution of costs.[29] The increase in the amount of waste dealt with was actually greater than the tonnages indicate; changes in the density and general composition of refuse meant that the volume of waste had increased disproportionately and that its disposal was becoming more of a problem.[30] Correspondingly detailed figures were not collected for Scottish LAs, but their Local Financial Returns allow the extraction of some data on public cleansing and show a similar picture of rising costs: there, £4.3 million was spent in 1953, and £9.55 million in 1965, each year amounting to 3.2% of total LA spending, a substantially greater proportion than for England and Wales.[31] We are not sure what accounts for this discrepancy, but are investigating this in our ongoing research.

These figures indicate considerable and significant economic activity, whether measured in terms of numbers served, weights handled, or sums spent. In fact, the actual figures must have been considerably higher than those quoted because virtually the whole of the UK's population was provided with regular services for waste removal and disposal, an operation whose scale appears to have been unrecognised nationally at the time and which seems to have been taken for granted both by its clients (the general public) and central government which partly contributed to its costs through the system of general Treasury grants to local authorities. The LAs themselves, on the other hand, were intimately involved with refuse and by no means unconcerned about it. On the basis of the records examined so far, refuse handling for them was a constantly problematic and expensive reality which occupied a high position in the hierarchy of municipal responsibilities and needed to be run in a decidedly business-like fashion if it were to function at even a minimally effective level.

The records for the City of Glasgow's Cleansing Department demonstrate the scale and nature of one of Britain's largest municipal refuse operations from the end of World War II until the major local government reorganisations in the mid-1970s.[32] During that period, domestic waste handling was still firmly the province of the public sector within a structure of local government, which was substantially unchanged from the 1930s. The dominant ethos remained throughout the period one of municipal responsibility for both the scale and standards of services provided. Archival material relating to the department gives an insight into its motivation and management, and also into the influences that shaped its changing attitudes to dealing with refuse. In addition, it sheds light on contemporary attitudes to environmental concerns and salvage, suggesting that they regularly impinged on municipal thinking, albeit in a form rather different to those displayed at the start of the twenty-first century.

Glasgow had one of the UK's oldest cleansing departments, and officials there consistently claimed that it had always been one of—if not the—largest and best organised in the UK.[33] Its origins dated back to 1800 when the Chief Officer of Police was charged with the duty of cleansing the streets by sweeping and removing refuse left on them, although the collection of domestic wastes was not involved.[34] In 1868, as a result of the new Public Health Act, a Municipal Cleansing Department was formally established, replacing a private contractor. In its first year it dealt with 140,240 tons of waste generated by a population of some 395,000, at a cost of just under £11,000, which represented a saving of one third on the contractor's charges for the previous year.[35] By the last financial year before World War II seventy years later, a dedicated workforce of over 600 handled over 423,000 tons from a population exceeding one million, at a total cost of £431,000.[36] The war, perhaps surprisingly, did little to alter the trend towards growth in amounts of waste generated and in costs, and by the time operations had adjusted to peacetime conditions in 1947-48, the scale of work had increased substantially. The Cleansing Department's total workforce then numbered nearly 2,000 with a budget of almost £990,000. That figure was exceeded only by education, health and welfare services and policing, indicating the substantial size and importance of public cleansing in the city's economy.[37] Twelve hundred of the Cleansing Department's workers were employed specifically on the collection and handling of 456,000 tons of assorted refuse, gathered from over 288,000 properties at a net cost of almost £580,000.[38]

The organisation for doing all this was substantial, professionally managed (although ultimately politically directed), and far ranging in its operations. The latter included not just the removal and treatment of refuse, but also what would now be labelled recycling activities. It had many of the characteristics of a large monopoly business enterprise, although the forces driving it were substantially different from the private sector. Between 1945 and the wholesale reorganisation of local government in the early 1970s, its operating costs (effectively its turnover) increased five-fold and the capital invested rose by a factor of 8.5, catering for a client base averaging over a million distributed across more than 300,000 households.[39] In Glasgow during this time, waste handling operations not only grew substantially, but also evolved in response to changes in the nature of waste and economic conditions which were outside its own area of control.

The department's remit (in common with any other similar municipal organisation) was to collect and dispose of the city's waste in an efficient manner within the budget set by the elected Council. That figure was determined largely by the extent and frequency of the collection service deemed essential, factors which in turn were determined partly by the public health considerations which were the province of the professional managers, and partly by the politically rooted impulse of the elected representatives to satisfy their constituents that they had used public funds prudently to provide a satisfactory service. Financing came chiefly from the general rate fund provided by locally levied rates based on household property values, and this was supplemented to a small extent by grants made by central government. The budget itself was derived from estimates produced by the department's staff and was frequently exceeded, with implications that we shall return to shortly.

Departures from projected expenditure resulted from unpredictable variables affecting the amount of refuse to be collected, none of which were actually susceptible to short-term control. Waste "arisings" from individual households were in effect spontaneous and unpredictable, and their generation was not subject to any practical limitation or regulation. They tended to increase yearly, both in weight and volume, although to a degree unknown in advance, and so the department was essentially reactive rather than proactive in dealing with them. Waste was constantly generated and, as a result, required prompt removal—it was an immediate and ever-present problem that created by far the largest part of the Department's expenditure. Bad weather, labour shortages or disputes, and (especially in the early post-war years) a lack of efficient disposal facilities all retarded collections and created excess costs through overtime payments or the temporary recruitment of extra workers.[40] There was no alternative to this because it was unacceptable to leave refuse uncollected for more than a short time without generating what were perceived as substantial risks to public health and a strongly negative public reaction.[41] Cost overruns—the equivalent of a loss in private sector business—eventually appeared as a deficit in the department's operating accounts and were ultimately met from the Rate Fund account, and eventually these additional expenditures were passed on to the public either in the form of increased rate levies or reduced levels of service elsewhere in the Council's municipal provisions. If expenditure was less than the budget in a given financial year, any surplus was absorbed back into the Rate Fund rather than being credited to the department for the following financial year.

This system of control would seem to imply a recipe for financial laxity with little incentive to run the department with increasing efficiency, especially given the fact that losses would be, to all intents and purposes, written off continually. Although it is hard to deny that there was a lack of incentive for improving efficiency, the overall financial and audit controls within local and central government acted to rein-in any tendencies towards departmental profligacy and the financial role of managers was principally to estimate costs as accurately as possible.[42] A concomitant skill required of the department's managers was to persuade the politically influenced councillors, who were practically equivalent to the board of directors in a company, to accept what were seen as the necessary costs to be incurred by the department in carrying out its duties, an area which involved not just current or "ordinary" expenditure but also the sometimes very large long-term capital costs involved in sustaining or extending the whole infrastructure of municipal waste handling.

The policies setting out the principles under which the service was run were determined by the elected Councillors—with guidance from the professional staff—and the main aim was to provide all residents with at least a weekly removal of their household waste.[43] This was very much the "public face" of waste disposal and, to the population, by far the most important aspect of it. Any shortcomings in collection standards were immediately apparent, something that taxed management considerably, especially when the service was still highly labour intensive during the 1940s and 50s. The work of collecting refuse bins was always heavy and frequently unpleasant, problems exacerbated by the city's heavy concentration of tenemented properties where, for logistical reasons, collections had traditionally been made at night. Nocturnal working was particularly unattractive, and finding and retaining workers was extremely difficult even when unemployment was running at a substantial level.[44] It was recognised that the only effective remedy was a fundamental revision in the organisation and methodology of collection, a course governed by an intermeshing complex of political, social, and technological factors that shaped the policies proposed by the professionals managing the service.

Political necessity meant that the elected members of the City Council had to provide an acceptable service to the large numbers of people in tenement properties; it was simply not possible to declare the operation uneconomic and abandon it, something that a commercial enterprise would almost certainly have done without hesitation. Here we see an essential difference between the public and private sector arising from differing objectives and systems of governance: accountability to the electorate on the Council's part, coupled with an ethos of social responsibility and professional pride amongst the practitioners, meant that, irrespective of difficulties, the work had to go on.

The apparently simple expedient of eliminating night collections in favour of day work was impossible because there was a chronic lack of capacity to deal with all the city's collections should they be concentrated into a normal eight-hour working day. The problem here lay in a bottleneck further up the system of waste handling that the city had constructed over the previous decades. After all, the collection of waste was simply the start of the process of waste disposal: refuse did not evaporate when bins were emptied into carts or lorries. In Glasgow it went on to be sorted, first by removal of inert material such as ash and clinker, and then for "salvage" (i.e. what we would now term recyclables), which was a source of some income generation for the department, after which the remainder was reduced by incineration to further inert ash and clinker. The latter substances were relatively easy to dispose of in the 1940s and 50s since they could be used as road bedding or to fill-in the disused quarries and mine works to which the city had access. Although the city was home to a number of incineration works, a further constraint on collection and disposal was that they had little storage capacity for holding refuse pending its incineration; and, because of the works' location in built-up areas, there was also no prospect for extending it. Thus, delivery times had to be staggered so that material was received gradually across a 24-hour cycle, another key reason for night collections.

The greatest proportion of domestic refuse was handled at one principal location in 1946.[45] This had been opened in 1928 and was a combined facility for separation, incineration of all un-reclaimed matter, and for the generation of electricity as a by-product of the high-temperature burning process. That had been economically viable before the war, with most of the current produced being sold to the city's own electricity supply department or used to charge the storage batteries of the Cleansing Department's own electrically powered vehicles. Post-war changes caused a major re-evaluation of the facility, however, partly because of what were described as "the very objectionable conditions" endured by the workforce and the population living around the plant. Apart from the regular and sustained emission of considerable amounts of smoke, large dust particles and soot, refuse awaiting combustion was "constantly catching fire" because restricted space necessitated its being held in close proximity to working incinerators. This not only endangered workers and threatened the whole works, but also added to the local air pollution, which appears to have bordered on the intolerable. Of equal significance was that the changing nature of the refuse being fed into the furnaces made incineration harder and more costly. The ash and clinker from domestic fires, which then made up a substantial proportion of household waste, had previously included partly burned coal residue, but it now reflected the poorer quality of domestic coal which tended to include a higher proportion of incombustible material, increasing the amounts of fuel needed to start and sustain the furnaces' burning cycle. The costs of running the works and generating electricity rose and the amount of current produced diminished, undermining the economics of production and emphasising the pressing need to reconsider the plant's future. A decision to rebuild the works, taken as early as 1946, was principally the result of commercial pragmatism, with environmental awareness playing what can best be described as a supporting, though doubtless significant, role.[46]

However, progress on rebuilding the works was slow, something compounded by the fact that the other smaller disposal plants available were inadequate to handle the city's wastes on their own. The solution adopted was to construct a completely new plant first, in addition to those already in use, which would eventually permit decommissioning and reconstruction of the older site. That decision, taken very soon after the one to reconstruct the existing facility, was part of a broader policy for dealing with anticipated long-term changes in the city's generation of waste that was expected to increase as a result of plans to build large numbers of new dwellings. These were meant to replace much of Glasgow's older municipally owned housing which was inadequate or worn-out. Exactly how long the project would take was not anticipated in 1946, and indeed was not acknowledged publicly even after the works was eventually operational in 1958: as late as 1968, it was claimed that work had begun only in 1955, ignoring the long saga of problems that had retarded the project for almost ten years.[47] The worst delays were actually outside the city's control—central government's reluctance to approve the scale and cost of the project caused a delay of six years, for instance, and another delay was caused by the reluctance of contractors to tender at all for the projected work, mainly because no schedule could be given in the absence of government approval.

The delays meant that various developments in other areas of waste production, handling, and management, as well as in business and the economy more generally, called into question the original assumptions that had underpinned the decision to build it. By the time the plant was fully working in late 1958 it was already clear that the generation of refuse was increasing at such a rate that a further incineration works would be needed, even after reconstruction of the older plant. Well before then, in 1956, plans were begun for another facility to cater for only for current demand, but also for what the head of the Cleansing Department described as "the vast new housing development" expected to take place in the City.[48] That project was as protracted as the earlier one; a site was not acquired until 1963 and the specification for the plant not produced until 1966. But this time part of the delay was caused by the realisation among planners that major changes in the nature of domestic refuse and the markets for recyclable materials were compromising the viability of the "separation and incineration" plants. In 1956, it had still been expected that the practice of separating collected refuse would continue, with salvaged materials such as ferrous and non-ferrous metals and textiles being baled and sold on to selected contractors, a labour-intensive operation which nevertheless generated a substantial net return. The design for the works initially followed those principles. By 1962, however, it was clear that income from salvage was falling away and showing little profit.[49] Despite this, a complete specification for issue to would-be contractors was virtually complete by the autumn of 1964, but it was not issued because it was believed necessary to reconsider how the works might best function in the light of emerging trends in waste creation and disposal.[50] In consequence, tenders were also invited for a plant designed to incinerate everything delivered without any attempts at prior separation, leading to further delays.

A decision was eventually made in 1967 to abandon all hand salvage work, and to confine separation to whatever ferrous metals could be recovered by magnets after incineration.[51] Everything that was delivered to the facility would be cremated there, even if material such as non-ferrous metals and cardboard would be lost. This had the attraction of reducing labour costs as well as simplifying the plant itself, which, in turn, produced significant capital cost savings. On the other hand, it increased the amount of material being burned and added substantially to the plant's exhaust emissions, although these were factors which at that time were of less immediate political or social concern than would later become the case. It was felt that filters to eliminate dust from the effluent would be a satisfactory solution to objections about pollution.[52]

The final decision to adopt a regime of cremation of all of the waste received in the plants was driven by operational considerations aimed at maximising throughput and minimising both capital and operating costs, considerations which would have been quite at home in any privately operated industrial environment. The aim of the Cleansing Department, though, was not to generate profits, but to minimise the growth in the expense of providing an essential, though ever more financially burdensome, social service. In 1946 its budget was just over £800,000, in 1956 £1.8 million, and in 1966 it was almost £3.4 million—a four-fold increase over a period which saw little inflation of prices. The value of all salvage sales had more than halved in relation to costs in those twenty years, falling from 10.6 percent of expenditure to 5.2 percent. Over 70 percent of the latter proportion actually came from a separate waste paper collection operation.[53] By 1967, whatever attachment there might have been for in-plant waste separation, its economic contribution to cost reduction had all but disappeared and its continuation was no longer justifiable in the context of efficient management, which was increasingly prioritising the disposal of an ever increasing amount of domestic refuse.

The City of Glasgow disappeared as a local government body in 1974. In its final year, the Cleansing Department employed over two and a half thousand workers and handled 304,135 tons of refuse with a budget of almost £8 million out of the city's total spending of £222.7 million—which is to say, 3.59% of the total. The cost of managing domestic waste was exceeded in non-revenue accounts only by the Police Service (£14.63 million) and by Education (£69.34 million), and these costs had remained substantially similar in relation to one another throughout the period after 1945. The preliminary examination of Glasgow's records, along with some of the national data for England, strongly suggest that municipal waste management bodies had much in common with large commercial enterprises, certainly in respect of their size and complexity of operations, as well as their ultimate duty of financial responsibility to supervisory bodies, although clearly those supervisory bodies had different objectives and composition than those in the private sector. The problems of policy formulation and labour supply and relations were clearly apparent, as well as the early recognition that domestic waste was an issue which had the potential to raise national issues concerning its disposal.

Conclusions

These early investigations have been both stimulating and challenging. They show that—at least in one case—municipal waste management was a convoluted operation that embraced many issues that we face today, although of course contemporary perceptions came from a different perspective. And they suggest that it may be necessary to revise at least some of our own initial conceptions of the industry's structure and operations prior to the emergence of an active commercial sector in domestic waste management. So far, we have only looked at the public sector governance of domestic waste handling, but even here we see as an important issue the question of whether Glasgow was typical of municipal practice or whether the authority was actually *sui generis*, and—if so—then why? The bulk of the work is still to do on the UK case, and we have yet to begin sustained work on the German one. Apart from making comparative studies of a selected sample of UK local authorities, we have yet to examine the effect of central government's role in shaping the nature of municipal waste control through legislation and political encouragement. The important issue of how private enterprise came to be melded into the "big picture" during the 1980s—in particular in relation to the privatisation question—has also still to be addressed, as has the matter of how public and private governance affected both the qualitative and quantitative performance of collecting and disposing of refuse. How, for instance, did commercial operators reconcile the need to discharge socially essential tasks with attaining efficient economic performance? To what degree were companies reliant on recruiting managers from the public sphere in order to extend their operations? Was the organisation of the burgeoning private waste management industry modelled on a municipal model, or was it derived from existing business practices?

There are, at this point, more questions than answers, and the mechanics of answering them may also be complicated because the commercial sector has been characterised by the fluidity of its constituent population: mergers and take-overs have been commonplace, companies have vanished entirely or their identities and records subsumed into others. Much documentation is likely to have been lost, although individuals' oral evidence may partly bridge the gaps. We are, though, encouraged by results of the research carried out so far, which indicate that the business-historical and comparative approach will provide new and fruitful historical perspectives on a pressing contemporary problem at the intersection of political economy, corporate governance, consumer society, and the environment.

Notes

[1] This a slightly revised version of a manuscript originally completed in January 2008. The project described here is now largely completed, and the results will appear shortly in a monograph. Written by Raymond G. Stokes, Roman Köster and Stephen Sambrook, *Making Waste Pay: A History of the Household Refuse Business in Britain and Germany, 1945 to the Present* is in press with Cambridge University Press

[2] BDE (Bundesverband der deutschen Entsorgungswirtschaft), "Zahlen und Daten zur Entsorgungswirtschaft 2005" (available at: www.bde.org).

[3] D. Hall, "Waste Management Companies in Europe." Report Commissioned by the European Federation of Public Service Unions (available at: www.psiru.org).

[4] Key Note, *Waste Management Report 2010*. Ch. 4 (available at: www.keynote.co.uk).

[5] Carl A. Zimring, *Cash for Your Trash: Scrap Recycling in America* (New Brunswick, NJ: Rutgers University Press, 2005).

[6] "Constructing the Waste Management Business in the United Kingdom and West Germany, 1945 to the Early 1990s": ESRC Project Reference RES-062-23-0580.

[7] See, for instance, H. Lanier Hickman. *American Alchemy: The History of Solid Waste Management in the United States* (Santa Barbara, CA: Forester Press, 2003).

K. A. Gourley, *World of Waste: Dilemmas of Industrial Development* (London: Zed Books, 1992). Gottfried Hösel, *Unser Abfall aller Zeiten. Eine Kulturgeschichte der Städtereinigung* (Munich: Jehle Verlag, 1987).

[8] Martin Melosi, *Garbage in the Cities: Refuse, Reform and the Environment* (College Station, TX: A&M University Press, 1981); Joel A. Tarr, *The Search for the Ultimate Sink: Urban Pollution in Historical Perspective* (Akron OH: University of Akron Press, 1996); Susan Strasser; *Waste and Want: A Social History of Trash* (New York: Henry Holt, 1999); and Zimring *Cash for Your Trash*.

[9] David E. Nye, *Consuming Power: A Social History of American Energies* (Cambridge MA: MIT Press, 1998), p. 223; Michael Brower and Warren Lyon, *The Consumer's Guide to Effective Environmental Choices* (New York: Three Rivers Press, 1999), p. 5.

[10] Matthew Gandy, *Recycling and the Politics of Urban Waste* (NY: St. Martin's Press, 1994).

[11] See, for instance, John Scanlan, *On Garbage* (London: Reaktion, 2005); John F.M. Clark, "The Incineration of Refuse is Beautiful: Torquay and the Introduction of Municipal Refuse Destructors," *Urban History*, Vol. 34: 4 (2007), pp. 254-76; John Scanlan, "In Deadly Time: The Lasting On of Waste in Mayhew's London," *Time & Society*, Vol. 16: 2/3 (2007), pp. 189-206; Timothy Cooper, "War, Waste and the Rediscovery of Recycling, 1900-1950," *Historical Research*, Vol. 81: 214 (2008), pp. 710-731; Mark Riley, "From Salvage to Recycling – New Agendas or Same Old Rubbish?," *Area*, Vol. 40: 1 (2008), pp. 78-89; Timothy Cooper, "War on Waste? The Politics of Waste and Recycling in Post-war Britain, 1950-1975," *Capitalism Nature Socialism*, Vol. 20: 4 (2009), pp. 53-72.

[12] N. Fuchsloch, "Metamorphosen oder Euphemismen? Vom Wandel der Abfälle zu Wertstoffen," *Technikgeschichte,* Vol. 68 (2001), pp. 373-394; and M.A.J. Park, "Von der Müllkippe zur Abfallwirtschaft. Die entwicklung der Hausmüllentsorgung in Berlin (West) von 1945 bid 1990," PhD. Berlin: TU-Berlin, 2004.

[13] Raymond G. Stokes, "The Ecological Burden of the Past: Environmental Aspects of German Unification," paper presented at symposium on *The Impact of German Unification: Multiple Perspectives*, Houghton Michigan: Michigan Technological University, 1991; Raymond G. Stokes, *Constructing Socialism: Technology and Change in East Germany, 1945-1990* (Baltimore: John Hopkins University Press, 2000).

[14] G. Hösel, *Unser Abfall aller Zeiten. Eine Kulturgeschichte der Stadtereiniging* (Munich: Jehle Verlag, 1987).

[15] Melosi, *Garbage in the Cities*; Robert Millward, *Private and Public Enterprise in Europe: Energy, Telecommunications and Transport, 1830-1990* (Cambridge: Cambridge University Press, 2005).

[16] Eunomia Research & Consulting Ltd, *Towards Sustainable Waste Management Practice* (Edinburgh: SNIFFER, 2002), p. 9; Volrad Wollny, "Verpackungsabfallverwertung in Grossbritannien" (available at: www.wip.tu-berlin.de).

[17] Gandy, *Recycling and the Politics of Urban Waste*; Strasser, *Waste and Want.*

[18] C3 Consultants Ltd., "Research Capacity within Scotland Relating to Key Subsectors of the Environmental Goods and Services Sector," Final Report: Scottish Enterprise (2006).

[19] See, for example, Alfred D. Chandler, *Shaping the Industrial Century: The Remarkable Story of the Modern Chemical and Pharmaceutical Industries* (Cambridge MA: Harvard University Press, 2005).
[20] Millward, *Private and Public Enterprise in Europe.*
[21] Edith T. Penrose, *The Theory of the Growth of the Firm*, 3rd ed. (Oxford: Oxford University Press, 1995); O. E. Williamson, "The Modern Corporation: Origins, Evolution, Attributes," *Journal of Economic Literature,* Vol. 19 (1981), pp. 1537-68; Alfred D. Chandler, *Scale and Scope: The Dynamics of Industrial Capitalism* (Cambridge MA: Belknap Press, 1990).
[22] Lewis Herbert, *The History of the Institute of Wastes Management 1898-1988* (Northampton: IWM, 1988). Ch. 1 provides an overview.
[23] See Ministry of Health, *Public Cleansing: Refuse Collection and Disposal; Costing Returns* (London: HMSO, 1938) Table VIII, p. 18ff. Costs also appeared in the Ministry of Health's contemporary annual *Local Government Financial Statistics* (London: HMSO) but only expressed in terms of rate in the pound of LAs' expenditure.
[24] Dougall Meston, *The Public Health Act 1936* (London: Sweet & Maxwell Ltd, 1937) p. xxvii.
[25] Melosi, *Garbage in the Cities*; William P. McGowan, "American Wasteland: A History of America's Garbage Industry, 1889-1989," *Business and Economic History,* Vol. 24: 1 (1995), pp. 155-163.
[26] Ministry of Housing and Local Government, *Public Cleansing Costing Returns* (London: HMSO, published annually after 1953) See, for example, 1952-53, Table 9, col. 5.
[27] *Ibid.*, Table 9, columns 17-22, grand totals.
[28] D. S. Lees and W. Appleyard, *Local Expenditure and Exchequer Grants* (London: Institute of Municipal Treasurers and Accountants, 1956).
[29] *Public Cleansing Costing Returns* 1964-65, Table 8, p. 27.
[30] *A Review of Public Cleansing in Glasgow from 1868 to 1968* (Glasgow: Corporation of the City of Glasgow, 1968). p.11 provides some examples.
[31] Great Britain, Scottish Office, *Local Financial Returns* (London: Scottish Office, 1965). Table "Rate fund services." The returns give no explanation for the difference.
[32] These are held by the Mitchell Library, Glasgow, as Cleansing Department, Annual Reports, collection reference DTC/7/3/1, subsequently cited as DTC.
[33] *A Review of Public Cleansing in Glasgow from 1868 to 1968* (Glasgow: Corporation of the city of Glasgow, 1968) chapters 1, 2 and 4 give an overview of the pre-1945 period.
[34] Ibid. p. 5.
[35] DTC 7/3/1 (1), Annual Report 1868/69.
[36] DTC/7/31 (5), Annual Report 1938/39.
[37] Health and Welfare was the term that replaced "Public Health" in the City's accounts. It included, amongst other things, domiciliary midwifery and nursing, health visiting, vaccination, and residential accommodation for the aged and infirm.
[38] Mitchell DTC 7/3/1 (5) Annual Report 1947/48.

[39] Extracted from DTC 7/3/1 Annual Reports for financial years 1945/46 to 1971/72.
[40] DTC 7/3/1 Annual Reports, Superintendent's comments, financial years 1945/46 to 1951/52.
[41] *Ibid.*
[42] Tony Byrne, *Local Government in Britain, Fourth Edition* (Harmondsworth: Penguin Books, 1986), pp. 216-215.
[43] *100 Years of Public Cleansing*. p. 7.
[44] DTC 7/3/5/1 (5) 1951/52, Superintendent's comments, p. 3.
[45] *100 Years of Public Cleansing*. Chapter 4 provides background material and quotations for this section, and DTC 7/3/1 (5) Annual Reports 1945/46 et seq. comment on the condition of the works and progress towards replacement.
[46] DTC 7/3/1 (5) Annual Report 1945/46.
[47] *100 Years of Public Cleansing*.
[48] DTC 7/3/1 (5) Annual Report 1955/56, Superintendent's comments.
[49] DTC 7/3/1 (5) Annual Report 1962/63, p. 6.
[50] DTC 7/3/1 (5) Annual Report 1963/64, p. 5 and Annual Report 1965/66, p. 5. Although the financial year ended on 31 May the Report was not drafted until the early Autumn.
[51] DTC 7/3/1/ (5) Annual Report 1966/67, p. 4.
[52] *Ibid.*
[53] Figures extracted from DTC 7/3/1 (5) Annual Reports for those years.

NOTES ON CONTRIBUTORS

FREDRIK ALBRITTON JONSSON is an Assistant Professor of History at the University of Chicago. He is the author of *Nature's Commerce: the Environmental Context of the Scottish Enlightenment 1750-1820* (Yale University Press, 2013).

JOHN F. M. CLARK is Lecturer in Modern History and Director of the Institute for Environmental History at the University of St Andrews. He is the author of *Bugs and the Victorians* (Yale University Press, 2009), as well as numerous papers on the history of the environment, science, and medicine.

STEVEN CONNOR is Grace 2 Professor of English at the University of Cambridge. His recent books include *The Book of Skin* (Reaktion, 2004), *Fly* (Reaktion, 2006), *The Matter of Air: Science and Art of the Ethereal* (Reaktion, 2010), and *Paraphernalia: The Curious Lives of Magical Things* (Profile Books, 2011).

TIMOTHY COOPER is Senior Lecturer in history at the University of Exeter (Tremough Campus). He was research fellow at the AHRC Research Centre for Environmental History at St. Andrews between 2004 and 2006 where he worked on the history of recycling. His research now focuses on the historical and ideological uses of waste and the environmental transformations it enables. He has had articles published in journals such as *Historical Research*, *Capitalism Nature Socialism* and *Environment and Planning A*.

MAURA COUGHLIN is Assistant Professor of Visual Studies at Bryant University in Smithfield, Rhode Island. Her current research project, "The Visual Culture of Loss and Recuperation on the Brittany Coast, 1880-1914," looks at intersections of ecology, ethnology and mourning in visual and material rural culture.

ALBERTO DUMAN is an Italian artist and university lecturer based in London whose practice is centred on the city in all its possible conceptualizations as the ultimate loci to contemplate, understand and intervene upon our contemporary global condition. He is currently working at Middlesex University, London.

HARVIE FERGUSON was Professor of Sociology at the University of Glasgow until his retirement in 2011. His research and publications have focused on the intellectual and experiential character of modernity, broadly conceived. His books include *Self-Identity and Everyday Life* (Routledge, 2009), *Phenomenological Sociology: Insight and Experience in Modern Society* (Sage, 2006), and *Modernity and Subjectivity: Body, Soul, Spirit* (University of Virginia Press, 2000).

REX FERGUSON is Lecturer in English Literature at the University of Birmingham. He is author of *Criminal Law and the Modernist Novel: Experience on Trial* (Cambridge University Press, 2013).

JENNIFER GABRYS is Senior Lecturer in the Department of Sociology at Goldsmiths, University of London. She is the author of *Digital Rubbish: A Natural History of Electronics* (University of Michigan Press, 2011).

EDWARD J. K. GITRE is Lecturer in History at Seattle University and Associate Fellow of the Institute for Advanced Studies in Culture at the University of Virginia. His research focuses on the modern history of the United States, with particular interests in its intellectual, cultural, and religious aspects. He holds a PhD in history from Rutgers University, as well as two Masters degrees, one in theology and the other in European cultural history. He is currently working on a monograph titled "America Adjusted: Conformity, Boredom, and the Modern Self, *c.*1920–1980."

PIERS J. HALE is Assistant Professor of the History of Modern Science in the Department of the History of Science at the University of Oklahoma. His research explores the changing relationship between biology and politics in Anglo-American society from the nineteenth century to the present. He is currently writing a book about the ways in which changing conceptions of evolution have shaped the possible worlds that people have imagined from the publication of Darwin's *Origin of Species* to the present.

JAIMEY HAMILTON is Assistant Professor of Art History at the University of Hawaii at Manoa where she teaches on topics related to contemporary visual culture, transnationalism, and consumer subjectivity. Currently, she is working on a book about contemporary sculptural practices that appropriate everyday materiality. It is tentatively titled “The Canned Universe.”

WILLIAM KUPINSE is Associate Professor in the Department of English at the University of Puget Sound, where he teaches British literature, ecocriticism, and creative writing. He has published essays on waste in the work of several modernist writers, including H. G. Wells, James Joyce, and G. V. Desani.

JILL R. PAYNE lectures in the School of Sociology, Politics and International Studies at the University of Bristol. Her research focuses on environment and development in Britain and sub-Saharan Africa.

STEPHEN SAMBROOK is a Research Fellow at the Centre for Business History in Scotland at the University of Glasgow. He is working on the ESRC project “Constructing the Waste Management Business in the United Kingdom and West Germany, 1945 to the early 1990s.”

JOHN SCANLAN teaches in the Department of Sociology at Manchester Metropolitan University. He was Project Officer at the AHRC Centre for Environmental History at the University of St Andrews between 2004-07. His books include *On Garbage* (Reaktion, 2005) and *Memory: Encounters with the Strange and the Familiar* (Reaktion, 2013).

RAY STOKES is Professor of Business History in the Department of Economic and Social History at the University of Glasgow, where he is also Director of the Centre for Business History in Scotland. His publications include a co-authored history of BASF Corporation, *German Industry and Global Enterprise. BASF: The History of a Company* (Cambridge University Press, 2004). He is currently working on two co-authored books: a history of waste management in Germany and the UK after 1945; and a history of the international industrial gases industry since 1895.

WILLIAM VINEY is a Research Fellow in the Centre for Medical Humanities and Department of English Studies at the University of Durham. He is the author of *Waste: A Philosophy of Things*, which is forthcoming from I.B. Tauris.

Index

A

acedia, 223-24
aesthetic fatigue, notion of, 2, 11
aesthetic sensibility, 7-8, 13, 14, 32-40, 88, 92, 94, 112-4, 122, 134, 154, 159-160, 166-167, 169, 173-174, 177-181, 183, 195, 208, 211, 222, 227-228, 241, 252, 261
affluence, 125, 190-203, 230, 252, 265, 300
After London; or, Wild England (Richard Jefferies, 1885), 153-55
air pollution, 12, 71-77, 311
Aiton, William, 18, 24, 27
"Angel of History" (Walter Benjamin), 72, 153
Angyal, Andras, 237-38
Arendt, Hannah, 142
Arman (Armand Pierre Fernandez), 13, 173-186
Association for the Protection of Rural Scotland (APRS), 36, 40

B

Bakhtin, Mikhail, 244
Ballard, J. G., 151, 230
Banham, Reyner, 45, 49, 51, 58
Barr, Robert, 77
Barthes, Roland, 283
Bataille, Georges, 13, 80, 105, 133, 177, 185, 244
Baudelaire, Charles, 210, 227, 236
Baudrillard, Jean, 59, 63, 182, 185
Bauman, Zygmunt, 120, 130, 291
Bégout, Bruce, 60
Benjamin, Walter, 72, 153, 161, 163, 210, 211, 229, 231, 233
Ben Lomond (place), 38
Bernays, Edward, 127, 129, 130
Black, Joseph, 24
Blade Runner (film, 1982), 146
Blair Drummond (place), 20, 22
Bloch, Ernst, 220, 230, 231
boredom, 13, 14, 203, 227-236, 243, 245-247
Boyle, Robert, 75-76
Brennan, Teresa, 6
Brewer, Daniel, 139
British Aluminium Company, 35
Burnet, Thomas, 18
Burning Man Festival (Nevada), 51, 52, 53
Burton, Robert (*Anatomy of Melancholy*), 224, 225, 230

C

Californian state of mind, 49, 53
Capitalism (agrarian), 21-22, 25-27
capitalism, 6, 12, 14, 21, 27, 82-95, 115, 120, 159, 173, 176, 184-86, 211, 225, 226
Casino (film, 1995), 64
Clarion (newspaper), 83
Cézanne, Paul, 13, 159, 163-69
Chadwick, Edwin, 123
Clean Air Act (1956, UK)
cleanliness, 120, 123-124, 134, 253, 259-260, 268
Cochrane, Archibald, 20, 23
Cold War, 37, 40, 54, 57, 202, 203
Colloredo-Mansfeld, Rudi, 277
Commons Preservation Society, The, 254
Connor, Steven, 155
consumption, 3, 6, 13, 79, 81, 82, 90, 95, 119, 121-123, 125-130,

133, 143, 159, 161, 165, 167, 169, 173-86, 190-91, 208, 211-13, 221, 231, 255, 266, 274-280, 282, 284, 288, 290, 292-293, 301, 302
Corbin, Alain, 161, 162
Countryside Commission for Scotland, 39
Courbet, Gustav, 160, 163, 164, 165

D
Dale, David, 22
Darling, Fraser, 19
Darwinism, 100 (see also 'Social Darwinism')
Davis, Erik, 52
Davis, Mike, 58, 59, 61
Death Valley (California-Nevada), 51, 52
Debord, Guy, 176-77, 181-83
decadence, 213-15
Decline and Fall of the Roman Empire, The (Gibbon), 147
Decline of the West, The (Spengler), 147
DeGroot, Gerard, 55, 57
Denton, Sally, 50, 62
Depression (economic, 1930s), 193-94, 196
Derrida, Jacques, 138,
desert (human culture and), 47, 49, 50, 51
destruction, 61, 63, 127, 130, 144, 148, 175, 182-185, 203
Dickens, Charles, 76
Dillon, Lindsay, 3
disgust, 14, 207, 221-223, 226, 227, 236-247
disposal practices, 2, 5, 54, 55, 74, 108, 120, 131-133, 212, 251-52, 254, 274-93, 298-299, 300, 303-07, 309-14
Douglas, Mary, 167, 239, 251
Doyle, Arthur Conan, 78
Duchamp, Marcel, 177

E
Eastgate Building (Harare, Zimbabwe), 3
ecologies of remembering and forgetting, 8
Edgeworth, Maria, 226
emptiness, 47-48, 72, 75, 222, 230, 233-34, 236
energy-use, 2, 3-7, 11, 19, 32-33, 37-38, 40-42, 52, 56, 63, 84, 113, 125, 173, 195, 197, 208, 282, 284, 299
Eliade, Mircea, 52
Elias, Norbert, 244
Engler, Mira, 32
Ether, 79-80
Everest, Kelvin, 145
excess, 51, 59, 60-63, 73, 79-80, 85, 121, 129, 173, 176-77, 179, 181, 183-86, 221, 225, 226, 242-43

F
Ficino, Marsilio, 224
Fitzgerald, F. Scott, 13, 207-08, 210, 213, 215-16
Fladager, Vernon, 278
fog, 76-77, 225, 234, 236
forgetting, 8, 10, 12, 45-46, 48, 53, 55, 56, 57, 58, 62, 64, 130, 151, 201
Foster, Sir Norman, 4
Foucault, Michel, 51, 244
Fox, William, 50, 58
Friends of the Earth, 40
Frederick, Christine, 126-27, 129, 130
Fresh Kills (landfill site, New York), 3
Freud, Sigmund, 125, 241, 244, 246
Futurist Architecture (Antonio Sant'Elia, 1914), 150

G
Gandy, Joseph, 13, 140-41
Gandy, Matthew, 300
Gambler, The (Dostoevsky), 61

Gattaca (film, 1997), 10
Glacken, Clarence, 47
Glasgow Cleansing Department, 307-15
Glen Affric (place), 36-37, 39
Glen Nevis (place), 37, 40
Gould, Peter C., 95
Great Gatsby, The (Fitzgerald), 14, 206-16
Greenpeace, 40
green politics, 12, 40

H

"hard" and "soft" wastes (Michel Serres), 7
Hatherley, Owen, 150
Hawkins, Gay, 279
Heart of Darkness (Conrad), 210
Hebrides, 17, 19, 27
Heidegger, Martin, 5, 45, 46, 62, 73, 75, 143, 232-34, 235, 246
Heller-Roazen, Daniel, 48
Hetherington, Kevin, 286
Higgs, Mary, 261
Highlands (Scotland), 17, 19, 32-43
"Highland myth," 34-35
Hilton, Matthew, 255
House_n (intelligent home, MIT), 4-5
Hugo, Victor, 164
Hume, David, 21
Huyssen, Andreas, 160
Huxley, Thomas, 81, 85-88, 100
hydroelectricity, 12, 32-33, 35-41
Hygëia (Thomas Beddoes), 121, 122, 127
Hygeia (Benjamin Ward Richardson), 120, 123-24, 134-35
hygiene, 120, 122-24, 128, 239-40, 252, 263, 265, 267-68

I

incineration, 288, 300, 311-13
Industrial society, 7, 11-12, 19, 22, 25, 28, 32-34, 39, 41-42, 54, 71, 76, 78, 81-95, 107-09, 121, 125, 129, 168, 174-75, 182, 236-37, 254, 286, 289, 290, 298-99, 301-02, 313
imperfection, 9-10
improvement, 11-12, 17, 22, 23, 27, 36, 39, 147, 160
In the Country of Last Things (Paul Auster, 1988), 147
intelligent systems, 5-6
Invisible Cities (Italo Calvino), 121, 130-31, 274
Irigaray, Luce, 73, 74, 79
Isaiah (Book of), 19-20

J

Jardim Gramacho (landfill site, Brazil), 3
Jasper, David, 47
Jenner, Mark, 251
Joad, C.E.M, 256
Jullien, François, 8
junk bonds, 2

K

Kant, Immanuel, 214, 215, 227, 228, 241-42
Kaprow, Allan, 180
Keep Britain Tidy Group, 265-67
Kierkegaard, Søren, 226, 231, 232, 234-37, 246
Kirby, Peter Wynn, 7-9
Klein, Norman, 57
Klein, Yves, 174-76
Kolnai, Aurel, 242-43
Kristeva, Julia, 220, 244-45
Kubler, George, 11,
Kuletz, Valerie, 51, 54, 57

L

Lamby Way (Cardiff), 119, 131-35
landfill, 3, 13, 119, 130-37, 285, 287, 300
Las Vegas, Nevada, 46, 47, 50
leisure, 5, 58, 119, 132, 185, 226, 230, 232, 255-56

Leopold, Aldo, 38
Liebig, Augustus, 124
Life (magazine), 190-91, 195, 196, 197, 198, 199, 202
Lightfoot, John, 17, 20, 26
liminal spaces, 12, 46, 51, 52, 59
Litter Act (1958, UK), 253, 266
Litter in the Countryside (Countryside Commission, 1971), 266
Loch Lomond, 38, 40
London, 72, 76-78, 124, 151-155
Lord Kames (Henry Home), 20, 21-22, 27
Los Angeles, 48, 50, 52, 58
Lowenthal, David, 143

M

Macaulay, Thomas, 152
McKibben, Bill, 6
McWilliams, Carey, 49
Malthus, Thomas, 81
Manson, Charles, 52
Manzini, Ezio, 283-84
Marx, Karl, 246, 292
Masco, Joseph, 51, 54, 57
Matless, David, 33, 34, 255
melancholy, 166, 203, 224-26, 229-30
Melosi, Martin, 299
Mendelssohn, Moses, 241
Metropolitan Public Gardens Association (London), 254
Miller, William L., 240-41
Millet, Jean-François, 160, 165
Modernity, 1, 6-15, 24, 64, 72, 99, 102, 107, 119-120, 127, 130, 135, 159-61, 164, 167, 173, 186, 198, 208, 211, 222-23, 225, 226, 228-30, 232, 235-37, 239, 243, 246, 247, 251
Mojave Desert, 12, 45
Montaigne, Michel de, 225, 230
Moravia, Alberto, 235
Morris, Roger, 50, 62
Morris, William, 12, 82-4, 87-95
Mr. Britling Sees It Through (Wells), 103
Muir, John, 38

N

Napoleonic Wars, 21, 24
National Trust, 254
National Trust for Scotland (NTS), 36, 40, 41
Nellis Air Force Base, Nevada, 54
Nevada Test Site (NTS), 54-58, 64
New Lanark, 22
The New Machiavelli, The (Wells), 108-14
Nietzsche, Friedrich, 215, 226
Nineteen Eighty-Four (Orwell), 147
North of Scotland Hydro-electric Board (NSHEB), 35-38
nuclear waste, 54-57
Nussbaum, Martha, 246
Nye, David, 207

O

Odes (Horace), 145
Ohmann, Richard, 211
On the Creation (Philo), 47
Origin of Species (Darwin), 101
Orlando, Francesco, 47
Otter, Chris, 251-252
"Ozymandias" (Percy Bysshe Shelley), 144-46, 148-51, 154-55

P

packaging, 174-75, 178, 186, 212-13, 253, 255, 263-64, 266, 267, 278-79, 282-86, 289, 301
Pascal, Blaise, 231-32, 234
Pawley, Martin (*Garbage Housing)*, 4
Pessoa, Fernando, 235
Pindar (Greek poet), 47
Planet of the Apes (film, 1968), 138, 146,
Poison Belt, The (Arthur Conan Doyle), 78-80

poverty and waste, 3, 85, 88, 254, 255
preservation, 20, 22, 39, 130, 147, 252, 254-58, 262-65, 268
Presley, Elvis, 62, 63
progress, 9-12, 17, 32, 36, 41, 54, 85-87, 102, 106, 108-09, 113, 121, 123, 125-27, 134, 141-50, 193, 225, 230, 254, 262, 279, 304, 312

Q

Queen Victoria, 34

R

Rauschenberg, Robert, 174, 180
recycling, 4, 14, 53, 120, 211, 267, 275, 285-292, 300-01, 303, 306, 308
Reno, Nevada, 46
resource use, 1, 5-7, 11, 13, 24, 33, 35, 42, 32-41, 63, 71, 73, 74, 81, 86, 88, 93, 108, 120-121, 124-25, 127, 133, 144, 161, 266-67, 276, 303
Ricoeur, Paul, 138, 238
Riegl, Alois, 149
Ritvo, Harriet, 254
Robert, Hubert, 13, 139-141
Rogers, Heather, 289
Romanticism (and landscape), 32-33, 34, 32-42
Rothman, Hal, 61, 64
Rozin, Paul, 238, 239
Ruskin, John, 83, 254

S

Sandburg, Carl, 191, 192, 193, 194, 196
Scheler, Max, 246
Scott, Sir Walter, 34
Scottish Enlightenment, 17, 19, 21
Sebald, W. G., 151
Sekula, Alan, 289
Seurat, Georges, 231-32
Serres, Michel, 7, 74, 141
Signs of Danger (van Wyck), 55, 56
Silent Spring (Rachel Carson), 40
Simmel, Georg, 154, 210
Skilton, David, 152
Smith, Adam, 21, 22, 24, 85
Smithson, Robert, 151, 159, 168-69
social Darwinism, 81, 84-87, 103, 105
Society for Checking the Abuses of Public Advertising (SCAPA), 251, 257, 262
Solnit, Rebecca, 53, 57
Sorrows of Young Werther (Goethe), 226
Speer, Albert, 149
Spencer, Herbert, 12, 84, 85, 86, 87, 90
Spigel, Lynn, 4
Stalker (film, 1979), 146
Steedman, Carolyn, 154
Steichen, Edward, 191, 193, 194, 195
Strasser, Susan, 212, 299,
suburban life, 13, 108, 115, 175, 199-203
sustainability, 1, 4-7, 9, 15, 50, 60, 99, 108-09, 111, 114, 120, 122, 266, 302

T

Tarr, Joel A., 299,
Tinguely, Jean, 174, 184, 185
Terminator (films, 1884; 1991), 146
Text-Book of Biology (Wells), 100, 106
Time Machine, The (Wells), 99, 100, 146-48
Thomson, David, 49, 52
Thompson, Michael, 209
Thoreau, Henry David, 38,
"throwaway society", 264, 277
time consciousness, 2, 5-9, 11, 15, 45, 53-57, 61-64, 72, 74, 119, 120, 124, 138-156, 183, 185,

221-22, 224-25, 226, 227, 229, 230
Toffler, Alvin, 274, 285
Tomkins, Silvan, 238
Tono-Bungay (Wells), 99, 103, 104, 105, 106, 108-11,
Turner, Victor, 51

U

Ukeles, Mierle Laderman, 131
Unna, Percy, 39
urban planning, 12, 119, 134-35, 165
utopia, 10, 12, 49, 50, 89, 91, 106, 115, 119, 121, 122-25, 129, 131, 134-35, 166, 190-91, 198, 202

V

Van Gogh, Vincent, 13, 159, 163-69
Veblen, Thorstein, 46
"Viva Las Vegas" (song, Elvis Presley), 63

W

Walker, John, 17, 19, 20, 21, 23, 24-26, 27
waste (idea of), 1-11, 18-21, 32-35, 45-47, 72-75, 81-88, 99-101, 140-42, 172-73, 208-09, 220-222, 251-52
waste management, 3, 12, 14, 74, 119-20, 132, 298-305, 314-15,
wastelands, 3, 7, 11, 12, 17-21, 27, 32, 38-39, 41, 45-46, 51, 53-54, 56, 60, 64, 107-09, 111, 115, 159-61, 168, 207
wastes (bodily), 10, 74, 163, 239-40
War of the Worlds (Wells), 99, 101, 102
Washburn, Katherine, 148
Waste Land (2010 film), 3
Waste Makers, The (Vance Packard), 121, 125, 127, 128, 129, 274
Weaver, William, 253
Weber, Max, 228
Weinrich, Harald, 45-46, 57
Wells, H.G., 12, 99-115, 146
White, Richard, 81, 86, 88,
wilderness, 19, 32, 34, 38-40, 41, 51-52, 59, 72, 151, 256
Woodward, John, 18
World Resources Institute, 276
World War II, 13, 32, 36, 190

Y

Yates, Frances, 45
Young, David, 21

Z

Zabriske Point (film, 1970), 51
"zero waste," idea of, 3-4, 120, 122
Zzyzx, Nevada (place), 48